ENERGY BUDGET IN THE HIGH ENERGY UNIVERSE

ENERGY BUDGET IN THE HIGH ENERGY UNIVERSE

Proceedings of the International Workshop

Kashiwa, Japan 22 – 24 February 2006

editors

Katsuhiko Sato • Junji Hisano

The University of Tokyo, Japan

World Scientific

NEW JERSEY • LONDON • SINGAPORE • BEIJING • SHANGHAI • HONG KONG • TAIPEI • CHENNAI

Published by

World Scientific Publishing Co. Pte. Ltd.
5 Toh Tuck Link, Singapore 596224
USA office: 27 Warren Street, Suite 401-402, Hackensack, NJ 07601
UK office: 57 Shelton Street, Covent Garden, London WC2H 9HE

British Library Cataloguing-in-Publication Data
A catalogue record for this book is available from the British Library.

ENERGY BUDGET IN THE HIGH ENERGY UNIVERSE
Proceedings of the International Workshop

ISBN-13 978-981-270-010-0
ISBN-10 981-270-010-2

Printed in Singapore by B & JO Enterprise

Preface

The International Workshop on "Energy Budget in the High Energy Universe" was held February 22-24, 2006 at Institute of Cosmic Ray Research, Kashiwa campus of the University of Tokyo. This workshop is the fourth in the series of international workshops of the 21th Century COE program "Quantum Extreme Systems and Their Symmetries" of the University of Tokyo (for a detailed description of the project, visit, http://bilbo.phys.s.u-tokyo.ac.jp/coe21/index_e.htm).

Aim of the Workshop was to discuss of our understanding in the non-thermal, high energy Universe. The existence of materials with very high specific energies, much exceeding the local virial temperature, is best represented by cosmic rays, of which the origin has long been a mystery. Recent astrophysical observations in X-ray, gamma-ray, neutrino and high energy cosmic ray experiments, in conjunction with theoretical studies, have been revealing various new aspects of the High Energy Universe, including promising candidates for the cosmic-ray acceleration sites. However, each approach has its own advantage and limitations in proving the whole view of the issue. Joint efforts by experimentalists and theorists in various related fields are essential in our deeper understanding of the issue. In this Workshop, we discussed to what extent we understand the fluxes, the spectra and the maximum energies of the cosmic radiation produced in various astrophysical sites.

This proceedings includes about 30 reviews by distinguished researchers and 21 contributed papers by active researchers. We hope this proceedings is useful for scientists in this field, and contributes further understanding of the High Energy Universe.

This workshop was supported by the National Astronomical Observatory of Japan (NAO) and the Inter University Program of the Institute for Cosmic Ray Research, the University of Tokyo (ICRR). Finally the editors would like to express deepest gratitude to the 21th Century COE program "Quantum Extreme Systems and Their Symmetries" of the University of Tokyo, NAO and ICRR for their support on behalf of all participants.

October, 2006 Katsuhiko Sato and Junji Hisano

CONTENTS

Contributions

AGASA RESULTS AND THE STATUS OF TA

M. FUKUSHIMA *
Institute for Cosmic Ray Research,
University of Tokyo,
Kashiwanoha 5-1-5, Kashiwa
Chiba, 277-8587 Japan
E-mail: fukushima@icrr.u-tokyo.ac.jp

The Akeno Giant Air Shower Array (AGASA) had observed 11 events of Extremely-High Energy Cosmic Rays (EHECRs) with energies exceeding 10^{20}eV. The energy of these cosmic rays is beyond the Gleisen-Zatsepin-Kuzmin (GZK) cutoff expected by the interaction of EHE protons with the Cosmic Microwave Background (CMB). New experiments with much larger acceptance and improved energy determination, Telescope Array (TA) experiment and Pierre Auger observatory are being built to confirm the existence of EHECRs and to understand their origin.

1. Super-GZK Cosmic Rays

The AGASA reported that the rate of observing EHECRs exceeding 10^{20} eV was consistent with a continued spectrum with a power law of $E^{-2.7}$ and the expected GZK-cutoff structure[1] was not observed. The Fly's Eye air fluorescence telescope also reported an event with 3×10^{20} eV in 1994[2].

High energy astronomical objects such as the active galactic nuclei and radio galaxies were searched as a possible origin of such EHECRs, but none were found in the arrival direction of these events within 100 Mpc of our galaxy[3]. More distant origins may be considered, but only if a special mechanism to allow a longer propagation of EHECRs to take place, for example the violation of special relativity[4] or the EHE neutrinos as the carrier of such energy[5].

It was therefore conceived that super-GZK ($E > 10^{20}$ eV) cosmic rays may be generated by the decay of super-heavy particles in the nearby uni-

*On behalf of Telescope Array collaboration

verse. Particles with energy $> 10^{20}$ eV are easily produced if the mass of the particle is at the Grand Unification scale. Such super-heavy particles may be surviving as a relic of the Big Bang or presently generated by the decay of topological defects[6]. An abundant generation of EHE gamma rays and neutrinos, in place of protons and nuclei, is characteristic in the decay of such particles.

2. Overview of TA

The Telescope Array (TA) was thus proposed in 2000[7] in order to investigate the origin of super-GZK cosmic rays by employing a large array of fluorescence telescopes with ~100 times larger acceptance than AGASA.

The HiRes experiment, however, presented an energy spectrum indicating the existence of the cutoff in the 27th ICRC and the result was later published in 2003[8]. The HiRes spectrum was composed of two monocular spectra each obtained by a single telescope. A preliminary stereo spectrum was presented in 2005 at the 29th ICRC[9]; it exhibits a clear cutoff but the E^3 multiplied flux below $10^{19.5}$ eV is larger than the monocular flux by a factor of ~1.5.

With contradictory results appearing from AGASA and HiRes, it became urgent to understand the experimental bias in the energy and the acceptance determination of two experiments. The construction of full TA is thus deferred, and we decided to build a composite detector at first with AGASA-like ground array and TA fluorescence telescopes[10]. We call it as phase-1 TA (ph-1 TA). We expect that simultaneous measurement of the same EHECRs by two detectors will reveal systematics of two methods and will guide us to a reliable determination of the primary energy and the acceptance.

The phase-1 TA consists of a large plastic scintillator array and 3 stations of air fluorescence telescopes overlooking the array from periphery as shown in Fig.1. The ground array has an aperture of ~1200 km^2 sr, which is an order of magnitude larger than that of AGASA. The fluorescence telescope will have a stereoscopic aperture of ~300 km^2 sr at 10^{20} eV with 10% duty factor. The telescope will also supply information on the primary particle species by measuring the longitudinal shower profile. It will be built in the West Desert of Utah, 140 miles south of Salt Lake City (lat. 39.3°N, long. 112.9°W, alt. ~1400 m).

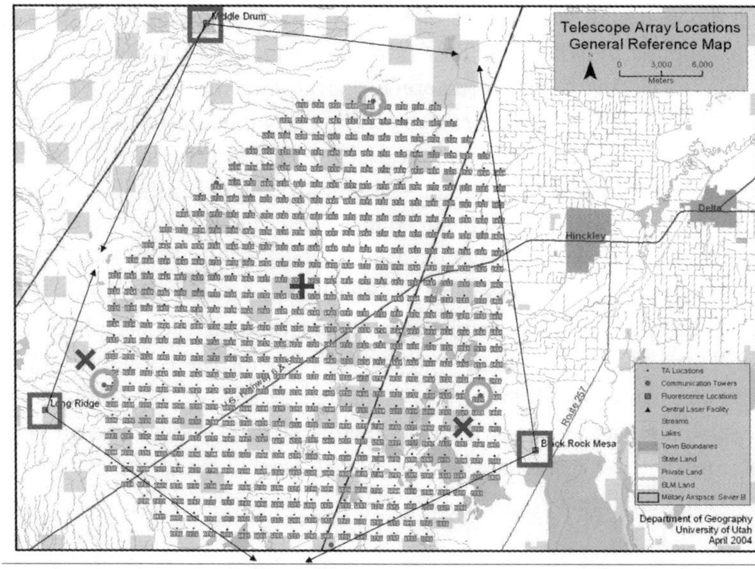

Figure 1. Detector Arrangement of ph-1 TA. The surface detector locations are indicated by small numbers. The fluorescence telescope stations are marked by square boxes.

3. Ground Array of ph-1 TA

The ground detector consists of 576 plastic scintillators deployed in a grid of 1.2 km spacing. It covers the ground area of $\sim$760 km^2. Approximately 80% of them will be on the Federal land, $\sim$10% on the state trust land and the rest on privately owned land. The detection (trigger) efficiency is $\sim$100% for cosmic rays with energies more than $10^{19.5}$ eV with zenith angle less than 45°.

The counter is composed of two layers of plastic scintillator overlaid on top of each other. The scintillator (CI, CIMS-G2) is 1.2 cm thick, 3 m^2 large and is read out by 96 wave length shifter fibers installed in a groove on the surface. The fiber (Kuraray, Y-11(200)M) has a diameter of 1 mm and a length of 5 m. Both ends of the fiber are optically connected to the photomultiplier (Electron Tube 9124B). A passage of cosmic ray muon gives $\sim$15 photo-electrons in average. Two layers are used for the coincidence measurement, for the muon calibration trigger and for extending the dynamic range by setting different PMT gains for two counters.

The signal from each PMT is continuously digitized with a 12-bit flash ADC with 50 MHz sampling. When both of the PMTs record more than 1/3 of the muon signal, wave forms of ~2.5 μs duration are stored with a time stamp supplied by the GPS. This rate of local buffering is less than 1 kHz. The relative timing between remotely separated counters will be controlled with better than ±20 ns accuracy by the GPS, which is sufficient to supply good resolution for the determination of the arrival direction.

When one of the PMT signals exceeds a trigger threshold of 3 muons, its timing is recorded in a local trigger list. The content of the list is transmitted to a branch DAQ board by the wireless LAN at 1 Hz. The list may contain less than 100 events for normal counters. The branch DAQ board is installed on a communication tower built at the periphery of the array. Three towers of ~15 m high will be built for the communication up to ~20 km. An air shower event is identified by the branch DAQ firmware by requiring clustered hits with a good coincidence timing. The air shower event rate will be less than 1 Hz when at least 3 adjacent counters are required in coincidence.

Figure 2. One of the Deployed Surface Detectors.

When an air shower trigger is generated in a branch DAQ board, a command is broadcasted to all counters and relevant counters storing the event with good coincidence timing respond by transmitting the wave form data to the branch DAQ board. The data are then transmitted to a central DAQ system via tower to tower wireless communication and stored in a mass storage. We employ a commercially produced wireless transmitter with the maximum speed of 11 Mbps using 2.4 GHz spread spectrum technology. The dead-time-less DAQ operation is aimed with the high transmission speed together with a large buffering memory at each counter.

One of the counters test-deployed to the field in December 2004 is shown in Fig.2. The total electrical power consumed by the PMT, ADC, GPS and LAN is approximately 7 W and is locally generated by the solar panel of ~120 W capacity (Kyocera KC-120J, see Fig.2). Behind the panel will be a heat-insulated enclosure containing a backup battery (12V, ~65 Ah and deep cycle) and all the electronics. A communication antenna is fixed at the top of 3.3 m tall mast. The total weight of the counter is less than 250 kg, such that it can be easily deployed by helicopter without disturbing the wilderness environment.

4. Fluorescence Telescope

Twelve reflecting telescopes are installed at each station and cover the sky of $3° - 34°$ in elevation and 108° in azimuth looking toward the center of the ground array (see Fig.3). The field of view of each telescope is 18.0° in azimuth and 15.5° in elevation. A spherical dish of 6.8 m^2 is composed of 18 hexagonal mirrors with a radius of curvature of 6067 mm. The direction of each mirror is individually adjustable and a spot size of less than 20 mm in diameter is realized at the focal plane (2960 mm). The mirror is made by 10.5 mm thick high thermal resistivity glass (Schott Borofloat) and is aluminum coated by vacuum deposition. The surface of the aluminum is protected by producing a ~50 nm thick anodization layer.

The air shower image is detected by a mosaic PMT camera on the focal plane. A set of 16×16 PMTs (Hamamatsu 6234) with a hexagonal window is used for one camera. Each PMT covers $1.1°\times1.0°$ patch of the sky. A UV transmitting glass filter (Schott BG3, 4 mm thick) is attached in front of each PMT for blocking the night sky background in the visible light range. The whole camera is assembled in a chassis with a window made by a UV transparent plexiglass.

Negative high voltage is applied to the PMT by a bleeder circuit using zener diodes to ensure a stable operation under high night sky background. The high voltage is individually adjustable for all PMTs. With a PMT gain

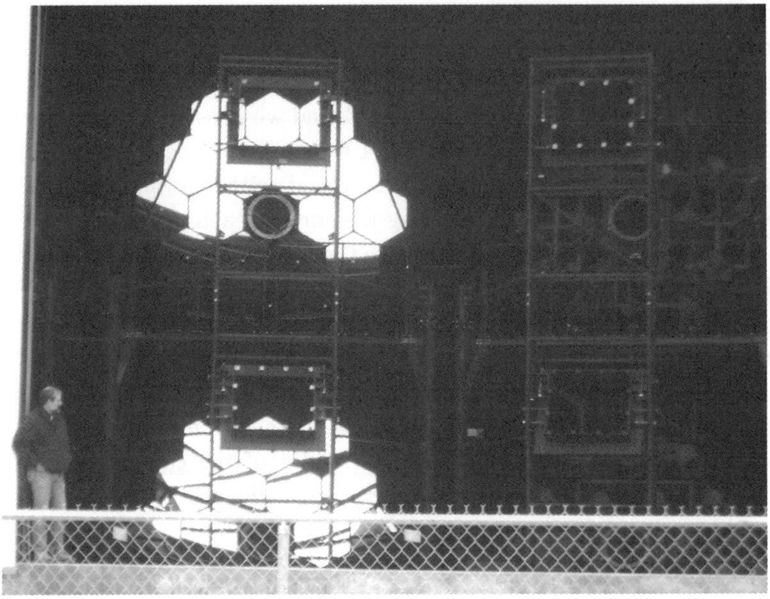

Figure 3. Fluorescence Telescopes of TA

of $\sim 10^5$, a linearity of up to 32 k photoelectrons in 100 ns is achieved. A signal from the PMT is amplified by a factor of 50 by the pre-amplifier and is sent to a Signal Digitizer and Finder (SDF) with 25 m twisted pair cable. The SDF module receives the signal with a shaping filter and digitizes it with a 12-bit, 40 MHz FADC. Consecutive 4 samplings are added by the following FPGA. A trace of fluorescence signal is searched in pipeline at the FPGA employing a sliding sum algorithm for every 25.6 μs of the time window. The dc component from the night sky background is estimated every 1 ms and is subtracted. The SDF is a 9U VME module and 16 channels are mounted in one module.

The result of the "hit" search by the SDF is reported to a Track Finder (TF) in the same VME crate and an air shower track is searched in one camera. A track is found when 5 or more than 5 adjacent PMTs are fired. A looser track definition is applied to a camera-crossing event. The results of all TF modules are concentrated to a Central Trigger Decision (CTD) module and the decision of data acquisition is made. The wave form data stored in the SDF memory are read out to "a camera PC" in parallel and a complete event is subsequently built from the camera PCs by Ethernet.

The calibration of telescope sensitivity and the measurement of atmospheric correction are essential for obtaining an accurate cosmic ray energy. They are described in references[11].

5. Prospects

The phase-1 TA is being built by the collaboration of Japanese and American physicists. The group consists of physicists who have been working in AGASA, HiRes and HEP experiments in the US and Japan. The Japanese fund for ph-1 TA was approved in 2003 by the Grants-in-Aid for Scientific Research (Kakenhi) of Priority Areas. The US group has submitted a proposal for matching fund to the NSF in 2005. The US proposal includes a construction of TALE, a Low Energy extension of TA down to 10^{17} eV, to investigate the modulation of CR composition and spectrum expected by the galactic to extra-galactic transition of CR origins. The infrastructure of TA and TALE in Utah is also the responsibility of the US group.

As of December 2005, a total of 370 surface detectors were produced, of which 18 were test-deployed in the field in 2004. We plan to build communication towers and deploy the rest of counters into the field when the land use permit by the Bureau of Land Management (BLM) is granted, which is expected in February, 2006.[a] The first fluorescence station was built and 12 telescope frames were installed, of which two were equipped with mirrors and a test observation was made with prototype camera and electronics in July, 2005. We expect to complete the construction of TA in April, 2007 and start taking data.

The Pierre Auger group is constructing a large hybrid experiment in Argentina with 1600 water tank detectors. The construction will be complete by the end of 2006. The group presented the first EHECR spectrum at the 29th ICRC in August, 2005 using an exposure already larger than what AGASA had accumulated in 13 years of operation. There was no event exceeding 10^{20} eV. The group considers, however, premature to conclude the existence of GZK-cutoff because the present systematic error of energy determination is estimated to be 50% at 10^{20}eV. The Auger group calibrated the ground array energy estimator $\rho(1000)$, the muon density 1000 m away from the shower center, by the measurement of shower energy from the fluorescence telescope. The extrapolation of the calibration from the lower energy, where most of the hybrid events were collected, caused the major part of the systematic error.

---

[a]The BLM grant was obtained in May 2006.

The construction of ph-1 TA will be finished a few months after the Pierre Auger is completed in Argentina. The acceptance of Auger ground array is ~4.5 times larger than that of ph-1 TA assuming the same zenithal acceptance. The scintillator of TA counts the number of penetrating charged particles and it is dominated by the electrons which outnumber the muons by an order of magnitude. The water tank of Auger on the other hand is more sensitive to the penetrating high energy muons rather than the soft electrons which stop near the surface of the water tank and do not generate as many Cherenkov photons.

The energy measurement of ph-1 TA therefore is less sensitive to the unknown particle composition of primary cosmic rays and the details of hadronic interactions at EHE, whereas its sensitivity for determining the primary particle species using the muon content is severely limited. The identification of EHE gamma rays and neutrinos will be difficult. It is our belief that the characteristic features of ph-1 TA, the sampling of electromagnetic shower energy, the unique calibration of fluorescence generation and the measurement in the Northern Hemisphere, will make an essential contribution to the understanding of the intricate problem of GZK cutoff.

References

1. K.Greisen, *Phys. Rev. Lett.* **16**, 748 (1966); T.Zatsepin and V.A.Kuzmin, *JETP Lett.* **4**, 178 (1966).
2. D.J.Bird et al., *Astrophys. J.* **424**, 491 (1994).
3. Y.Uchihori et al., *Astropart. Phys.* **13**, 151 (2000); G.Sigl et al., *Phys. Rev.* **D63**, 081302 (2001).
4. H.Sato and T.Tati, *Prog. Theo. Phys.* **47**, 1788 (1972); S.Coleman and S.L.Glashow, *Phys.Rev.* **D59**, 116008 (1999).
5. T.J.Weiler, *Astropart. Phys.* **3**, 303 (1999).
6. V.Kuzmin and I.Tkachev, *JETP Lett.* **68**, 271 (1998); V.Berezinsky, P.Blasi and A.Vilenkin, *Phys. Rev.* **D58**, 103515 (1998); K.Hamaguchi, I.Izawa, Y.Nomura and T.Yanagida , *Phys. Rev.* **D60**, 125009 (1999); V.Berezinsky, *Nucl. Phys. Proc. Suppl.* **81**, 311 (2001).
7. The Telescope Array Project: Design Report, July, 2000.
8. T.Abu-Zayyad et al., *Phys. Rev. Lett.* **92**, 151101 (2004); T.Abu-Zayyad et al., *Astropart. Phys.* **23**, 157 (2005).
9. R.W.Springer et al., *Proc. of 29th ICRC, Pune* **7**, 391 (2005).
10. M.Fukushima et al., *Proc. of 28th ICRC, Tsukuba* **2**, 1025 (2003); S.Kawakami et al., *Proc. of 28th ICRC, Tsukuba* **2**, 1033 (2003); F.Kakimoto et al., *Proc. of 28th ICRC, Tsukuba* **2**, 1029 (2003).
11. H.Tokuno et al., *Proc. of 29th ICRC, Pune* **8**, 221 (2005); T.Yamamoto et al., *Nucl. Instr. and Methods* **A488**, 191 (2002).

Results from the High Resolution Fly's Eye Experiment

Charles C. H. Jui, for the HiRes Collaboration
Department of Physics, University of Utah Salt Lake City, Utah, USA
E-mail: jui@physics.utah.edu

The High Resolution Fly's Eye (HiRes) Experiment has been in operation in monocular mode since 1997. The HiRes results on the cosmic ray spectrum are consistent with the GZK Suppression at $10^{19.8}$ eV and observes an ankle structure at $10^{18.5}$ eV. Composition studies of Hires stereo data shows a predominantly light composition in the energy range $10^{18.0} - 10^{19.3}$ eV. We also report on the result of the proton-air cross section measurement from the tails of the X_{MAX} distribution. Various anisotropy studies have yielded null results. However, an apparent correlation between HiRes stereo events and BL-Lac objects has been reported.

Keywords: HiRes; cosmic rays; Proceedings; World Scientific Publishing.

1. Introduction

The High Resolution Fly's Eye experiment is located on the U.S. Army Dugway Proving Ground in the West Desert of Utah. The project is a collaboration between the University of Utah, Columbia University, Rutgers University, University of New Mexico, University of Montana,the Los Alamos National Laboratory (LANL), University of Tokyo, and the Institute for High Energy Physics (IHEP) in Beijing, China. The experiment itself consists of two fluorescence detector stations placed 12.6 km apart. The HiRes-1 site comprises 22 mirrors covering $3° - 17°$ in elevation, and has been in operation since 2007. The HiRes-2 site began routine observations at the end of 1999. It consists of 42 mirrors which view elevation angles in the range $3° - 31°$.

Each of the 64 HiRes mirror units employ the same basic optical design. Ultra-violet (*fluorescence*) light, emitted in the wake of extensive air showers, are collected using a $\sim$ 2 m diameter spherical mirror (3.72 m^2 effective collection area) onto a 16×16, hexagonal *cluster* of photo-multiplier tubes (PMTs) Each PMT pixel covers a 1° cone in the sky. Each mirror unit thus covers about 16° in azimuth, and about 14° in elevation angle. The 1° pixel size gives HiRes a factor of five improvement in resolution over the original Fly's Eye Experiment.[1] The resulting improvement in sensitivity gives HiRes an order-of-magnitude increase in the aperture over the Fly's Eye. Additional details of the detectors components and readout electronics can be found elsewhere[2,3]). We give an overview of the HiRes results to date in the following sections.

2. Energy Spectrum

The HiRes experimental design was optimized to measure the spectrum, composition, and anisotropy of ultrahigh energy cosmic rays at energies $> 10^{18}$ eV. The combined monocular spectrum from HiRes-1 and HiRes-2 was first published in 2004.[4] Figure 1 shows an updated spectrum, which clearly shows two spectral features expected from attenuation from the cosmic microwave background: (a) the GZK suppression (from photo-pion production)[5] at $\sim 10^{19.8}$ eV and the associated *pile-up* just below, and (b) the dip (*ankle*) structure[6] at $\sim 10^{18.5}$ eV. The location of the latter is in excellent good with the stereo spectrum observed by the original Fly's Eye Experiment.[1,7]

As quantitative test of the significance of these features, the spectrum was fitted to power laws. A single unbroken power law model results in a poor fit with a $\chi^2/dof = 162/39$. Allowing for two independent power indices, the fit finds a (floating) break at $10^{18.63}$ eV with a $\chi^2/dof = 68.2/37$. Finally, as shown in Figure 1, a fit allowing for three independent power indices finds breaks at $10^{18.63}$ eV and $10^{19.75}$ eV, and yields an excellent $\chi^2/dof = 34.7/35$. Moreover, a linear extrapolation of the middle segment of fit in the figure predicts 44.9 events whereas only 14 were seen, giving a chance probability of $\sim 10^{-7}$. This calculation takes into account the small overlap in the Hires-1 and HiRes-2 monocular data sets.

3. Composition and Cross-Section Measurements

The shower maximum slant depth (X_{MAX}), from the HiRes stereo data, was used in a study of UHE cosmic ray composition for energies $> 10^{18}$ eV. The average X_{MAX} vs. energy has been published.[8] These are shown, with the measurements at lower energies from the HiRes/MIA hybrid, in Figure 2. The figure also shows model predictions for proton and iron. The HiRes results, spanning the energy range of $10^{18.0} - 10^{19.3}$ eV, indicate a constant and predominantly light composition. The logarithmic slope $dX_{MAX}/d\log E$, known as the elongation rate, is essentially parallel to the

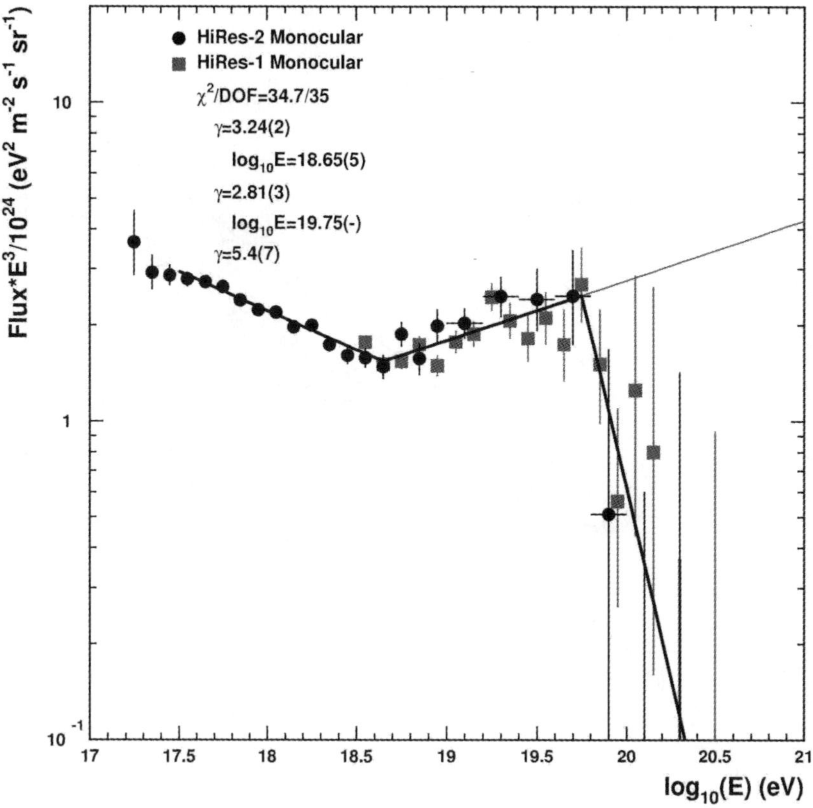

Fig. 1. The HiRes monocular spectrum (6/2005) showing clearly a dip (ankle) structure at $\sim 10^{18.5}$ eV consistent with previous measurements, and a suppression at $\sim 10^{19.8}$ eV consistent with the GZK Effect.[5] The plot also shows a power-law fit with three free indices and two floating breaks. The results of the fit confirms the presence of the dip and the suppression.

model lines. This is another indication of constant composition. For completeness, Figure 2 also includes the result of the HiRes/MIA hybrid,[9] which shows a transition from predominantly heavy composition at $\sim 10^{17}$ eV to a predominantly light one by 10^{18} eV.

The stereo observation of HiRes offers the advantage of redundancy. The difference between the measurements from the two sites can be used to check the accuracy of the Monte Carlo simulations, especially in its ability to model the detector resolutions. Figure 3 shows the relative difference

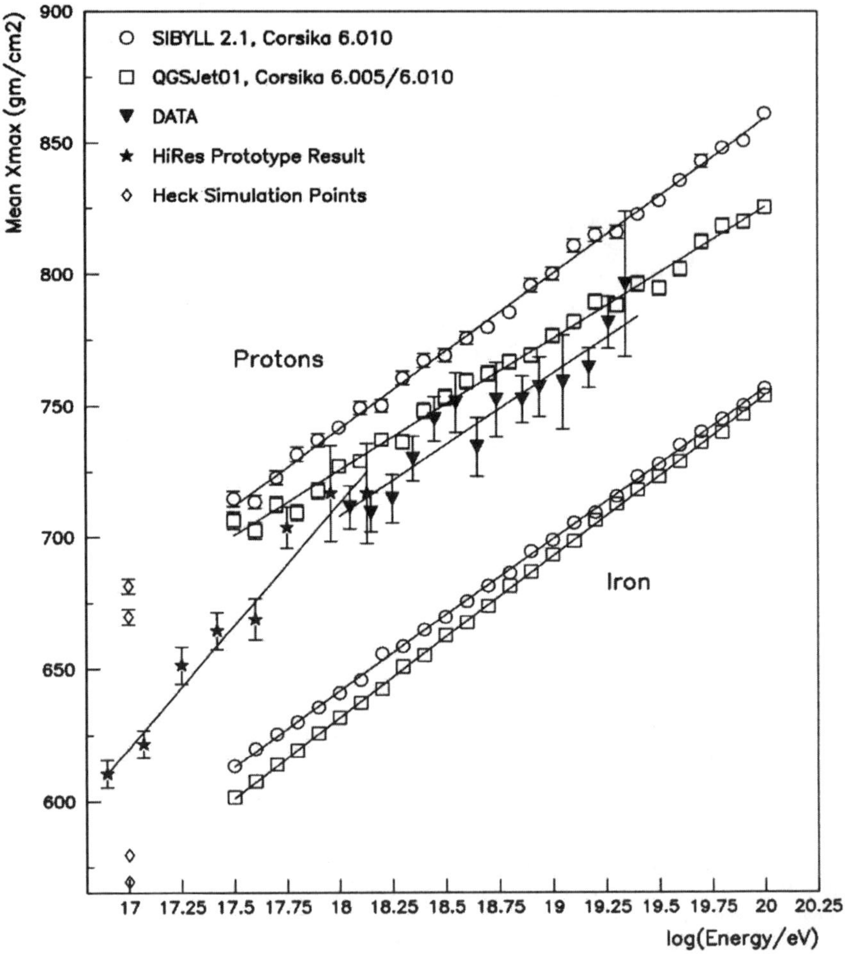

Fig. 2. HiRes[8] (stereo data) and HiRes/MIA[9] (hybrid data) composition results: Average X_{MAX} vs. $\log E$. Model predictions are also shown. These results are consistent with a transition from heavy to light composition from $\sim 10^{17}$ eV to 10^{18} eV, and a constant, light composition above 10^{18} eV.

between the X_{MAX} values measured from HiRes-1 and HiRes-2 detectors, shown for both data (left) and Monte Carlo (right). The vertical axes in these graphs are in log scale. The two histograms have the same width. This excellent agreement serves to validate the X_{MAX} resolution of ~ 30 g/cm^2 obtained from simulations.[8]

Given the predominantly light (protonic) composition, the the expo-

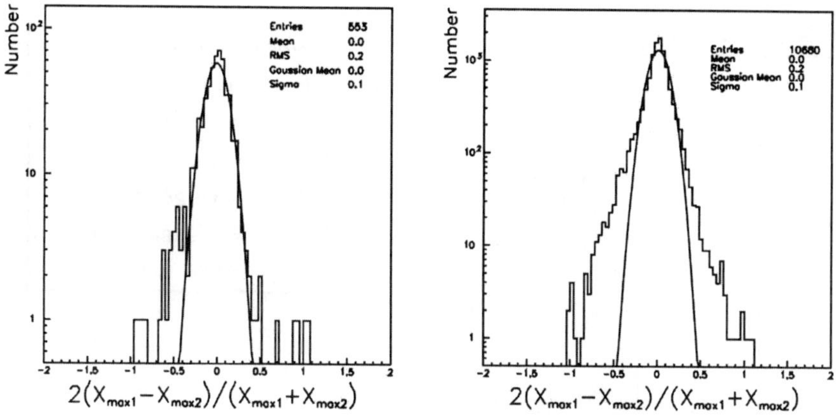

Fig. 3. Distributions of the "pull" (relative difference) between the X_{MAX} measured from HiRes-1 and that from HiRes-2. Data is shown on the left and Monte Carlo to the right. These exhibit excellent agreement and serve to validate the X_{MAX} resolution of ~ 30 g/cm^2 obtained from simulations[8]

nential tail of the X_{MAX} distribution was used to measure the inelastic proton-air cross-section. This analysis uses a deconvolution technique in a reverse Monte-Carlo fit to determine λ_{p-air}, the interaction depth of the primary cosmic rays in the atmosphere. To minimize the effect of the possible presence of heavy components, only the region $X_{MAX} > 700$ g/cm^2 is used in the fit. The result for the HiRes stereo data (up to Jan. 2004) is shown in Figure 4. From $\lambda_{p-air} = 52.7$ cm^2/g, a proton-air cross-section of 456 ± 17(stat.)+39(sys.)−11(sys.) mb was obtained at $E = 10^{18.5}$ eV. Details of this analysis can be found elsewhere.[10]

The proton-air cross-section has been converted to a proton-proton cross-section.[11] This result is shown in Figure 5 with accelerator-based pp data selected by M. Block using a robust sieve method.[11] The latter was also fitted to a parametric model[11] and shown in Figure 5. The extrapolation of this curve passes within the overall error bars of the converted HiRes measurement. The figure also shows the results from the Fly's Eye experiment and the old Akeno Array, renormalized with a k-factor of 1.32 obtained from CORSIKA simulations (using both QGSJet and Sybill).[11]

4. Anisotropy

A number of papers have been published by HiRes on both small and large-scale anisotropy using the HiRes-1 monocular data. They include searches for point-sources[12] and cross-correlation with the AGASA doublets and

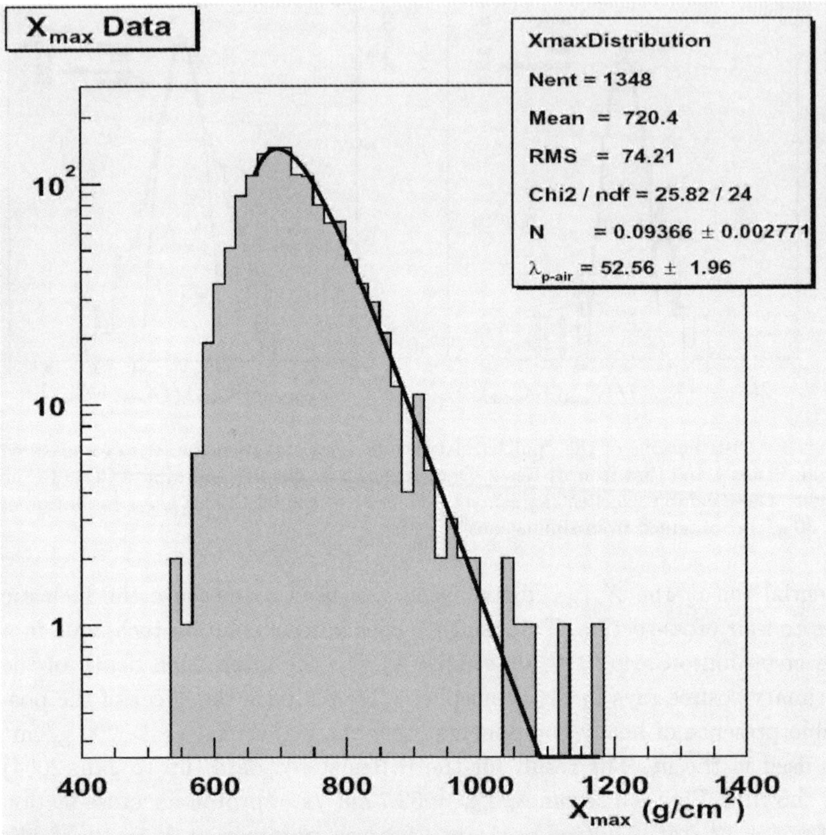

Fig. 4. Measured X_{MAX} distribution and the result of the de-convolution fit to its exponential tail. A value of $\lambda_{p-air} = 52.7$ cm^2/g was obtained, which corresponds to a proton-air cross-section of 456 ± 17(stat.)+39(sys.)−11(sys.) mb.

triplet.[13,14] No significant excesses were found in any of these. Null results were also found in searches for dipole enhancements in the direction the Galactic Center, Centaurus A, and M87.[15] Moreover, HiRes did confirm AGASA's report of small-scale clustering.[16,17]

Anisotropy studies were also performed on stereo data, which has better than 0.6° angular resolution. We have reported that the distribution of stereo events above 10^{19} eV is consistent with isotropy at all small angular scales.[18] While one stereo event was found to overlap with the AGASA triplet, a Monte Carlo study of 10^4 isotropic datasets yielded 47 that gave the same coincidence with equal or greater statistical significance, which

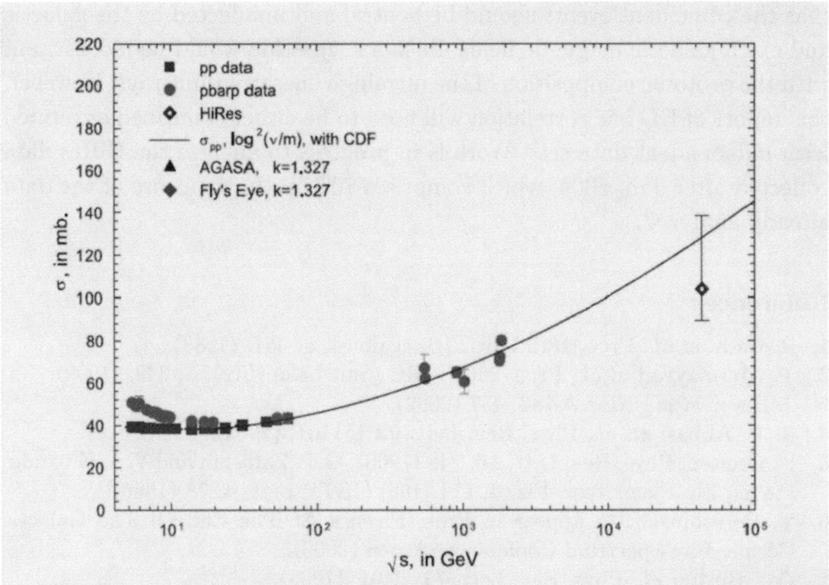

Fig. 5. The HiRes proton-air cross-section converted to a pp cross-section.[11] The HiRes data point is shown with an extrapolation of lower-energy pp data,[11] which gives a prediction that lies within the error bars of the HiRes result. Also shown are the Fly's Eye and Akeno results re-normalized using a k factor of 1.32 obtained from CORSIKA simulations (with QGSJet and Sybill).[11]

corresponds to a $\sim 2.5\sigma$ fluctuation. Moreover, the chance probability of $\sim 5 \times 10^{-3}$ does not account for the statistical penalty associated with the *a posteriori* nature of cuts used by AGASA, which would further reduce, but in an inestimable way, the significance of the apparent overlap.

Finally, an apparent overlap between HiRes stereo events and BL-Lacertae (BL-Lac) objects (with magnitude $m < 18$) was reported by Gorbunov *et al.*[19] Using a binned analysis, the authors compared the positions of HiRes stereo events (collected before Jan. 2004), above 10^{19} eV, extracted from previous HiRes publications, to 156 identified "BL" objects found in the Veron Catalog.[20] They observed 11 HiRes events to lie within 0.8° of "BL" objects and estimated a chance probability for this coincidence to be $\sim 10^{-3}$. A subsequent independent study by HiRes using a maximum-likelihood method (without binning) was able to reproduce the correlation, with an estimate for the chance probability of $\sim 10^{-4}$.[21]

A peculiar feature of this Result is that the correlation is consistent with the HiRes stereo angular resolution. At face value, this would suggest that the coincident events should be neutral and unaffected by the galactic and extra-galactic magnetic fields. Such a suggestion would be inconsistent with the protonic composition of the ultrahigh energy cosmic rays. However, this report of BL-Lac correlation will need to be either confirmed or refuted with independent data sets. Work is in progress to analyze the HiRes data collected after Jan. 2004, which comprises 70% of the exposure of the data already analyzed.

References

1. R. Cady *et al.*, Proc. 18th ICRC (Bangalore), **9**, 351, (1983).
2. T. Abu-Zayyad *et al.*, Proc. 26th ICRC (Salt Lake City), **5**, 349, (1999).
3. J. Boyer *et al.*, NIM **A482**, 457 (2002).
4. R. U. Abbasi *et. al.*, Phys. Rev. Lett. **92** 151101, (2004).
5. K. Greisen, Phys. Rev. Lett. **16**, 748 (1966); G.T. Zatsepin and V.A. K'uzmin, Pis'ma Zh. Eksp. Teor. Fiz. **4**, 114 (166) [JETP Lett. **4**, 78 (1966)].
6. V. Berezhinskii, to appear in Proc. Physics At The End Of The Galactic Cosmic Ray Spectrum Conference, Aspen (2005).
7. D.J. Bird *et al.*, Phys. Rev. Lett. **71**, 3401, (1993).
8. R. U. Abbasi *et. al.*, Astrophys. Journal **622**, 910-926, (2005).
9. T. Abu-Zayyad *et al.*, Phys. Rev. Lett. **84**, 4276, (2000).
10. K. Belov for the HiRes collaboration, Nucl. Phys. B (Proc. Suppl.) **151** (2006) 197-204.
11. M. Block, to appear in Proc. Physics At The End Of The Galactic Cosmic Ray Spectrum Conference, Aspen (2005).
12. R. U. Abbasi *et. al.*, submitted to Astroparticle Phys.
13. M. Teshima *et al.*, Proc. of the 28th ICRC (Tsukuba, 2003) 437.
14. R. U. Abbasi *et. al.*, submitted to Astroparticle Phys.
15. R. U. Abbasi *et. al.*, Astroparticle Phys. **21** (2004) 111-123.
16. M. Takeda *et al.*, Astrophys.J. **522** (1999) 225-237
17. R. U. Abbasi *et. al.*, Astroparticle Phys. **21** (2004) 111-123.
18. R. U. Abbasi *et. al.*, Astrophys. J. **610** (2004) L73-76.
19. D. S. Gorbunov, P. G. Tinyakov, I. I. Tkachev, and S V. Troitsky, JETP Lett. **80** (2004) 145-148; Pisma Zh.Eksp.Teor.Fiz. **80** (2004) 167-170
20. Veron-Cetty,M.-P., and Veron, P. 2000, A Catalogue of Quasars and Active Nuclei (9th ed.; Garching: ESO) A&A, **374**, (2001) 92.
21. R. U. Abbasi *et. al.*, Astrophys. Journal **636** (2006) 680-684.

ASTROPHYSICAL ORIGINS OF THE HIGHEST ENERGY COSMIC RAYS

SUSUMU INOUE

National Astronomical Observatory of Japan,
2-21-1 Osawa, Mitaka, Tokyo 181-8588, Japan
E-mail: inoue@th.nao.ac.jp

Theoretical aspects of potential astrophysical sources of the highest energy cosmic rays are discussed, including their energy budget and issues on particle escape and propagation. We highlight the possibility of heavy nuclei originating from cluster accretion shocks. The importance of X-ray and gamma-ray signatures in addition to neutrinos as diagnostic tools for source identification is emphasized.

1. Introduction

Several decades after their discovery, the origin of ultra-high energy cosmic rays (UHECRs), cosmic particles with energies 10^{18}-10^{20} eV and above, remain one of the biggest mysteries in astrophysics. Many issues contribute to the difficulty of the problem. For experimentalists, the extremely low event rates necessitate detector facilities with huge effective area in order to obtain reliable results. The relevant energies far exceed those of terrestrial experiments, making the determination of basic observables such as particle energy and composition dependent on interaction models with large uncertainties. On the theoretical side, conceiving viable ideas for the production of UHECRs with conventional physical mechanisms in known astrophysical objects is a great challenge. The unavoidable yet uncertain influence of Galactic and intergalactic magnetic fields on UHECR propagation pose further complications.

However, great advances are expected in the coming years with the advent of new generation facilities such as the Pierre Auger Observatory and the Telescope Array, as well as future projects such as EUSO. Combined with crucial complementary information from neutrino, X-ray and gamma-ray observatories, the solution of the mystery could be within sight soon. This article discusses selected theoretical topics in this exciting field, focusing on the astrophysical aspects.

2. Issues on UHECR propagation

We first touch upon some issues concerning the propagation of UHECRs, the basics of which have been well reviewed elsewhere (e.g. Ref.1). The observed global isotropy in the arrival directions strongly suggest that UHECRs are of extragalactic origin. If UHECRs are protons, photopion interactions with cosmic microwave background (CMB) photons must induce severe energy losses at $\gtrsim 7 \times 10^{19}$ eV for propagation distances $\gtrsim 30$ Mpc. Unless the sources lie much nearer, a spectral ("GZK") cutoff is expected above these energies. Whether or not there is actual evidence for this in the observed UHECR spectrum is a matter of controversy at the moment, and will not be discussed here.

Here, we call attention to the possibility that the highest energy UHECRs are composed mainly of heavy nuclei such as iron. For nuclei, the dominant energy loss process above 10^{19} eV during intergalactic propagation is photodisintegration and pair production interactions with photons of the far-infrared background (FIRB) and the CMB. Evaluations based on the most recent determination of the FIRB show that the energy loss distance for iron nuclei at 10^{20} eV is $\gtrsim 100$ Mpc, somewhat larger than that for protons [2]. Observationally, the nuclear composition of UHECRs, particularly at the highest energies, is very uncertain. Important clues come from fluorescence measurements of shower maximum depths by the HiRes experiment, indicating that the composition changes from heavy-dominated below 10^{18} eV (presumably of Galactic origin) to light-dominated above this energy [3]. However, above $\sim 3 \times 10^{19}$ eV the statistics run out and there is no information currently available. Even for the lower energy range, other methods give rather different results [4], and systematic uncertainties in interaction models and atmospheric optical attenuation remain a serious concern [5]. In any case, at present, it is quite viable that UHECRs at $\gtrsim 3 \times 10^{19}$ eV are predominantly heavy nuclei originating from extragalactic sources. This picture will be elaborated on in Sec.4.

Another crucial issue that is often underestimated is deflection by intergalactic and Galactic magnetic fields, which can strongly affect the arrival directions of UHECRs and also act to lengthen their effective propagation distance. The strength and distribution of magnetic fields in the intergalactic medium and at high latitudes in the Galaxy are very poorly known, both observationally and theoretically. Faraday rotation measurements of distant radio sources give only upper limits in the nanogauss range for intergalactic fields on average, subject to assumptions on the field reversal scale [6]. Realistically, whatever their origin, intergalactic fields can be expected to have some correlation with the distribution of large scale structure. Attempts to model this through cosmological simulations haven given different results depending on the input physics and numerical methods [7,8]. The effect of Galactic magnetic fields is also highly model-dependent [9]. Definitive answers may not come until the operation of the Square Kilometer Array (SKA), slated to undertake an "all-sky" survey of rotation measures toward $> 10^7$ background sources with typical angular separations $\sim 90''$, out to different redshifts beyond $z \sim 3$ [10]. Until such information becomes available, we cannot rule out the possibility that deflections by intervening magnetic fields are significant even for the highest energy CRs.

3. Candidate astrophysical sources of UHECRs

A minimum requirement imposed on astrophysical sources of UHECRs is the ability to magnetically confine particles of the requisite energies. For particles with energy E and charge Z, this implies the condition $(R/\mathrm{pc})(B/1\mathrm{G}) \gtrsim (E/10^{20}\mathrm{eV})/Z$ between the system's size R and magnetic field B. Only a select few types of objects are known to meet this criterion, among them the jets of radio-loud active galactic nuclei (AGNs), gamma-ray bursts (GRBs), and clusters of galaxies [11]. Notwithstanding other candidates, in what follows, we focus on these three as representative types of potential UHECR sources. The actual maximum energy that can be attained under different circumstances must be evaluated case by case by comparing the timescales for particle acceleration, usually that for the first order Fermi mechanism in shock waves, against the timescales for limiting processes such as source lifetime, particle escape, adiabatic or radiative energy loss, etc.

Equally important is the available energy budget. Fig.1 shows estimates of the kinetic energy output averaged over the universe as a function of redshift z due to AGN jets, GRB explosions and accretion onto clusters, which should be proportional to their cosmic ray output. The plotted quantity is differential per unit z, $dE_{\mathrm{kin}}/dz = (dt/dz) \int L(dn/dL)dL$, where L is the kinetic luminosity per object and dn/dL is the z-dependent luminosity function. For AGN jets, we have made use of the observed radio luminosity function along with the observed correlation between the radio and jet kinetic luminosities of radio galaxies [12]. GRBs were assumed to occur each with kinetic energy $E_{GRB} = 10^{53}$ erg at a rate that follows the star formation history and matches the log N-log S distribution observed by BATSE [13] (note that this estimate is roughly independent of the beaming factor). The three curves each for AGNs and GRBs correspond to different evolutionary assumptions at the highest z, with only small differences at low z. Cluster accretion will be discussed in Sec.4.

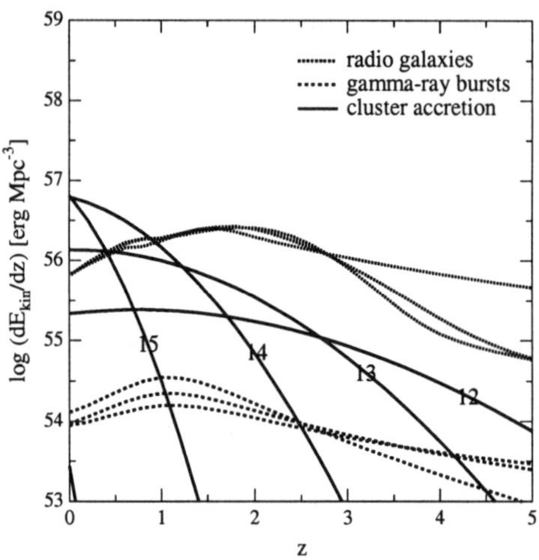

Figure 1. Energy budget of candidate UHECR sources. See text for details.

The results at low z can be compared with the observed energy density of UHECRs, which is $\simeq 10^{-19}$erg cm^{-3} $\simeq 10^{54}$erg Mpc^{-3} at 10^{19} eV. It is apparent that whereas AGN jets and cluster accretion shocks have comfortable margins to accommodate the energetics of UHECRs, GRBs, with an energy budget 2-3 orders of magnitude less, require a very high efficiency of energy conversion into UHECRs, a fact that has already been noted [14].

For both AGN jets and GRBs, different locations along the outflow can be potential UHECR production sites, since one expects $B \propto R^{-1}$ under the naive assumption that the ratio of magnetic to kinetic energy is constant. One candidate in AGNs is the inner jet region with $R \sim 10^{16}$-10^{17} cm and $B \sim$ 0.1-1 G, known through observations of blazars to be a site of particle acceleration, perhaps due to internal shocks. Estimates show that the maximum proton energy should be limited by photopion interactions with low frequency internal radiation to somewhat below 10^{20} eV [15]. Furthermore, efficient conversion to neutrons may be necessary to allow the particles to escape the jet without suffering adiabatic expansion losses, and contribute to UHECRs by decaying back to protons outside. A more promising site may be the hot spots of powerful radio galaxies, termination shocks where large-scale jets are decelerated by the external medium, with $R \sim 10^{21}$ cm and $B \sim 1$ mG. Here the maximum energy may reach 10^{21} eV, limited by escape [16]. However, note that these particles must further traverse the cocoon of shocked, magnetized jet material in order to completely escape the system and constitute UHECRs, an issue that has not been examined in detail. A further possibility that may be worth exploring is acceleration by the bow shocks being driven into the ambient gas by the expansion of the cocoon [17].

In order to verify an AGN origin, detailed analysis of observed UHECR arrival directions and cross correlations with source catalogs will undoubtedly be essential. However, given the uncertainties in intervening magnetic fields (Sec. 2), it will also be highly desirable to have some means to pinpoint individual sources through characteristic, UHECR-induced signatures of secondary neutral radiation. Although neutrinos provide a clear earmark of high energy hadrons, currently planned neutrino detector facilities are unlikely to be able to resolve individual AGNs [18]. Thus, distinctive electromagnetic signals will also be extremely valuable. For radio galaxy hot spots, synchrotron emission from UHE protons can produce nonthermal X-rays and could be distinguished from other processes such as electron inverse Compton through multiwavelength observations [19]. If the medium surrounding the source is sufficiently magnetized, as would be the case for AGNs inside clusters, UHE protons propagating diffusively in the source's vicinity can lead to diffuse gamma-ray emission from photomeson-triggered cascades that may be detectable by current and upcoming instruments [20].

Potential locales for UHECR acceleration in GRBs include internal shocks, external reverse shocks, and external forward shocks, believed to be the emission sites of the prompt X-rays and gamma-rays, optical flash and radio flare, and the radio to X-ray afterglow, respectively [21]. The external forward shock could be disfavored since the ultrarelativistic shock velocity implies that particles are scattered mostly in the weak magnetic fields of the upstream region, but loopholes exist [22]. While this is irrelevant for the mildly relativistic internal and external reverse shocks, a different problem is that for the particles to escape the acceleration site without significant losses, neutron conversion may be required, as with AGN inner jet regions [23,14]. This entails an inevitable reduction in efficiency, which cannot be too severe in view of the tight energy budget, as discussed above.

Although there is some hope that a rare, nearby GRB might be detected as an isolated source of high energy neutrinos, this will not the case for the majority of bursts. Thus, as with AGNs, photon signatures of UHECR production will be important. We have conducted a detailed study of this issue for internal shocks utilizing a comprehensive Monte Carlo code that includes all relevant processes, and found that in some cases, unique features such as synchrotron radiation from muons and protons may be observable by GLAST (Asano & Inoue, in preparation; Asano et al., this volume).

4. Nuclei from cluster accretion shocks as UHECRs

Cluster accretion shocks are candidate UHECR sources that have not received as much attention. Below we briefly summarize our recent work on this subject, leaving more details to a forthcoming paper (Inoue, Sigl, Miniati and Armengaud, in preparation).

In the currently standard picture of hierarchical structure formation in the CDM cosmology, all massive clusters of galaxies should be surrounded by strong accretion shocks, as a consequence of continuing infall of dark matter and baryonic gas [24]. Such shocks should be interesting sites of particle acceleration, and have been proposed as sources of UHECRs [17,26]. For clusters of mass M, the rate of gas kinetic energy flow through accretion shocks can be estimated as $L_{acc} \simeq 3 \times 10^{46} (M/10^{15} M_\odot)^{5/3}$ erg/s [25]. This can be combined with the Press-Schecter mass function to evaluate the energy output averaged over the universe for clusters of different M as shown in Fig.1, where $\log M$ is labeled on each solid curve. Note that due to the hierarchical nature of structure formation together with the nonlinear nature of gravity, $dE_{\rm kin}/dz$ reaches maximum at $z = 0$, with ample room to supply the UHECR energy budget.

However, realistic estimates show that the maximum energy $E_{\rm max}$ for protons falls short of 10^{20} eV by 1-2 orders of magnitude [26,25]. A fiducial cluster of $M = 2 \times 10^{15} M_\odot$ has shock radius $R_s \simeq 3.2$ Mpc and shock velocity $V_s = (4/3)(GM/R_s)^{1/2} \simeq 2200$ km/s. The uncertain shock magnetic field B_s is taken to be a parameter in the range 0.1-1μG. The timescale for shock acceleration is $t_{acc} = (20/3)(Ec/ZeBV_s^2)$, assuming the Bohm limit for scattering by magnetic irregularities as inferred for supernova remnant shocks. To be compared are the energy loss timescales for photopair and photopion interactions with the CMB, as well as the shock lifetime t_s, taken to be the dynamical time $t_s = (4/3)R_s/V_s \simeq 1.9$ Gyr, which also roughly equals the limiting escape time from the acceleration region. As is clear in Fig.2, for protons $E_{\rm max} \sim 10^{18}$-10^{19} eV. It is unlikely that B_s is much larger, since $B_s = 1\mu$G already amounts to 5 % of the gas thermal energy at R_s, and there is also some observational support for $B_s \sim 1\mu$G [27].

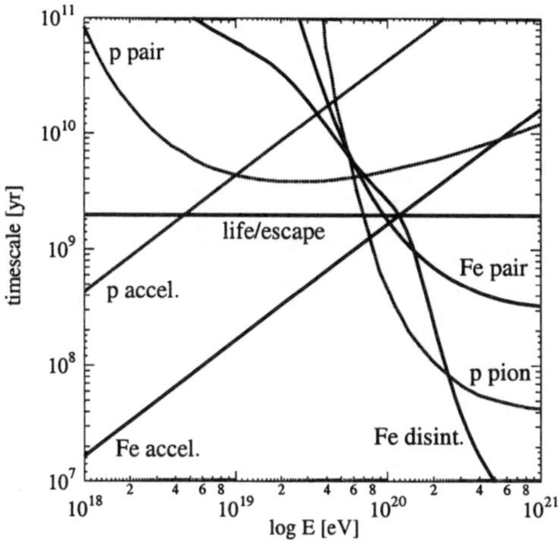

Figure 2. Comparison of timescales for shock acceleration, energy loss by photopair/photopion/photodisintegration interactions, and shock lifetime/escape limit when $B_s = 1\mu$G, for protons and Fe nuclei.

However, heavy nuclei with higher Z have correspondingly shorter t_{acc}, and Fe may be accelerated up to 10^{20} eV in the same conditions, notwithstanding energy losses by photodisintegration with the FIRB and CMB (Fig.2). Heavy elements are known to exist in the intracluster gas with $\sim$ 0.3 solar abundance. Both observations and theory suggest that they are also present in the gas accreting onto clusters with $\gtrsim$ 0.1 solar abundance [28].

In order to explore whether nuclei from cluster accretion can provide a viable picture of UHECR origin, we undertake detailed propagation calculations of UHE nuclei in realistic model distributions of sources and intergalactic magnetic fields that trace large-scale structure, as in Ref. 29. The source density is taken to be $n_s = 10^{-6}\text{Mpc}^{-3}$, appropriate for massive clusters. A fraction f_{CR} of the accretion luminosity L_{acc} is injected into cosmic rays with energy distribution $\propto E^{-p}\exp(-E/E_{\max})$, where $E_{\max}$ is determined for each element by comparing timescales as in Fig.2. A crucial assumption is the elemental composition at injection, which we take to have the same relative abundances at fixed energy/nucleon as that of the observed Galactic cosmic rays at GeV energies for elements heavier than He, as in Ref.30. The abundances of these elements with respect to p and He is given by the metallicity of the gas Z_{gas}. The calculation follows the trajectories of all particles in the assumed intergalactic magnetic field, including secondary nuclei arising from photodisintegration. The case of negligible magnetic fields is also considered.

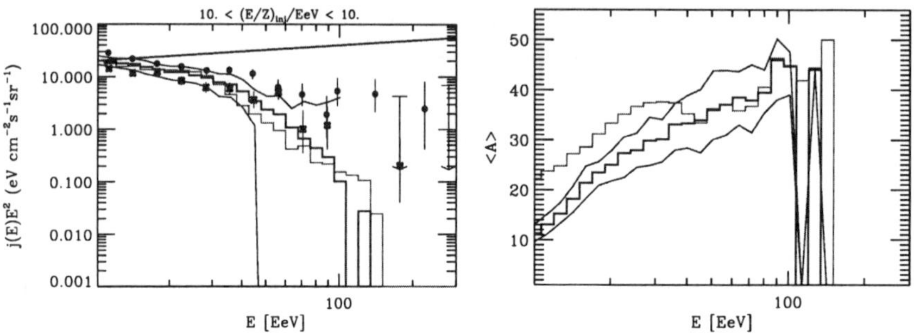

Figure 3. Left. The energy spectrum at Earth for $E = 10^{19}$-3×10^{20} eV expected in the cluster accretion shock scenario for $p = 1.7$ and $Z_{\rm gas} = 0.1$. The histograms show the cases with (thick) and without (thin) intergalactic magnetic fields, and the thin curves represent the model variances for the former case. The straight line illustrates the injection spectral index. The current data for HiRes (squares) and AGASA (circles) are also shown. Right: As with the left panel, but for the mass composition at Earth.

Fig.3 shows the resulting spectra and composition at Earth for $p = 1.7$ and $Z_{gas} = 0.1$, which are quite consistent with current observations. This value for p is naturally expected at the high energy spectral end in acceleration theories that take into account the nonlinear CR back reaction [31]. Normalization to the observed spectrum fixes f_{CR}, which is $\simeq 0.03$ for the case with intergalactic magnetic fields, and $\simeq 0.005$ for the case without. The low value of f_{CR} may reflect the inefficient escape of CRs from the system, which is quite conceivable considering that the shock downstream region is inside the cluster, and that escape may require transverse diffusion away from large-scale filament structures, or perhaps the occurrence of a major merger disrupting the system.

The "GZK" cutoff in the spectra, together with the rapid increase in the fraction of heavy nuclei above $\simeq 3 \times 10^{19}$ eV are clear predictions of the scenario that can be tested in the near future with data accumulated by Auger and TA. A further prediction is the characteristic anisotropy, which should become significant at small multipoles with a sufficient number of events, resulting from the dominance of a few nearby sources. Important information on intervening magnetic fields may also come from the anisotropy.

In this picture, photopion interactions at the source are unimportant, and neutrinos may only arise from the decay of photodisintegrated neutrons, with quite low flux. However, we may look forward to very unique signatures in X-rays and gamma-rays. Protons accelerated to 10^{18}-10^{19} eV in cluster accretion shocks should efficiently channel energy into photopairs of energy 10^{15}-10^{16} eV, which then emit synchrotron radiation in hard X-rays and inverse Compton radiation in TeV gamma-rays. Fig.4 displays the predicted spectra for a Coma-like cluster (see Ref.25 for more details). The detection prospects are very promising for Cerenkov telescopes such as HESS, and hard X-ray observatories such as Suzaku and the future NeXT mission. Photopair production by nuclei may also be efficient and induce further important signals that are worth investigating (Fig.2).

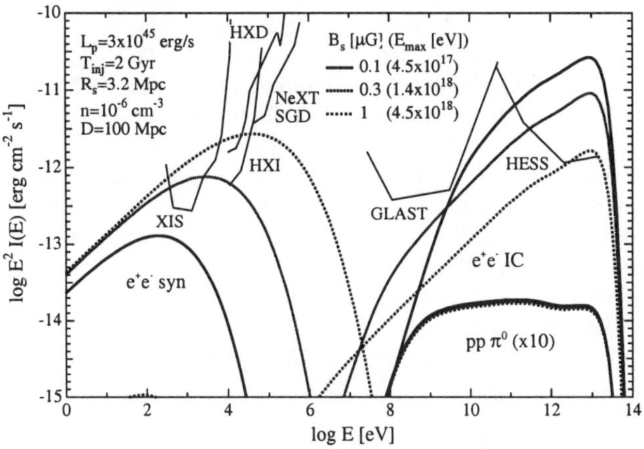

Figure 4. Spectra of UHE proton-induced photopair emission from the accretion shock of a Coma-like cluster, for B_s =0.1, 0.3 and 1 μG. The sensitivities for a 1 degree extended source are overlayed for HESS, GLAST, Suzaku XIS+HXD, and NeXT HXI+SGD.

Who is the true culprit behind the UHECR mystery: clusters, AGNs, GRBs or some other source? The combined effort of upcoming CR, neutrino, X-ray, gamma-ray as well as radio observations may bring us close to the answer in the near future.

Acknowledgments

The author thanks F. Aharonian, E. Armengaud, F. Miniati, G. Sigl and N. Sugiyama for past and ongoing colllaborations.

References

1. P. Bhattacharjee and G. Sigl, *Phys. Rep.* **327**, 109 (2000).
2. F. W. Stecker and M. H. Salamon, *Astrophys. J.* **512**, 521 (1999).
3. R. U. Abbasi et al., *Astrophys. J.* **622**, 910 (2005).
4. M. Ave et al., *Astropart. Phys.* **19**, 61 (2003).
5. A. A. Watson, astro-ph/0408110.
6. P. P. Kronberg, *Astron. Nachr.* **327**, 517 (2006).
7. G. Sigl, F. Miniati and T. A. Ensslin, *Phys. Rev.* D **68**, 43002 (2003); *Phys. Rev.* D **70**, 43007 (2004).
8. K. Dolag, D. Grasso, V. Springel and I. Tkachev, *JCAP* **1**, 9 (2005).
9. H. Takami, H. Yoshiguchi and K. Sato, *Astrophys. J.* **639**, 803 (2006).
10. B. M. Gaensler, *Astron. Nachr.* **327**, 387 (2006).
11. A. M. Hillas, *Ann. Rev. Astron. Astrophys.* **22**, 425 (1984); D. Torres and L. A. Anchordoqui, *Rep. Prog. Phys.* bf 67, 1663 (2004).
12. S. Inoue and S. Sasaki, *Astrophys. J.* **562**, 618 (2001).
13. C. Porciani and P. Madau, *Astrophys. J.* **548**, 522 (2001).
14. V. Berezinsky, A. Gazizov and S. Grigoreva, *Phys. Rev.* D **74**, 43005 (2006).
15. K. Mannheim, R. J. Protheroe and J. P. Rachen, *Phys. Rev.* D **63**, 3003 (2001).
16. P. L. Biermann and P. A. Strittmatter, *Astrophys. J.* **322**, 643 (1987); J. P. Rachen and P. L. Biermann, *Astron. Astrophys.* **272**, 161 (1993).
17. C. A. Norman, D. B. Melrose and A. Achterberg *Astrophys. J.* **454**, 60 (1995).
18. F. Halzen, astro-ph/0602132.
19. F. A. Aharonian, *Mon. Not. R. A. S.* **332**, 215 (2002).
20. S. Gabici and F. A. Aharonian, *Phys. Rev. Lett.* **95**, 251102; E. Armengaud, G. Sigl and F. Miniati, *Phys. Rev.* D **73**, 83008 (2006).
21. E. Waxman *Phys. Rev. Lett.* **75**, 386 (1995); M. Vietri *Astrophys. J.* **453**, 883 (1995).
22. Y. A. Gallant and A. Achterberg, *Mon. Not. R. A. S.* **305**, L6 (1999); M. Vietri, D. De Marco and D. Guetta *Astrophys. J.* **592**, 378 (2003).
23. K. Asano, *Astrophys. J.* **623**, 967 (2005).
24. D. Ryu et al., *Astrophys. J.* **593**, 599 (2003).
25. S. Inoue, F. A. Aharonian and N. Sugiyama, *Astrophys. J.* **628**, L9 (2005).
26. H. Kang, D. Ryu and T. W. Jones, *Astrophys. J.* **456**, 422 (1996); H. Kang, J. P. Rachen, P. L. Biermann, *Mon. Not. R. A. S.* **286**, 257 (1997).
27. L. Feretti and D. M. Neumann, *Astron. Astrophys.* **450**, L21 (2006); M. Johnston-Hollitt and R. Ekers, astro-ph/0411045.
28. F. Nicastro et al. *Nature* **433**, 495 (2005); R. Cen and J. P. Ostriker, astro-ph/0601008.
29. E. Armengaud, G. Sigl and F. Miniati, *Phys. Rev.* D **72**, 43009 (2005).

30. D. Allard et al., *Astron. Astrophys.* **443**, L29 (2005); G. Sigl and E. Armengaud, *JCAP* **10**, 16 (2005).
31. H. Kang and T. W. Jones, *Astrophys. J.* **620**, 44 (2005).

COSMIC RAYS AND MAGNETIC FIELDS IN LARGE SCALE STRUCTURE OF THE UNIVERSE

HYESUNG KANG
Department of Earth Sciences,
Pusan National University,
Pusan, 609-735, KOREA
E-mail: kang@uju.es.pusan.ac.kr

Diffuse radio emission from galaxy clusters indicates the presence of GeV electrons and microgauss level magnetic fields in the intracluster medium (ICM). Nonthermal emission due to Inverse Compton scattering of relativistic electrons off the cosmic microwave background radiation (CMBR) has been also detected in several clusters. Considering that most acceleration mechanisms preferentially accelerate more protons than electrons, we can deduce that the energy contained in CR protons could be even more substantial. We review the possible astrophysical sources that can inject and accelerate CRs, and generate and amplify magnetic fields in the ICM. From both observational and theoretical grounds, we conclude that the energy budget of CR protons, magnetic fields, and turbulence is each order of 10 % of thermal energy, while CR electrons may contain less than 1 % of thermal energy.

1. Introduction

Cosmic rays (CRs) and magnetic fields in astrophysical plasmas are closely related, since astrophysical gas is most often ionized and heated by shocks, and strongly magnetized. Most astrophysical shocks are so-called collisionless shocks which form in a tenuous plasma via electromagnetic viscosities, *i.e.*, collective electromagnetic interactions between the particles and the underlying irregular magnetic fields [28]. Hence the existence of magnetic fields or self-generation/amplification of magnetic fields by shocks, especially irregular component, are required for the shock formation process. The origin of the cosmic magnetic fields is one of outstanding problems in astrophysics. The generation of magnetic fields by Biermann battery mechanism [25] and the Weibel instability [29] at cosmic shocks on a cluster scale have been considered in addition to the amplification of seed primordial fields by dynamo actions.

According to the diffusive shock acceleration (DSA) theory, about $10^{-4} - 10^{-3}$ of incoming particles can become CRs and 10-50 % of shock kinetic energy is transferred to CRs at strong quasi-parallel shocks. Such particle acceleration has been observed directly at various kinds of shocks including interplanetary shocks, Earth's bow shocks, and supernovae remnant shocks [4]. On a galactic scale it is well known that CRs and magnetic fields are dynamically important components of the interstellar medium of our Galaxy with an approximate energy equipartition among different components, *i.e.*, $\varepsilon_{CR} \sim \varepsilon_B \sim \varepsilon_{therm} \sim \varepsilon_{CMBR} \sim 1\mathrm{eVcm}^{-3}$. Although the Galactic CRs are commonly believed to be accelerated mostly at supernova remnant shocks [3, 43], CR acceleration is probably important in all shock-heated cosmic plasmas.

There are increasingly more observational evidences that clusters of galaxies and large scale structures may contain a significant amount of CR electrons and magnetic fields embedded in the tenuous baryonic medium. Several theoretical explanations for origin of those CR electrons have been proposed, including re-acceleration of relic relativistic electrons by merger shocks, fresh injection/acceleration of CRs at merger shocks and accretion shocks, and secondary electrons generated by inelastic collisions of CR protons and the ICM [5, 16, 30, 32]. Galactic winds driven by supernova explosions [44] and jets from active galactic nuclei (AGN) [10, 24] are also known to inject CRs and magnetic fields into the ICM. Thus these nonthermal components are important in understanding the energy budget in the ICM and in large scale structure (LSS).

In order to recapitulate the progresses in both observational and theoretical studies of the origin and roles of nonthermal components in LSS, the 3rd Korean Astrophysics Workshop entitled as "International Conference on Cosmic Rays and Magnetic Fields in Large Scale Structure" was held in Pusan, Korea in 2004 (http://canopus/chungnam.ac.kr/kaw3). This review paper is mainly based on the papers presented at that meeting whose presentations can be found in the meeting website.

2. Observational Evidences of Nonthermal Components

2.1. *Signatures of Cosmic Rays in the Intracluster Medium*

CR electrons and magnetic fields reveal themselves via synchrotron radiation, often in radio for electrons with GeV energies in microgauss fields. There are now more than 30 clusters of galaxies with diffuse radio emission [16, 17]. Depending on its observed properties the diffuse radio emission is further classified as either a *radio halo* when its morphology is regular and typically centered on and resembling the X-ray emissivity, or as a *radio relic* when it is irregular and located at the periphery of the cluster. Radio halos are usually found in rich clusters with high ICM temperature and high X-ray luminosity. The clusters with a radio halo and/or a relic show some signatures of recent merger events and significant substructures, but no cooling flows. The strong correlation between the radio power emitted at 1.4 GHz and the cluster X-ray luminosity, and the similarity of X-ray and radio morphology of radio halo clusters indicate a possible connection between the non-thermal and thermal energy components. The radio emission is unpolarized down to a few percent, implying turbulent nature of magnetic fields [17].

In addition to diffuse radio emission, observations in EUV and hard X-ray have also revealed that some clusters possess excess radiation compared to what is expected from the hot, thermal X-ray emitting ICM [*e.g.*,6, 18, 41]. One mechanism proposed for the origin of this component is the inverse Compton (IC) scattering of cosmic microwave background photons by the same CR electrons emitting radio synchrotron. Also it has been suggested that significant fraction of the diffuse gamma-ray background radiation could originate from the same process [26,31]. The same mechanisms that are capable of producing these CR electrons may have produced CR protons. Collisions of CR protons in the ICM generate a flux of γ-ray photons through the production and subsequent decay of neutral pions. Also ultra-high energy protons ($10^{18} - 10^{19}$eV) interacting with CMBR may produce electron-position pairs, which emit synchrotron and IC radiation in hard X-ray and TeV gamma ray [19]. However, these γ-rays have not been positively detected from galaxy clusters yet [35, 36].

2.2. *Observed Magnetic Fields in the ICM*

The mean magnetic field in the interstellar medium of our Galaxy is about 5-8 μG. It is believed that $\sim 10\mu$G regular fields can be generated by large-scale dynamo due to galactic rotation in spiral galaxies. In galaxy clusters, however, magnetic fields are dominantly irregular and likely generated from small-scale turbulent dynamo. Magnetic fields in the ICM can be measured using a variety of techniques [8]. Observations of radio synchrotron emission and IC scattering of CMBR by CR electrons can provide a way to estimate the energy densities of CR electrons and magnetic fields in the ICM. Fluctuations in Faraday rotation measures (RMs) are used to estimate total field strength. For example, 1) studies of synchrotron relic and halo radio sources within clusters indicate 0.4-1 μG [17], 2) studies of IC X-ray emission indicate 0.2-1 μG [14], and 3) surveys of Faraday RMs of radio sources both within and behind clusters indicate 1-40 μG [9]. Estimates based on Faraday RMs tend to give higher field strengths than those from synchrotron and IC measurements. It was pointed out that in order to claim cluster-wide field measurements from RMs we must first rule out the effects local to sources, and that current estimates may be biased towards strong fields in central region of clusters [37]. If the cluster magnetic field is intermittent rather than volume filling, however, the discrepancy between the field strengths estimated from Faraday RMs and those estimated from radio halos can be compromised.

In the case of synchrotron radio halos, average magnetic field strength can be estimated using the equipartition assumption, $\varepsilon_{cr} = \varepsilon_B = B_{eq}^2/8\pi$:

$$B_{eq} \propto [(K+1)\frac{I_\nu}{(lf_v)}]^{1/(\alpha+3)}, \tag{1}$$

where I_ν is the synchrotron intensity, K is the ratio of proton to- electron number density, l is a path length, f_v is a volume filling factor, and α is synchrotron spectral index [1]. Adopting the standard equipartition condition where equal density in CR protons and electrons, and $f_v = 1$, the field energy density for radio halos is typically $\varepsilon_B = 0.01 - 0.1 \mathrm{eVcm}^{-3}$ [17]. However, the parameters, K and f_v are rather uncertain and need to be considered carefully. According to Beck & Kraus (2005), for example, $K = 40 - 100$ for CR populations accelerated by shocks, $K \sim 100$ for MHD turbulence acceleration and $K = 100 - 300$ for secondary electrons. The analysis of observational data of several supernova remnants (SNRs) by Völk and collaborators and the proton-to-electron ratio of Galactic CRs also suggest $K \sim 100$ [2]. Also the volume filling factor f_v can be much smaller than one for highly intermittent fields where magnetic fields are concentrated in filaments.

3. Diffusive Shock Acceleration

Shocks are ubiquitous in astrophysical environments and CRs are diffusively accelerated at shocks by the first-order Fermi process [11]. For parallel shocks, in which the ambient magnetic field is aligned with the shock normal, the applicability of DSA theory is now fairly well established [22, 28]. Also it has been quite successful in explaining many aspects of the CR population, such as the nearly power-law spectrum of CRs detected at the top of the Earth's atmosphere, the relation between the break in the power-law around the $\sim 10^{15}$ eV knee energy to the maximum energy of CRs achievable in SNRs, and the nonthermal, power-law electron populations deduced from the radio synchrotron observations of SNRs [2, 4].

Plasma simulations of quasi-parallel shocks have shown that the particle velocity distribution has some residual anisotropy in the local fluid frame due to the incomplete isotropization during the collisionless shock formation process and so some particles can stream back upstream of the shock [28]. Streaming motions of high energy particles against the background fluid generate strong MHD Alfvén waves upstream of the shock, which in turn scatter particles and lead to the strong amplification of magnetic fields [27]. These particles can be accelerated further to higher energies via Fermi first order process. Hence the nonthermal particles are natural byproducts of the collisionless shock formation process and they are extracted from the shock-heated thermal particle distribution (thermal leakage injection) [22, 28]. With 10^{-4}–10^{-3} of the particle flux passing through the shock injected into the CR population, a significant fraction (up to 60%) of the kinetic energy of strong quasi-parallel shocks can be converted into CR protons and the nonlinear feedback to the underlying flow can be substantial [22, 23, 43]. For quasi-perpendicular shocks in which the mean magnetic field direction is perpendicular to the flow direction, thermal leakage injection is inefficient, since the transport of low energy particles normal to the average field direction is suppressed. Due to low injection rate the CR acceleration is much less efficient at perpendicular shocks, compared to parallel shocks.

On the other hand, electrons whose gyro-radius is much smaller than ionic gyro-radius need some additional processes to bridge the gap between the thermal electron population and the relativistic region. This pre-acceleration (electron injection) is not well understood, so we do not have a quantitative model for the electron acceleration yet. An empirical model that can fit observational data imply the CR proton-to-electron ratio of $K \sim 100$ [2].

4. Astrophysical Sources of Nonthermal Components in the ICM

In galaxy clusters there are several possible sources of CRs and magnetic fields including termination shocks of galactic winds driven by supernova explosions, jets of radio galaxies, merger shocks, structure formation shocks, and turbulence [12].

4.1. *Galactic Winds*

In order to explain the observed metal abundance of the ICM, $Z = 0.1 - 0.3Z_\odot$, a typical rich cluster should have gone through $N_{SN} \sim 10^{12}$ SNe which inject $E_{GW} \sim 10^{63}$ erg via SN-driven galactic winds [44]. Assuming a canonical CR acceleration efficiency of 10-30 %, the termination shocks of galactic winds can deposit the CR energy of $E_{CR} \sim 10^{62}$ ergs into the ICM. These particles are further accelerated by the gravitational contraction and merger/accretion shocks. Galactic winds also drag out the galactic magnetic fields of μG strength into the ICM, depositing magnetic energy $E_B \sim E_{CR}$ into the ICM and resulting in the cluster fields of $0.1 - 1\mu$G.

4.2. *Radio Galaxy Jets*

Radio jets from active galactic nuclei (AGN) are known to inject CRs and magnetic fields as well as photons into the ICM [10, 24]. The efficiency of energy conversion of the black hole (BH) infall energy is $\eta_{photon} \sim \eta_B \sim \eta_{CR} \sim 0.1$ [24]. So the injected energy from a BH with an average mass $M_{BH} \sim 10^8 M_\odot$ is $E_{nonthermal} \sim 0.2 M_{BH}c^2 \sim 4 \times 10^{61}$ ergs. With the average density of BH, $\rho_{BH} \sim 2 \times 10^5 M_\odot \mathrm{Mpc}^{-3}$, AGNs provide sufficient nonthermal energies to magnetize the IGM to $B \sim 0.1 - 1\mu$G level and to fill the ICM with CR protons and electrons. Large scale gravitation infall, merger shocks, and turbulences may provide further amplification of B and acceleration of CRs.

4.3. *Major Mergers*

In major mergers of clusters the subclusters ($M_{cl} \sim 10^{15} M_\odot$) collide at velocities of ~ 2000 km s^{-1}, releasing gravitational binding energies of $\gtrsim 10^{64}$ ergs. Shocks associated with these mergers dissipate energies of $\sim 3 \times 10^{63}$ ergs [40], most of which go into the thermal energy of the intracluster medium. These shocks have have speeds similar to typical SNR shocks in Sedov stage, but have lower Mach numbers (*i.e.*, $M_s \lesssim 3$) due to high preshock temperature in the ICM. According to the DSA calculations and observations of SNRs, typically order of 10% of the shock kinetic energy is transferred to CR protons, while ~ 0.1 % is deposited into CR electrons. So $E_{CR,p} \sim 0.1 E_{therm} \sim 10^{62}$ ergs. Fresh injection and acceleration CR particles at cluster merger shocks were studied by Miniati (2002) and Gabici & Blasi (2003). Merger events are expected to induce turbulent motions in the ICM with energies $E_{turb} \sim 0.2E_{therm}$, leading to turbulent acceleration of CRs and turbulent dynamo of magnetic fields [*e.g.*,7, 13].

4.4. *Structure Formation Shocks*

During the structure formation in the universe large-scale collisionless shocks are produced by flow motions associated with the gravitational collapse of nonlinear structures. Estimated speed and curvature radius of these shocks could be as large as $\sim 3 \times 1000$ km s^{-1} and ~ 5 Mpc, respectively. These shocks not only heat the ICM to a few to 10 KeV, but also can accelerate the CR ions and electrons to very high energies [3, 20, 21, 22, 33]

External shocks form around outermost surfaces that encompass nonlinear structures, so they are by nature accretion shocks that decelerate the never-shocked intergalactic gas. *Internal shocks* are produced within the nonlinear structures by infall of previously shocked gas to filaments and knots, during subclump mergers, as well as by chaotic flow motions. External shocks have high Mach numbers of up to $M_s \sim 100$ due to low temperature of the accreting gas. Internal shocks, on the other hand, have mainly low Mach numbers of $M_s \sim$ a few, because the gas inside nonlinear structures has been previously heated by shocks and so has high temperature. However, internal shocks are more important in energetics, because of higher preshock density. As a result, thermalization of gas and acceleration of cosmic rays occurred mostly at internal shocks. For internal merger shocks of $M_s \lesssim 3$ the energy transfer to CRs should be ~ 10 % of the shock kinetic energy at each shock passage, with an associated CR particle fraction of 10^{-3}. The spectrum of CR protons should be steeper than $N_p(E) \propto E^{-2.5}$ at internal shocks. On the other hand, external shocks typically propagate into the low density intergalactic medium, so the amount of kinetic energy passed through accretion shocks is small. Since the CR acceleration is the most effective for strong shocks ($M_s \gtrsim 5$), the shock flow should be significantly modified by the CR pressure. The accelerated proton spectrum may not be a simple test-particle power-law of $N_p(E) \propto E^{-2}$ because of nonlinear feedback of CR pressure at such strong shocks [39].

The CR acceleration at structure formation shocks imply $E_{CR,p} \sim (0.1 - 0.5)E_{thermal} \sim 10^{62}$ ergs and $E_{CR,e} \lesssim 0.01E_{thermal} \sim 10^{61}$ ergs with the uncertainties of a factor of ~ 2. Geometry of magnetic field lines (*i.e.*, dominantly quasi-parallel vs. quasi-parallel) and pre-acceleration processes of electrons (*i.e.*, the proton-to-electron ratio, K) are unknown elements that affect these estimates most.

4.5. *Turbulence*

Flow motions associated with LSS formation, *i.e.,* gravitational infalls and and hierarchical mergers, are expected to drive turbulent motions in the ICM. Random vortices generated at oblique accretion shocks and behind the wakes driven by galaxies and merging clumps develop into turbulences. Hydrodynamic simulations of LSS formation showed that the ICM becomes turbulent with $v_{turb} \sim 400$ km s^{-1} and eddy scale, $l_{eddy} \sim 50-500$ kpc [34]. Alfvén waves are generated by the fluid turbulence through the Lighthill mechanism. Via wave-particle resonant scattering the turbulent energy is channeled mostly into CR protons, but CR electrons can be accelerated by the same process. Studies of turbulent acceleration of CR protons and electrons via resonant Alfvén waves have been quite successful in explaining observed characteristics of radio halos [7, 13].

Recent MHD simulations for the evolution of turbulences and magnetic fields in the ICM showed that the fluctuation dynamo generates μG level random magnetic fields during the epoch of cluster formation and major mergers [42]. Turbulences generated by subclusters of $M_{sub} \sim 3 \times 10^{13} M_{\odot}$ have a characteristic velocity, $v_{turb} \sim 300$ km s^{-1}, resulting in intermittent fields of $B \sim 2-4\mu G$ with a coherent scale of 20-30 kpc and with a volume filling factor $f_v \sim 0.1-0.2$. Magnetic field energy is likely in equipartition with turbulent energy, $\epsilon_B \sim \epsilon_{turb}$ [42].

From turbulent acceleration and dynamo, the energy budget among different components is roughly $E_{CR,p} \sim E_{turb} \sim E_B \sim 0.1 E_{thermal}$.

4.6. *Primary vs. Secondary Electron Models*

The origin of radio emitting electrons in clusters with radio halos on $\sim$ Mpc scale calls for some continuous, *in situ* reacceleration, since the electron cooling time scale is much shorter than the cluster dynamical time. *Primary* electrons are directly accelerated by merger shocks or by the ICM truculence driven by major mergers. Seed particles are likely relic electrons injected by previous activities such radio jets, galactic winds, and structure formation shocks. It is often argued that cluster *radio relics* are due to shock acceleration, while cluster *radio halos* are produced by turbulent acceleration [40]. The primary electron model is consistent with the fact that diffuse radio halos and relics are found in clusters with some signatures of recent merger events. *Secondary electrons*, on the other hand, are produced by inelastic collisions of CR protons with thermal protons through production and decay of charged pions ($p + p \rightarrow p + p + n\pi$, $\pi^{\pm} \rightarrow e^{\pm}$). In this model the CR protons are accelerated by various shocks or by turbulence in the ICM. So the model does not rely on unknown electron acceleration processes.

5. Summary

There are growing observational evidences indicating that the intergalactic medium in large scale structure is strongly magnetized with ~0.1-1 microgauss fields, and that the intracluster medium may contain significant amounts of CR ions and electrons in addition to hot thermal plasma. In this paper we review several astrophysical sources of CRs and magnetic fields in galaxy cluster environments, all of which can inject $E_{CR} \sim E_B \sim 10^{61} - 10^{62}$ ergs into the ICM.

- Galactic winds driven by supernovae provide CRs and magnetic fields as well as metals into the ICM. Termination shocks of galactic winds are the main acceleration sites [44].
- AGNs inject order of 10 % of black hole infall energy in the form of nonthermal components via jets [24]. Hot spots formed near termination shocks of radio jets can accelerate CR protons to ultra-high energies.
- Hierachical mergers induce shocks and turbulence in the ICM. CRs are accelerated by merger shocks (Fermi 1st order) and by turbulence (Fermi 2nd order) via resonant scattering with Alfvén waves [40]. Magnetic fields can be amplified by turbulent dynamo [42].
- Accretion shocks and merger shocks induced by gravitational collapse during structure formation can generate magnetic field via Weibel instability and Biermann battery mechanism and accelerate CRs by diffusive shock acceleration [25, 29, 20]. The gravitational infall motions along the filaments/sheets stretch and amplify magnetic fields [38].
- Radio emitting electrons could be secondary electrons from inelastic collisions of CR protons with the ICM gas. These CR protons are accelerated by various shocks listed above and the have much longer life time than CR electrons [12].
- Most acceleration processes work more efficiently for protons than for electrons, leading to the proton-to-electron ratio, $K \sim 100$ [1, 2].

Obviously a combination of several processes could operate together in a specific cluster, depending on the dynamical state and AGN and starburst activities. AGN input and galactic winds could have been the most dominant sources at high redshifts, while merger shocks and turbulent acceleration might be more recent, on-going processes. Although different sources predict slightly different energy partition among CR protons, electrons, magnetic fields, turbulence and thermal gas, the energy budget of the ICM can be summarized approximately as follows:

$$E_{CR,p} \sim (0.1 - 0.5) E_{thermal} \tag{2}$$

$$E_{CR,e} \lesssim 0.01 E_{thermal} \tag{3}$$

$$E_{turb} \sim E_B \sim 0.1 E_{thermal} \tag{4}$$

This work was supported by the Korea Research Foundation Grant funded by Korea Government (MOEHRD, Basic Research Promotion Fund) (R04-2004-000-100590).

References

1. R. Beck and M. Kraus , *Astron. Nachr.* **326**, 414 (2005).
2. E.G. Berezhko and H. J. Völk, *A&A*, **427**, 525 (2004).
3. V. Berezinksy *et al.* , *ApJ*, **487**, 529 (1997).
4. R. D. Blandford, and D. Eichler, *Phys. Rept.*, **154**, 1 (1987).
5. P. Blasi, *Jour. of Korean Astro. Soc.*, **37**, 483 (2004).
6. S. Bowyer *et al.* , *ApJ*, **605**, 168 (2004).
7. G. Brunetti *et al.* , *MNRAS*, **350**, 1174 (2004).
8. C.L. Carilli, and G.B. Taylor, *ARAA*, **40**, 319 (2002).
9. T. E. Clarke, *Jour. of Korean Astro. Soc.*, **37** 337 (2004).
10. S. Colgate *et al.* , *Phys. Plasmas*, **8**, 2425 (2001).
11. L. O'C. Drury, *Rept. Prog. Phys.*, **46**, 973 (1983).
12. T. Ensslin, *Jour. of Korean Astro. Soc.*, **37** 439 (2004).
13. Y. Fujita *et al.* , *ApJ*, **584**, 190 (2003).
14. R. Fusco-Femiano *et al.* , *ApJ*, **602**, L73 (2004).
15. S. Gabici, and P. Blasi, *ApJ*, **583**, 695 (2003).
16. G. Giovannini, and L. Feretti, *Jour. of Korean Astro. Soc.*, **37** 323 (2004).
17. L. Feretti *et al.* , *Jour. of Korean Astro. Soc.*, **37** 315 (2004).
18. M. Henriksen, and D. Hudson, *Jour. of Korean Astro. Soc.*, **37** 299 (2004).
19. S. Inoue *et al.* , *ApJL*, **628**, L9 (2005).
20. H. Kang *et al.* , *ApJ*, **456**, 422 (1996).
21. H. Kang *et al.* , *MNRAS*, **286**, 257 (1997).
22. H. Kang, and T.W. Jones, *ApJ*, **620**, 44 (2005).
23. H. Kang, and T.W. Jones, *Astroparticle Phys.*, **25**, 246 (2006).
24. P. P. Kronberg, *Jour. of Korean Astro. Soc.*, **37**, 501 (2004).
25. R. M. Kulsrud *et al.* , 1997, *ApJ*, **480**, 481 (1997).

26. A. Loeb, and E. Waxmann, *Nature*, **405**, 156 (2000)
27. S. G. Lucek, and A. R. Bell, *MNRAS*, **314**, 65 (2000).
28. M. A. Malkov, and L.O'C. Drury, *Rep. Progr. Phys.*, **64**, 429 (2001).
29. M. V. Medvedev *et al.* , *ApJL*, **642**, L1 (2006).
30. F. Miniati *et al.* , *ApJ*, **542**, 608 (2000).
31. F. Miniati, F., *MNRAS*, **337**, 199 (2002)
32. F. Miniati, F., *Jour. of Korean Astro. Soc.*, **37**, 465 (2004).
33. C. Norman *et al.* , *ApJ*, **454**, 60 (1995).
34. M. L. Norman, and G. L. Bryan, *Lect. Notes in Phys.*, **530**, 106 (1999).
35. O. Reimer *et al.* , *ApJ*, **588**, 155, (2003).
36. O. Reimer, *Jour. of Korean Astro. Soc.*, **37**, 307 (2004).
37. L. Rudnick, *Jour. of Korean Astro. Soc.*, **37**, 329 (2004).
38. D. Ryu *et al.* , *A&A*, **335**, 19 (1998).
39. D. Ryu *et al.* , *ApJ*, **593**, 599 (2003).
40. C. L. Sarazin, *Jour. of Korean Astro. Soc.*, **37**, 438 (2004).
41. C. L. Sarazin and R. Lieu, *ApJL*, **494**, L177 (1998),
42. K. Subramanian *et al.* , *MNRS*, **366**, 1437 (2006).
43. H. J. Völk *et al.* , *A&A*, **433**, 229 (2005).
44. H. J. Völk and A. M. Atoyan, *Astroparticle Phys.*, **11**, 73 (1999).

CHARACTERIZATION OF MICROWAVE CONTINUUM EMISSION FROM UHECR EXTENSIVE AIR SHOWERS

B. T. STOKES, N. G. LEHTINEN, P. W. GORHAM, AND G. S. VARNER
University of Hawai'i at Mānoa
Department of Physics and Astronomy
Honolulu, HI 96822, USA

We describe our investigative work observing microwave emission from ultra-high energy cosmic ray (UHECR) extensive air showers. This work has consisted primarily of RF observations made at accelerator facilities and the deployment of an atmospheric detector in Honolulu, Hawai'i. While this technique appears promising, further verification will be needed in the form of coincident observations with an already accepted method for UHECR detection

1. Introduction

The origin and nature of the ultra-high energy cosmic rays (UHECRs) remains one of the enduring mysteries of experimental particle astrophysics. In spite of well over four decades of observations of 10^{20} eV UHECRs by many different experiments[1], we still do not have a confirmed astrophysical source for these particles, nor do we understand their composition in any detail, nor do we know how they propagate from their unknown sources to earth[2]. As the highest energy subatomic particles observed in nature, UHECRs must arise from the most energetic phenomena in our universe– one of the most compelling reasons to redouble our efforts. Their study is thus crucial to understanding the nature of acceleration processes that can attain energies some seven orders of magnitude higher than is currently achievable in the laboratory[3,4,5].

While there is general agreement among the different experiments as to the global properties of the UHECR spectrum, there is still significant disagreement and uncertainty on absolute flux scales and on some of the fundamental questions of UHECR research. The two primary techniques of UHECR observation, ground-based coincidence arrays and optical fluorescence detectors both suffer from tangible limitations. In the case of ground arrays, only a single slice of EAS longitudinal development can be observed.

This means that estimates of primary particle energy and composition require extrapolation via model-dependent estimates, which often disagree depending on the model used. While the optical fluorescence method enables one to observe longitudinal as well as transverse shower development, it is highly constrained by the fact that it can only work on clear, moonless nights. This leads to a net duty cycle of 5-10% and requires using large numbers of expensive, fragile photomultiplier tubes in remote observation stations[6]. Furthermore, because the highest energy events are observed at increasingly large distances, even small fluctuations in atmospheric aerosol contamination can have substantial effects on energy estimation.

An air shower dissipates virtually its entire energy budget through ionization, producing an initial tenuous plasma with an electron temperature of order 10^5 K or more. The ionization and subsequent de-excitation of molecular nitrogen in the N_2^{+*} 1N and 2P states leads directly to the optical N_2 fluorescence now observed. The hot air shower plasma cools rapidly on nanosecond time scales, distributing its thermal energy through collisions with the neutral molecules, primarily N_2, which has the largest cross section and number density. This rapid cooling process leads to additional excitation of rotational, vibrational, electronic valence, and other modes of kinetic energy distribution among molecules, many of which can also lead to subsequent emission. In turn, the hot electrons themselves, while producing this excitation, can produce their own emission, such as continuum bremsstrahlung emission, or recombination radiation. The fraction of total radiated energy in optical fluorescence, compared to the total available energy budget for secondary radiation, is very small, leaving much possible radiative energy still unaccounted for. Yet the possibilities for observing secondary air shower plasma emission other than optical fluorescence have not yet been explored in any detail. We have already been exploring in detail these alternatives in possible microwave emission; in this document we propose to take this exploration to the next level.

To investigate the possibility of other channels for EAS observations we propose to search for EAS microwave molecular bremsstrahlung radiation (MBR)[7], using the Air-shower Microwave Bremsstrahlung Experimental Radiometer (AMBER) array. MBR detection could potentially yield advantages comparable to those of optical fluorescence without the shortcomings associated with weather and limited duty cycle. By observing MBR, one is observing an EAS from the same perspective as with optical fluorescence via energy-loss processes that are closely related to the excitation of molecular nitrogen that leads to air fluorescence. However, observations can occur 24 hours per day, and at the microwave bands of interest there is virtually no attenuation due to atmospheric contamination from aerosols or clouds. Furthermore, commercially designed microwave reception equipment can be easily weatherproofed, and future arrays would most likely be able to employ off-the-shelf satellite television components, taking advantage of the tremendous economy of scale in wireless and satellite television technology. Following validation of the technique in coincidence with an existing EAS installation, MBR detectors could be potentially deployed as standalone UHECR observatories.

2. Accelerator beam tests

In two accelerator experiments designed to measure MBR from shower ionization regions, the total emission substantially exceeded the expectations of simple MBR. The results appear to lend support to the theory, while at the same time indicating that additional emission mechanisms, or suprathermal extensions to the MBR, may also be operative.

2.1. *AWA INCOBREMS*

In June 2003, the INCOBREMS experiment was performed at the Argonne Wakefield Accelerator (AWA). For this experiment, energetic γ-rays were produced by bombarding several thicknesses of tungsten target with a 12 MeV electron beam with pulse charges of 0.1-1 nC. The γ-rays were then used to ionize air in a 1 m^2 copper anechoic Faraday chamber equipped with C, Ku, and Ka-band antennae, producing the equivalent ionization of a PeV electromagnetic shower inside the Faraday chamber. Figure 1 shows a schematic view of the general layout in the two experiments.

In this experiment, both incoherent and phase-stable (at least partially coherent) emission components were observed for tens of nanoseconds after beam passage. The total energy of the incoherent component was consistent with thermal MBR, but the excess power of the partially coherent emission was not predicted. Problems with beam scraping backgrounds necessitated complex background subtractions, leading to doubts about the reliability of the results, so another experiment with a cleaner beam was scheduled at SLAC.

2.2. *SLAC T471/E165*

In the following year, a similar experiment, T471, was performed at the Stanford Linear Accelerator Center. The configuration of this experiment

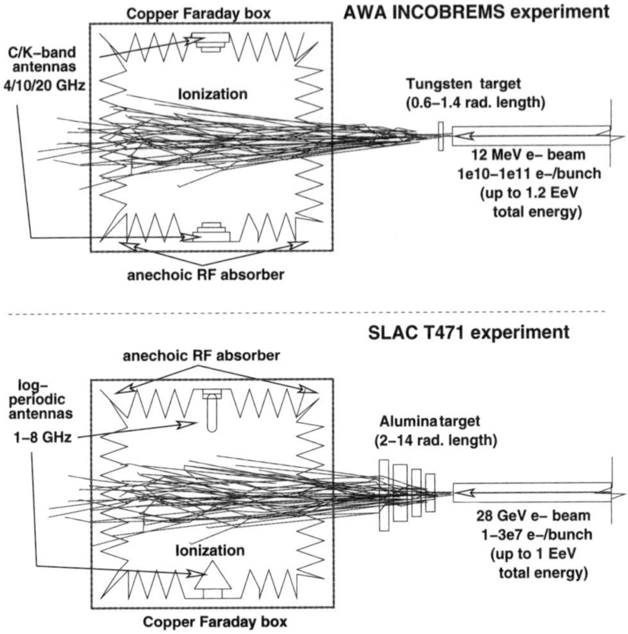

Figure 1. *Schematic of AWA INCOBREMS (top) and SLAC T471 (bottom) experiments, which used electron beams to shower in either Tungsten or alumina targets to produce ionization inside an anechoic Faraday chamber, observed by internal antennas.*

was largely the same as that of INCOBREMS, but additional precautions were taken against EMI and beam backgrounds, and verified in lab and beam calibration tests. This experiment was coordinated to be operated just downstream of the E165 FLASH experiment, which was used to do precise calibration of air fluorescence for the HiRes collaboration[8]. The SLAC T471/E165 experiments also used a precisely controlled, 28 GeV electron beam which was collided with a target consisting of 90% Al_2O_3 and 10% SiO_3 to make showers with varying particle number, from 0 to 14 radiation lengths of material.

While the analysis of this experiment is ongoing, these preliminary results indicate that once again a signal significantly in excess of the theoretical prediction was observed. However, as noted above, theoretical MBR predictions assume quasi-equilibrium Maxwellian thermal velocity electron populations, since this gives radiative balance in the analysis of the Einstein coefficients. Under non-equilibrium conditions, an electron population inversion in the ionized region is possible, and this can lead potentially to

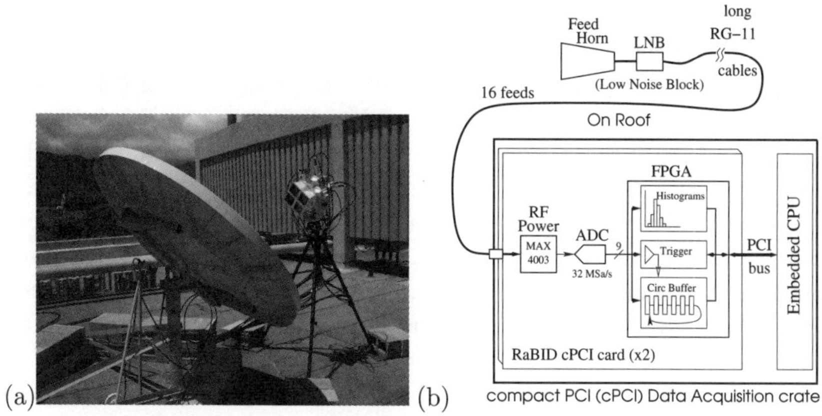

Figure 2. The AMBER System—(a) Prototype AMBER telescope and feed array on the roof of the physics building at UHM; (b) AMBER detector readout chain. The feed horn signals are amplified and down-converted in a Low Noise Block and then transmitted to a pair of RaBID cards for processing. See text for details.

stimulated emission[9]. Such inverted populations have been observed in discharge experiments in molecular nitrogen plasmas[10,11].

3. The AMBER system

We are currently developing a prototype of a system that could be used to search for detectable microwave emission from actual air showers. This system is built around a custom compact-PCI digitizer and data acquisition system, which we designate the Radio Bremsstrahlung Impulse Detector (RaBID). We have chosen the components and size of the prototype system such that it can be duplicated at low cost with mostly commercial parts. The proposed system, incorporating the RaBID prototype, is designated the AMBER for Air-shower Microwave Bremsstrahlung Experimental Radiometer. AMBER is currently operating on the rooftop of Watanabe Hall at the University of Hawaii at Mānoa (UHM) in Honolulu, Hawaii, pictured in Figure 2a. In its current configuration, the AMBER unit consists of a dual-band (C and Ku), dual-polarization feed horn array at the prime focus of a 1.8 m off-axis parabolic dish. The array is in a diamond-shaped configuration where each feed is $\sim 5.2^\circ$ from its nearest neighbor. Each feed produces four channels of signal which are amplified and down-converted in Low Noise Blocks (LNB) and then conveyed to the RaBID DAQ via RG11 coaxial cable, as shown in Figure 2b. The RaBID DAQ consists of a pair of

RaBID cards located inside a compact PCI (cPCI) crate, along with a cPCI CPU for data collection and logging. At the RaBID card input, the down-converted LNB outputs are measured with RF power monitor (MAX4003) chips, which provide output proportional to the received RF power, with approximately 70 ns integration time. This power level is sampled with a 16MSa/s ADC and processed inside a Field Programmable Gate Array.

These digitized samples are processed in 3 parallel paths: (1) 2048 sample C-band records are logged into a software histogrammer, which allows optimum threshold-riding with varying background (2) a trigger threshold is set based upon the histogram values; and (3) a circular buffer holds the samples in time sequence to be read out into the CPU upon detection of a trigger condition. In order to avoid biases in the triggering, each feed horn channel (of 12 total) is triggered separately, at minimum possible threshold, and the trigger times (corresponding to different transit times across the array field of view) are analyzed in the stored data. All sample times are recorded with respect to a common clock, which is synchronized to GPS via Network Time Protocol. An external trigger port (not shown) is available for forcing readout when observing in conjunction with another detector.

3.1. *Plan of Deployment*

We are currently proposing to fabricate and deploy a four-antenna AMBER array to coincide with the early stages of the scheduled deployment of the first Telescope Array (TA)[12,13] fluorescence detector in Delta, Utah in early 2006. At this point TA will consist of one $30° \times 120°$ fluorescence detector overlooking an array of 100 scintillator units.

References

1. J.Linsley, Phys. Rev. Lett. **10** 146 (1962).
2. J. W. Cronin, Nucl. Phys. Proc. Suppl. **138**, 465 (2005)
3. D. J. H. Chung, G. R. Farrar and E. W. Kolb, Phys. Rev. D **57**, 4606 (1998)
4. T. Stanev, Astrophys. J. **479**, 290 (1997)
5. F. W. Stecker and S. L. Glashow, Astropart. Phys. **16**, 97 (2001)
6. S. C. Corbato *et al.*, Nucl. Phys. Proc. Suppl. **28B**, 36 (1992).
7. R. O. Hundley, Tech. Rep. RM-3334-ARPA, The RAND Corporation (1962).
8. J. Belz *et al.* 2005, [arXiv:astro-ph/0507379].
9. G. Bekefi, *Radiation Processes in Plasmas* (Wiley, New York, 1966).
10. H. Singh & D. B. Graves, J. Appl. Phys. **87** 4098 (2000).
11. U. Kortshagen, I. Pukropski, & M. Zethoff, J. Appl. Phys. **70** 2040 (1994).
12. http://www.physics.utah.edu/~kai/ta.html
13. http://taws100.icrr.u-tokyo.ac.jp

COSMIC RAYS AT THE KNEE

T. K. GAISSER

Bartol Research Institute and Department of Physics and Astronomy, University of Delaware, Newark, DE, 19716 USA

Several kinds of measurements are combined in an attempt to obtain a consistent estimate of the spectrum and composition of the primary cosmic radiation through the knee region. Assuming that the knee is a signal of the high-energy end of a galactic cosmic-ray population, I discuss possible signatures of a transition to an extra-galactic population and how they might be detected.

Keywords: Cosmic Rays.

1. Introduction

The cosmic-ray spectrum extends from the sub-GeV region to at least 10^{11} GeV (Fig. 1.) Up to 100 GeV and somewhat higher, measurements with magnetic spectrometers flown above most of the atmosphere provide good momentum resolution along with identification of the charge and mass of individual primaries. Measurements with calorimeters continue to identify individual primaries to beyond 100 TeV, but with larger systematic uncertainties in the energy assignment. In the PeV region and beyond, the cosmic-ray intensity is too low for direct measurements; only indirect measurements of air showers from the ground are possible. Since the particles are not identified on an event-by-event basis, the energy spectrum derived from measurements of air showers is given as an "all-particle" spectrum, in terms of energy per particle rather than energy per nucleon. In the air-shower regime, identification of the primary mass is made in one of several indirect ways on a statistical basis, complicating the search for features in spectra of individual elements.

The gyroradius of a proton in a typical galactic magnetic field of 3 μGauss is about half a parsec at one PeV. The parsec scale is also typical of the size of structures in the interstellar medium driven by supernova explosions. For particles with higher energy and larger gyroradii, diffusion in the interstellar medium may become less efficient. Estimates of the

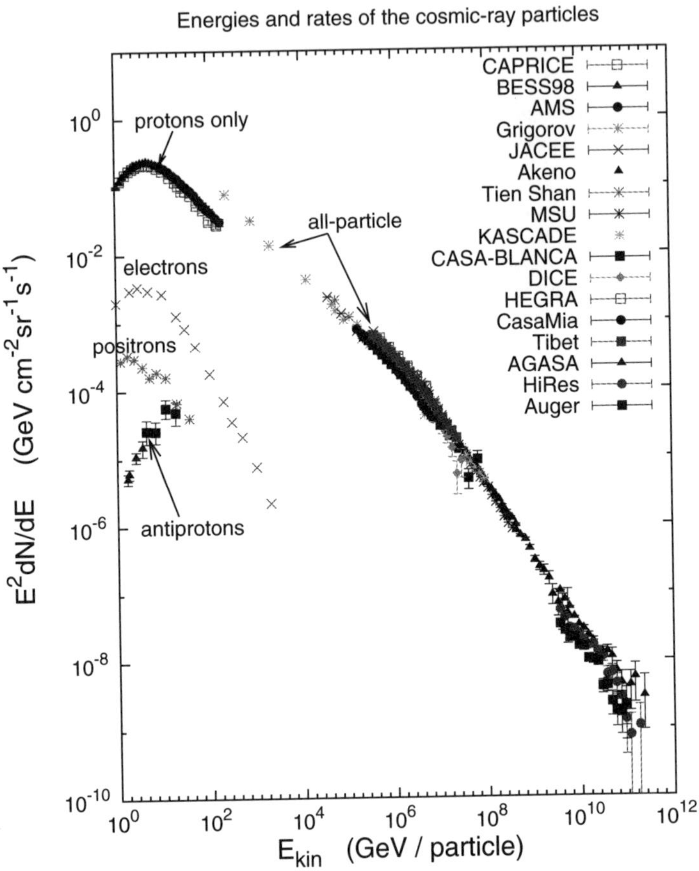

Fig. 1. Spectrum of cosmic rays.

maximum energy of particles accelerated at supernova shocks are around the same energy. It is therefore natural to guess that the knee of the cosmic-ray spectrum around 3 PeV has something to do with the end (or at least the beginning of the end) of the galactic cosmic-ray spectrum.

Peters[1] described the consequences for energy dependence of the primary composition if the spectrum is characterized by a maximum rigidity, R_c, which could be associated either with propagation or with acceleration (or both). The relation between rigidity and total energy is

$$R = \frac{Pc}{Ze}, \tag{1}$$

where $P\,c \approx E_{\rm tot}$ is the total energy of a nucleus of charge Z and mass A. If cosmic rays are classified by energy per particle, as is the case for air shower measurements, then the spectrum should steepen first for protons, then for helium, then for the the CNO group etc. Elsewhere[2] I have called this sequence the "Peters cycle". Several air shower measurements show some evidence that the spectrum becomes progressively enriched in heavy nuclei through the knee region. The clearest evidence for the Peters sequence for several groups of nuclei comes from analysis of the KASCADE experiment.[3]

2. Comparison of direct and indirect measurements

Data from direct measurements above 100 TeV are sparse, leaving a gap that is bridged by emulsion chamber data with low statistics below a PeV and by the threshold region of air shower experiments above 100 TeV. Apart from the technical problems of low statistics and systematic threshold effects, there is also the problem that air shower experiments do not identify individual primary nuclei. In addition, the efficiency of ground arrays depends strongly on primary mass in the threshold region. Nevertheless, it is possible to form a fairly consistent picture of the cosmic-ray spectrum up to the knee.

Figure 2 shows several measurements of the primary spectrum through the knee region. Up to 100 GeV there are good measurements of the spectra of protons and helium with magnetic spectrometers flown in spacecraft[4] and high-altitude balloons.[5,6] Using the AMS[4] and BESS[5] measurements, together with measurements of heavier nuclei at 10 GeV/nucleon,[7] I have converted the spectra of protons, helium, CNO, Ne-Si and Fe from energy per nucleon to total energy per particle and combined them to give the all particle spectrum at low energy (shown as the solid line in Fig. 2).

The highest energy direct measurements in which individual nuclei are identified directly are from emulsion chamber experiments. Measurements of protons and helium from RUNJOB[8] and JACEE[9] are shown in Fig. 2a. Preliminary data on protons and helium from the ATIC thin ionization chamber[11] (not shown here) are consistent with a smooth power-law extrapolaton between the BESS and AMS spectromter data at low energy and the RUNJOB data above 10 TeV. (The JACEE helium data are higher.) The solid line normalized to the all-particle spectrum at 100 GeV and extrapolated to 1 PeV with an $E^{-2.7}$ power law is consistent with the all-particle measurements of Grigorov[12] but about a factor 1.5 above the sum of the RUNJOB spectra at 100 TeV. (Using the higher JACEE measurement of helium would bring the emulsion chamber measurements into agreement

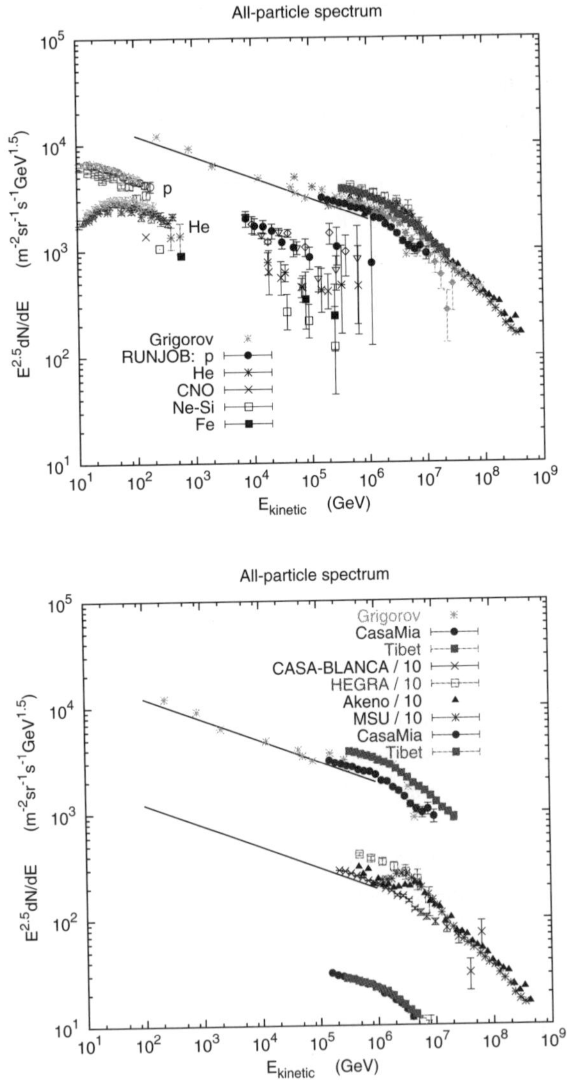

Fig. 2. All-particle spectrum. Upper panel illustrates an extrapolation from direct measurements at low energy to 1 PeV. Lower panel shows detail of air shower measurements in the knee region. The two lower sets of points are offset respectively by 1/10 and by 1/100 for clarity (see text).

with the $E^{-2.7}$ extrapolation at 100 TeV.) An external motivation for using the hard extrapolation from the spectrometer measurements comes from the Super-K measurement of the flux of atmospheric neutrinos, where the best fit to the data requires an even harder extrapolation of the primary spectrum above 100 GeV.[10]

Even with the hard $E^{-2.7}$ spectrum, the extrapolation of the direct measurements comes in somewhat below the air shower measurements. Moreover, the all-particle spectrum of RUNJOB[13] is somewhat below the lowest-energy air shower data, so there may be some systematic offset between the direct measurements and the measurements of air showers. There is also a slight correlation between hardness of the fitted spectrum and primary mass in this energy range.[14] There is, however, no sign[15] of the sharp steepening of the proton spectrum, below 100 TeV, which would be the case if the maximum rigidity accessible in most galactic cosmic-ray accelerators were around 100 TV or below as originally estimated for diffusive shock acceleration by expanding supernova remnants.[16]

3. Modelling the knee

Although there are systematic differences among the various air shower measurements in the knee region, all show that the spectrum steepens from -2.7 or slightly harder below 1 PeV to -3 above 10 PeV. Figure 2b shows several results. Two of the measurements[17,18] show evidence of structure in the knee region, while the others[19–22] do not. The figure also shows the spectra of two of the measurements (CASA-MIA and Tibet) with the energy assignments of the latter shifted down by a factor of 0.8, which may be taken as an indication of systematic uncertainties in energy assignment. The shapes of these two spectra agree well with each other.

Some authors[23] have suggested that a single source contributes significantly to the flux of cosmic-rays in the knee region. The possible offset between the air-shower measurements and the extrapolation of the direct measurements leaves room for such a possibility. Others[24,25] argue, however, that the overall smoothness of the spectrum indicates a single population, with a secondary acceleration mechanism boosting some of the cosmic rays accelerated by galactic supernova explosions to higher energy. Several authors[26–28] approach the problem from the high-energy end by modelling the highest energy spectrum as arising from a cosmological distribution of sources and subtracting the extra-galactic population from a model of the supernova-accelerated population to see if an extra, high-energy galactic source is required to fill in the energy region above the knee ("population B"[26]). This is an open question at present. (For a recent review and further references see .[29])

In any case it is interesting to estimate the power required to produce the high-energy end of the galactic cosmic-ray spectrum. To do so, I assume a model of galactic propagation with diffusion characterized by an equivalent leaky box model with a escape time given by

$$\tau_{\rm esc} \;=\; 2\times 10^{7}\,{\rm yrs}\times E^{-0.33} \qquad (2)$$

which applies for all energies. With this propagation model (motivated by lack of anisotropy at high enery), the source spectrum at low energy must be $E^{-2.37}$ to give the observed $E^{-2.7}$ spectrum inside the galaxy. Assuming a maximum rigidity of 1 PV for this low-energy component, with a composition as measured by an emulsion chamber experiment,[8] the assumed low-energy component can be subtracted from the observed spectrum. The total power required to account for the observed spectrum up to 1 EeV is then $\sim 2\times 10^{39}$ erg/sec. Such an estimate is obviousy very model dependent, but not unreasonably large compared to what might be available in individual galactic sources.

4. Primary composition from air shower measurements

What is needed to make progress is a precise knowledge of the energy-dependence of the major groups of nuclei (p, He, CNO, heavy) through the knee region. If there are several important groups of sources with different maximum rigidities then there should be a corresponding sequence of Peters cycles, perhaps characterized by different compositions. The transition to extra-galactic cosmic-rays would be characterized by a transition from heavy nuclei (from the highest galactic source) to the light component of a cosmological distribution of sources. A recent summary of composition measurements with air showers in the knee region is given in the rapporteur paper of Matthews at the last ICRC[30] (See also Swordy *et al.*[31]) Overall, the evidence suggests a change in composition toward hearvier primares at higher energy as expected. However, results of different measurements disagree in detail and the picture is not very clear. As noted in the introduction, the best evidence for the sequence of increasingly heavier groups of nuclei comes from the KASCADE experiment,[3] which uses the ratio of muons to electrons as a probe of primary composition.

Ratio of muons to electrons in the shower front is a classic probe of primary composition. At each interaction of the nucleons in an air shower, roughly 1/3 of the energy not retained by the projectile nucleon is transferred to the electromagnetic component of the cascade via $\pi^0 \to \gamma\gamma$ and

2/3 to charged pions. The charged pions either reinteract or decay depending on their energy and the depth (and hence the density) in the atmosphere where they are produced. Charged pions that interact contribute further to the electromagnetic component while those that decay feed the muon component of the air shower. Comparing nuclei of the same total energy, charged pions reach the energy at which they can decay earlier in the cascade for heavy primaries than for protons because the initial energy per nucleon is lower by E_{total}/A. As a consequence, the ratio of the muonic to the electromagnetic component of an air shower is larger for heavy primaries.

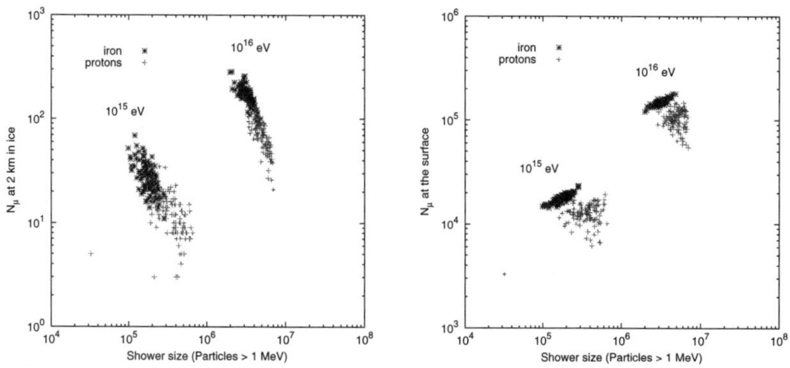

Fig. 3. Correlation between muons and shower size at the surface. Left panel, muons at 2 km in the ice; right panel, muons at the surface.

To measure this ratio with greatest sensitivity requires an observation near shower maximum so that the size of the shower is well correlated with the primary energy. In general, showers in the knee region and somewhat above are observed after shower maximum so that fluctuations in the relation between observed shower size and primary energy are important. KASCADE is a surface array near sea level so that these fluctuations are large.

Muons detected in KASCADE typically have energies of a few GeV. An alternative, realized with EASTOP-MACRO[32] and with SPASE-AMANDA,[33] is to sample the muon component with a deep underground detector and the electromagnetic component with an array on the surface above. Such a setup selects muons with sufficiently high energy at production to penetrate to the deep detector. The high-energy muons generally come more from the fragmentation region of phase space of the hadronic interactions in the shower, whereas than the low-energy muons reflect more the less well-understood central region. On the other hand, the multiplicity of high-energy muons is small so that fluctuations in muon number are more important. Some of these points are illustrated in Fig. 3 which compares N_μ vs. N_e for high-energy (> 0.5 TeV) muons and for low-energy muons in 10^{15} and 10^{16} eV showers simulated with CORSIKA. In general, the mass resolution is somewhat better for the low energy muons provided that the sampling is good enough to get a good measure of the muon number in individual showers, while the energy resolution is better for coincident events with high energy muons, particularly if the surface array is at high altitude. This complementarity was pointed out by Ralph Engel.[34]

It is interesting that analyses of both EASTOP-MACRO[32] and SPASE-AMANDA[33] suggest an increase in the mean primary mass in the decade of energy above a PeV, that is, through the knee region. As a muon detector, MACRO has the advantage of resolving muon tracks, whereas AMANDA reconstructs the light pool generated by the bundle, which gives a measure of the energy deposition of the muons. A limitation of both experiments is the relatively small sampling area of the deep muon detectors, which has to be accounted for in the comparison between data and simulations.

IceCube, with its surface component IceTop, now under construction at the South Pole[35] will have a much larger acceptance. Engineering data from the first year of operation of IceCube with four IceTop stations and one string demonstrate the ability of IceCube to reconstruct events with few nanosecond accuracy over distances of more than two kilometers.[36] Since January 2006 IceCube has been operating with 16 surface stations and 9 strings of detectors in the ice. Each string contains 60 digital optical modules (DOMs) evenly spaced in the clear ice between 1450 and 2450 meters below the surface. Each IceTop station consists of two ice Cherenkov tanks separate from each other by 10 meters and located 25 m from the top of the corresponding IceCube string. Each tank is instrumented with two of the same DOMs used in the ice. With a nominal grid spacing of 125 m, the acceptance of the current partial array is $\sim$ 1500 m^2sr, which allows detection of coincident events approaching 10^{17} eV. The acceptance of the full IceCube as a three-dimensional air shower array will be approximately 0.3 km^2sr, allowing measurement of coincident events up to an EeV. It is scheduled for completion in 2011 and will be operating in the meantime as new detectors are added.

5. Transition to extra-galactic cosmic rays

Analysis of measurements of energy-dependence of the depth of shower maximum with HiRes[37] suggest a change in composition from heavy to light as energy increases from 10^{17} to 10^{18} eV. Shower maximum is deeper in the atmosphere for protons than for heavy nuclei of the same total energy. The measured depth of shower maximum increases more rapidly than expected from model calculations, indicating an increasing fraction of protons as energy increases. A similar change was observed in analysis of the Fly's Eye data[38] but at a higher energy (above 10^{18} eV). It was recognized at the time that such a change of composition could be a signature of the transition from a population of galactic cosmic rays to an extra-galactic population. In view of the more recent data, the signature of a transition appears to occur at a lower energy (below 10^{18} eV). A discussion and references to cosmological scenarios in which the transition to extra-galactic cosmic rays occurs at relatively low energy is given by Hillas.[26]

Several new experiments are planned or in operation that can explore the energy region from the knee to the EeV region to overlap with the threshold region of the giant arrays Auger[39] and (formerly) AGASA.[40] The most advanced of these if KASCADE-Grande,[41] a large sea-level array that includes the original KASCADE as a subarray. IceCube is under construction as described in §4. Telescope Array[42] and its low-energy extension, TALE,[43] are under development in Utah. There is a propsed array of atmospheric Cherenkov detectors, TUNKA,[44] in Russia, and use of the radio technique for a large acceptance air shower detector is being explored.[45] Understanding the transition from a galactic to an extra-galactic population of cosmic rays is an interesting and important goal for the near future.

Acknowledgments

I thank Peter Niessen for the generating the data for Fig. 3 and Michael Hillas for enlightening discussions. This research is supported in part by the U.S. Department of Energy under DE-FG0291 ER 40626.

References

1. B. Peters, *Nuovo Cimento* **XXII**, 800 (1961).
2. T. K. Gaisser, *Proc. of the NATO Advanced Study Institute on Neutrinos and Explosive Events in the Universe* (Springer, 2005) pp. 3-31.
3. T. Antoni *et al.* (KASCADE), *Astropart. Phys.* **24**, 1 (2005).
4. J. Alcarez *et al.* (AMS), *Phys. Lett. B* **490**, 27 and **494**, 193 (2000).
5. T. Sanuki *et al.* (BESS), *Ap.J.* **545**, 1135 (2000).

6. M. Boezio *et al.* (CAPRICE), *Ap.J.* **518**, 457 (1999).
7. J. J. Engelmann *et al.*, *Astron. Astrophys* **233** 96 (1990).
8. A. P. Apanasenko *et al.* (RUNJOB), *Astropart. Phys.* **16**, 13 (2001).
9. K. Asakimori *et al.* (JACEE), *Ap.J.* **502**, 278 (1998).
10. Y. Ashie *et al.*, *Phys. Rev.* **D71** 112005 (2005).
11. H. S. Ahn *et al.* (ATIC), *Proc. 29th Int. Cosmic Ray Conf. (Pune)* **3** 57 (2005).
12. N. L. Grigorov *et al. Yad. Fiz.* **11**, 1058 (1970) and *Proc. 12th Int. Cosmic Ray Conf. (Hobart)* **2**, 206 (1971).
13. L. G. Sveshnikova *et al.* (RUNJOB), *Proc. 29th Int. Cosmic Ray Conf. (Pune)* **3**, 49 (2005).
14. M. Ichimura *et al.* (RUNJOB), *Proc. 29th Int. Cosmic Ray Conf. (Pune)* **3**, 21 (2005).
15. M. Hareyama *et al.* (RUNJOB), *Proc. 29th Int. Cosmic Ray Conf. (Pune)* **3**, 17 (2005).
16. P. O. Lagage & C. J. Cesarsky, *Astron. Astrophys.* **125**, 249 (1983).
17. M. Nagano *et al.* (AKENO), *J. Phys. G* **10**, 1295 (1984).
18. Yu. A. Fomin *et al.* (MSU), *Proc. 22nd Int. Cosmic Ray Conf. (Dublin)* **2**, 85 (1991).
19. M. A. K. Glasmacher *et al.* (CASA-MIA) *Astropart. Phys.* **10**, 291 (1999).
20. M. Amenomori *et al.* (Tibet), *Ap.J.* **461**, 408 (1996).
21. F. Arqueros *et al.* (HEGRA), *Astron. Astrophys.* **359**, 682 (2000).
22. J. W. Fowler *et al.* (CASA-BLANCA) *Astropart. Phys.* **15**, 49 (2001).
23. A. D. Erlykin & A. W. Wolfendale, *J. Phys. G* **27**, 1005 (2001).
24. W. I. Axford *Ap.J. Suppl.* **90**, 937 (1994).
25. H. J. Völk & V. N. Zirakashvili, *Proc. 28th Int. Cosmic Ray Conf. (Tsukuba)* **4**, 2031 (2003).
26. A. M. Hillas, *Proc. Cosmology, Galaxy Formation and Astroparticle Physics on the pathway to the SKA* Oxford (2006).
27. D. R. Bergman, *et al.* (HiRes) *Proc. 29th Int. Cosmic Ray Conf. (Pune)* **7**, 315 (2005).
28. R. Aloisio *et al.*, astro-ph/0608219.
29. A. M. Hillas, *J. Phys. G: Nucl. Part. Phys.* **31** R95 (2005).
30. J. Matthews, *Proc. 29th Int. Cosmic Ray Conf. (Pune)* **10**, 283 (2005).
31. S. Swordy *et al.*, *Astropart. Phys.* **18**, 129 (2002).
32. M. Aglietta *et al.*, (EASTOP-MACRO) *Astropart. Phys.* **20**, 641 (2004).
33. J. Ahrens *et al.*, (SPASE-AMANDA) *Astropart. Phys.* **21**, 565 (2004).
34. R. Engel, *Proc. Workshop on Physics at the End of the Galactic Cosmic Ray Spectrum* (2005).
35. J. Ahrens *et al.* (IceCube), *Astropart. Phys.* **20**, 507 (2004).
36. A. Achterberg *et al.* (IceCube), *Astropart. Phys.* (to be published) (2006) (astro-ph/0604450).
37. R. U. Abassi *et al.* (HiRes), *Astropart. Phys.* **23**, 157 (2005).
38. D. J. Bird *et al.* (Fly's Eye) *Phys. Rev. Lett.* **71**, 3401 (1993).
39. P. Sommers *et al.* (Auger) *Proc. 29th Int. Cosmic Ray Conf. (Pune)* **7**, 387 (2005).

40. M. Takeda *et al.* (AGASA), *Astropart. Phys.* **19**, 447 (2003).
41. A. Haungs *et al.* (KASCADE-Grande), astro-ph/0508286.
42. Y. Arai *et al.* (Telescope Array), *Proc. 28th Int. Cosmic Ray Conf. (Tsukuba)* **2**, 1025 (2003).
43. G. B. Thomson *et al.* (TALE), *Proc. 29th Int. Cosmic Ray Conf. (Pune)* **8**, 363 (2005).
44. N.M. Budnev *et al.* (TUNKA), *Proc. 29th Int. Cosmic Ray Conf. (Pune)* **8**, 255 (2005).
45. H. Falcke *et al.* (LOPES), *Nature* **435**, 313 (2005).

THE ANTI MATTER SPECTROMETER (AMS-02): A PARTICLE PHYSICS DETECTOR IN SPACE

ROBERTO BATTISTON†

Physics Department and INFN Section,
Perugia, 06123, Italy

On behalf of the AMS-02 Collaboration

AMS-02 is a space borne magnetic spectrometer designed to measure with accuracies up to one part in 10^9 the composition of Cosmic Rays near Earth. With a large acceptance (5000 cm^2 sr), an intense magnetic field from a superconducting magnet (0.7 T) and an accurate particle identifications AMS-02 will provide the highest accuracy in Cosmic Rays measurements up to the TeV region. During a three years long mission on the ISS AMS-02 will achieve a sensitivity to the existence of anti-He in the Cosmic Rays of one part in a billion as well as important informations on the origin of Dark Matter. We review the status of the construction of the AMS-02 experiment in preparation for the three years mission on the ISS.

1. Introduction

AMS is a particle physics experiment in space. The purpose is to perform accurate, high statistics, long duration measurements of the spectra of energetic (up to multi-TeV) primary cosmic rays in space. Some of the physics goals are:

- **Dark Matter:** There are many theoretical suggestions [1] that particles predicted by SUSY theories, for example the neutralino χ, are a component of the Dark Matter which constitutes one quarter of the mass of the universe. Collisions of dark matter in the galactic halo produce $\overline{p}$, e^+ and γ via:

$$\begin{aligned}\chi + \chi &\rightarrow \overline{p} + \ldots \\ &\rightarrow e^+ + \ldots \\ &\rightarrow \gamma + \ldots\end{aligned}$$

† Work partially supported by the grant I/021/05/0 of the Italian Space Agency (ASI)

The $\overline{p}$, e^+ and γ from these collisions will produce deviations from the smooth energy spectra. Therefore, the precision measurement of the $\overline{p}$, e^+ and γ spectra will enable AMS to establish whether SUSY particles are the origin of Dark Matter. There are also predictions that antideuterons can be produced from the collision of SUSY particles, at a level which AMS could detect [2].

- **Antimatter**: The strong evidence which supports the Big Bang theory of the origin of the universe requires matter and antimatter to be equally abundant at the very hot beginning [3]. The absence of sharp annihilation γ ray peaks excludes the presence of large quantities of antimatter within our local super cluster of galaxies. However there is no direct experimental evidence regarding the remainder (10^8) of the universe.

- **Cosmic rays**: AMS will collect ~ 10^9 nuclei and isotopes (D, He, Li, Be, B, C ...Fe). Among the interesting issues which will be studied are the accurate determination of the ratio of boron to carbon over a wide range of energies provides crucial information regarding the propagation of cosmic rays in the galaxy. In particular, the ratio of ^{10}Be (mean lifetime of 3.6 x 10^6 years) to the stable ^{9}Be will enable us to extend the low-energy measurements of the Ulysses satellite to higher energies and to provide important information on the understanding of cosmic ray propagation.

To perform a high accuracy measurement of the spectra of energetic charged particles in space, the AMS detector is based on experience and observations from experiments to study rare signals among intense backgrounds, such as the study of leptonic decays of vector mesons from $\gamma + N \rightarrow N + (\rho, \omega, \phi \rightarrow e^+e^-)$ at DESY [4] and the discovery of the J particle from $p + N \rightarrow J (\rightarrow e^+e^-) + ...$ [5] combined with the precision measurements made in the study of Z decays [6]. These experiments were successful because they have:

(a) Minimal material in the particle trajectory so that the material itself is not a source of background nor a source of large angle nuclear scattering;

(b) Many repeated measurements of momentum and velocity so as to ensure that background particles which had experienced large angle nuclear scattering from the detector itself be swept away by the spectrometer and not confused with the signal.

It was the strict adherence to these techniques that ensured that a background rejection of 10^{10} was indeed possible and made these experiments successful. AMS is designed following the same principles. Figure 1 shows the AMS-02 detector configuration for the International Space Station (ISS).

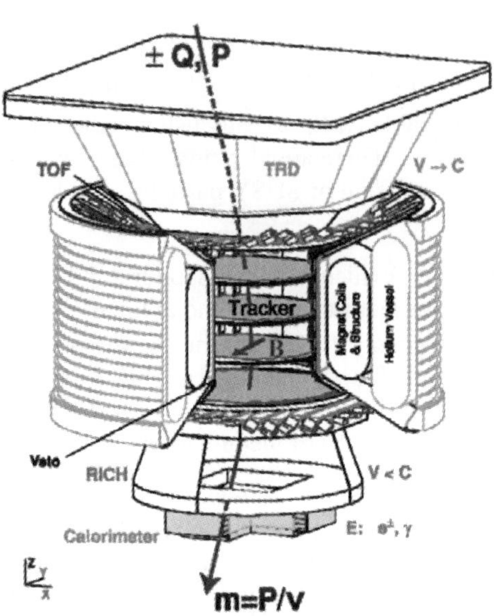

Figure 1: A TeV Detector in Space: AMS-02 on the Space Station.

AMS-02 will contain the following main components:

(1) A twenty layer Transition Radiation Detector (TRD) which identifies electrons and positrons with a measured rejection factor against hadrons of 10^3 to 10^2 from 1.5 GeV to 300 GeV.

(2) Four layers of Time of Flight (TOF) hodoscopes that provide precision time of flight measurements (~ 120 picoseconds) and dE/dX measurements.

(3) The superconducting magnet, which provides a bending power of $BL^2 = 0.86\ Tm^2$.

(4) Eight layers (6.45 m^2) of silicon tracker, which provide a proton rigidity (= momentum/charge) resolution of 20% at 0.5 TV and a helium (He) resolution of 20% at 1 TV and charge resolution of nuclei up to iron (Z=26).

(5) Veto, or anticoincidence, counters (ACC) which ensure that only particles passing through the magnet aperture will be accepted.

(6) A Ring Imaging Cerenkov Counter (RICH), which measures the velocity (to 0.1%) and charge |Z| of particles or nuclei. This information, together with the measurement of momentum in the

tracker, will enable AMS to unambiguously determine the mass of these particles and nuclei.

(7) A 3-D sampling calorimeter (ECAL) made out of 16.7 X_0 (radiation lengths) of lead and scintillating fibers which measures the energy of gamma rays, electrons and positrons and distinguishes electrons and positrons from hadrons with a rejection of 10^4 in the range between 1.5 GeV to 1 TeV.

(8) A system of two star trackers (AMICA) to allow the precise reconstruction of the origin of high energy gamma rays reconstructed in the detector.

Due to the limits in the available space, in the following sections we will discuss the characteristics of only a two among the subsystems of the AMS-02 apparatus, the Cryomagnet and the Silicon Tracker which are at the hearth of the magnetic spectrometer.

2. The Cryomagnet

The purpose of the superconducting magnet is to extend the energy range of our measurements of particles and nuclei to the multi-TeV region [7]. The magnet design was based on the following technical considerations:
(i) Experience in designing and manufacturing the AMS-01 magnet which was 10 times safer than stress limits allowed.
(ii) The result of many years of intensive R&D collaboration to design a magnet with the following properties :

- Identical field configuration to the AMS-01 magnet to maintain mechanical stability and follow NASA safety standards;
- Minimized heat loss (~100 mW) and minimized quench probability;

(iii) We have chosen to have the magnet built by experts from the Oxford Instruments R&D group, who has an excellent record to produce highly reliable magnets running in persistent mode without quench. This group has produced the 8 OSCAR magnets (2.36 Tesla) used in cyclotrons in Japan and England and which have operated for close to 30 years without any quench. This group also built the CLEO magnet at CORNELL and the CLAS torus at Jefferson Laboratory and the KLOE magnet in Frascati. All are large, high-field, special-purpose magnets which have operated for years without quench. To ensure that these experts are able to devote all their efforts to the construction of the AMS-02 magnet, we have supported a new company: Space Cryomagnetics Ltd., entirely staffed by the experts of the Oxford Instruments R&D group and entirely dedicated to the AMS-02 magnet.

Two magnets are being built. One is the flight magnet and the other is used for space qualification tests. The magnet has no magnetic field during the shuttle launch and landing and so there is no force among the coils, hence for the test magnet the coils are replaced by mass equivalents.
The magnet system, as shown in Figure 2, consists of superconducting coils, a superfluid helium vessel and a cryogenic system, all enclosed in a vacuum tank. Outside of the vacuum tank are supporting electronics, located in the Cryomagnet Avionics Box (CAB), valves and cabling. The vacuum tank is toroidal with inner diameter of 1.1 m, outer diameter of 2.7 m and a length of the central cylinder surrounding the tracker of 0.9 m. The magnet operates at a temperature of 1.8 K, cooled by superfluid helium stored in the vessel. It is launched at the operating temperature, with the vessel full of 2500 litres of superfluid helium. The magnet will be launched with no field, it will be charged only after installation on the ISS. Because of parasitic heat loads, the helium will gradually boil away throughout the lifetime of the experiment. After the project time of 3 to 5 years, the helium will be used up and the magnet will warm up and no longer be operable.

The coil system consists of a set of 14 superconducting coils arranged, as shown in Figure 2, around the inner cylinder of the vacuum tank. The coil set has been designed to give the maximum field in the appropriate direction inside the bore tube, while minimising the stray field outside the magnet.
The single large pair of coils generates the magnetic dipole field perpendicular to the experiment axis. The twelve smaller flux return coils control the stray field and, with this geometry, they also contribute to the useful dipole field. The magnetic flux density at the geometric centre of the system is 0.86 T. Table 1 summarizes some of the magnet parameters.

The superconducting wire was developed specifically to meet the requirements of the AMS cryomagnet [1]. The current is carried by tiny (22.4 μm diameter) filaments of niobium titanium (NbTi) which - given the magnetic flux and current densities within the coils – carries the current without resistance provided the temperature is kept below 4.0 K. Because pure NbTi has rather low thermal conductivity, it is prone to instability. This can be overcome if it is in intimate contact with a material which has a high thermal conductivity at the cryogenic operating temperature, such as a pure metal. For this reason, the NbTi filaments are embedded in a copper matrix, which is encased in high-purity aluminium. The copper is required for manufacturing reasons, but the aluminium is extremely conductive and much less dense, thus providing maximum thermal stability for aluminium weight. The diameter of the round copper strand at the centre is 0.76 mm, and the aluminium dimensions are 2.0 mm x 1.5 mm. The current density in the superconductor is 2300 A/mm^2 or 157 A/mm^2 including the aluminium.

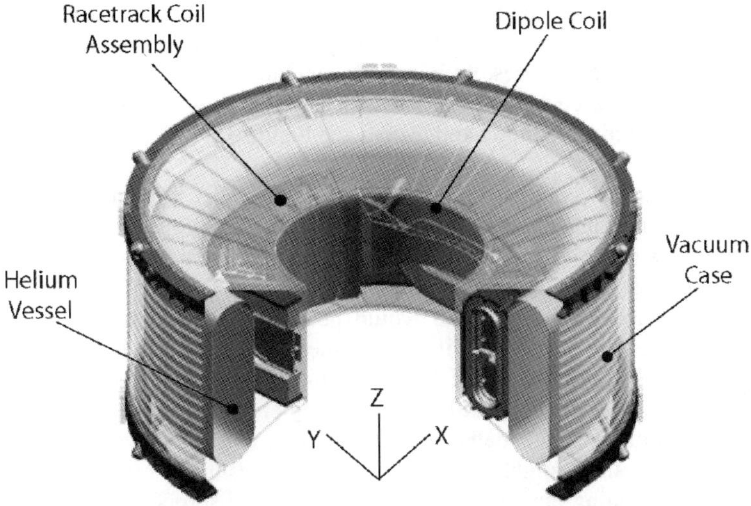

Figure 2: AMS-02 superconducting magnet layout.

Table AMS-02 Cryomagnet parameters.

Central Magnetic Field Bx (0,0) [T]	0.860
Dipole Bending Power [Tm^2]	0.862
Maximum Stray Magnetic Field at R=2.3 m [mT]	15.2
Maximum Stray Magnetic Field at Y=2.3 m [mT]	7.6
Maximum Stray Magnetic Field at R=3.0 m [mT]	3.9
Peak Magnetic Field on the Dipole Coils [T]	6.59
Peak Magnetic Field on the Racetrack Coils [T]	5.91
Maximum Torque in geomagnetic field [Nm]	0.272
Nominal Operating Magnet Current [A]	459.5
Stord Energy [MJ]	5.15
Nominal Magnet Inductance [H]	48

To manufacture the coils, the superconducting wire was first cleaned then insulated using 85 μm thick polyester tape. Each coil was wound separately onto an aluminium former from a single length of conductor before being impregnated under vacuum with epoxy resin. The impregnation process gives the coils mechanical integrity, and also provides electrical insulation between turns and layers. After completion, each coil is tested individually under conditions as representative as possible of flight, to test the integrity of the design and the quality of the build.

Each of the two larger (dipole) coils, which generate most of the useful field, has 3360 turns, and the 12 smaller (flux return) coils each have 1457 turns. The 14 coils are connected in series, with a single conductor joint between each pair of adjacent coils. The current when the magnet is operating is 459.5 A. All of the coils have been manufactured and tested (see Figure 3).

The coils are not coupled thermally. This means that a quench in one coil, leading to a rise in temperature, will not necessarily propagate to any of the others. If this was allowed to happen, the entire stored energy of the magnet (5 MJ) could be dissipated as heat in the coil which quenched. While the temperatures reached in that coil would not be dangerous, sufficient thermal stresses could be induced (by differential thermal contraction between parts of the coil) that the performance of the coil might be permanently reduced. For this reason, all the coils are constantly monitored by an electronic protection system. If the onset of a quench is detected in any coil, heaters are powered in the other coils to quench all 14 simultaneously. This distributes the stored energy between the coils, preventing any single coil from taking a disproportionate amount of energy which could otherwise result in degradation. The operation of these quench heaters is an important part of the testing and qualification procedure for the magnet coils.

Figure 3: Coil Assembly.

Each coil is subject to internal forces as a result of its own magnetic field. In general, these are burst forces, trying to expand the racetrack-shaped coil into a ring. These loads are in the plane of the coil, and are resisted by the former on which the coil was wound. In addition, each coil is attracted or repelled by all the other coils in the magnet. This leads to a relatively complicated load system on some of the coils, with forces perpendicular to the plane of the coil. The magnetic loads are quite large: the two dipole coils feel a net attraction to each other of around 250 tonnes. During individual coil testing, each coil is charged until some part of the winding is subject to the same force it will experience in flight.

The cold mass of the magnet is more than 2000 kg. This has to be supported from the experiment structure (in particular the vacuum tank), which is at ambient temperature (~270±40 K). The design of the support straps is therefore crucial, as they have to be able to carry the load without conducting significant heat across the large temperature gradient. Each consists of a pair of composite bands connected in parallel. One band is thin, with low stiffness and strength, and is permanently connected between the cold mass and the vacuum tank. The other band is much thicker and stronger, but possesses a passive disconnect feature. This means that it only forms a thermal path between the vacuum tank and the magnet during launch and landing. At other times (when it is not needed), differential thermal contraction between the bands and the removal of the high inertial load cause the disconnect to open, dramatically reducing the thermal conduction of the support. A total of 16 straps support the magnet from the vacuum tank. During normal operations on the ground or in space, only the low-stiffness band is engaged, and the heat conduction is very low (less than 3 mW per support). During launch, the high-stiffness band engages as well. The conducted heat load is much higher but because the launch takes only a few minutes the effect on the overall endurance of the system is not significant. The dual stiffness characteristic of the straps makes their behaviour non-linear and, as major structural components of the magnet system, they have been subject to special scrutiny and testing, particularly testing to failure to understand the safety margins.

The cooling of terrestrial superconducting magnets using liquid helium is a well-established technology, but there is much less experience of helium cryogenics in space. The cryogenic system for the AMS magnet combines technologies from terrestrial magnet cryogenics and space cryogenics to meet the particular challenges of the Space Shuttle launch and the environment of the ISS [8]. It maintains the magnet at a temperature of 1.8 K, under all operating conditions, for the duration of the experiment. Liquid helium (^{4}He) can exist in two forms. Normal liquid helium behaves in a conventional manner. But if it is cooled below 2.17 K, some of its properties change dramatically as it becomes a superfluid. In particular, its viscosity falls almost to zero, and its apparent thermal conductivity increases by many orders of magnitude.

The AMS-02 magnet is cooled by superfluid helium. There are two main reasons for this. Firstly, the specific latent heat and density of superfluid helium are both higher than in normal liquid helium. Since the amount of cryogen that can be carried is limited by the size of the helium vessel, this gives a useful endurance benefit (there is a greater mass of helium, and each kilogram has a higher cooling capability).

Secondly, in zero gravity there can be no convection currents. In normal liquid helium this can result in thermal stratification, making it difficult to ensure that all parts of the system are fully cold. In the superfluid state, however, the very high thermal conductivity makes it impossible for the helium to support large temperature gradients, so the system remains isothermal.

Any large cryogenic vessel has to be viewed as a potential safety hazard, particularly when it is in an enclosed space such as the payload bay of the Space Shuttle. Safety of the AMS magnet has to be assured in ground handling operations, during launch, on orbit and during landing. All cryogenic volumes, as well as the vacuum tank, are protected by burst discs to prevent excessive pressures building up in any fault conditions. Some of the burst discs have to operate at temperatures below 2 K, and these have been the subject of a special development and testing programme. In addition, extra protection is provided to mitigate the effect of a catastrophic loss of vacuum. This could be caused, during ground handling operations, by a serious rupture of the vacuum case. If the vacuum case had a large puncture, air could rush through the gap and condense on the surface of the helium vessel. This would result in rapid pressurisation and venting of the helium in the vessel. To slow down the rates of pressurisation and venting (making the pressure relief path smaller and more manageable) a 3 mm layer of lightweight cryogenic insulation will be applied to the outside surface of the helium vessel. Carefully constructed experiments have shown that this insulation reduces the heat flux to the superfluid helium by a factor of 8 following a sudden, total loss of vacuum. All parts of the AMS magnet system are subject to a battery of tests to ensure their quality and integrity, and their suitability for the mission.

Every one of the 14 superconducting coils will have been tested before assembly into the final magnet configuration. All twelve of the flux return coils have been tested successfully, and the first of the two dipole coils is undergoing final preparations for testing. A special test facility has been constructed which allows the coil to be operated under cryogenic conditions as close as possible to flight. In particular, the coil is mounted in a vacuum space, and is cooled by a thermal bus bar of the same construction as the one in the flight system.

A full-scale replica of the superfluid thermal bus bar system outlined above was designed and assembled. Heaters on the part of the thermal bus bar outside the helium vessel were then used to simulate the heat load due to the magnet coils being charged. Up to 6 W could be generated by the heaters and transferred by Gorter-Mellink conduction through the thermal bus bar to be dissipated in the vessel of boiling helium. If the heaters were used to generate more power, the Gorter-Mellink conduction broke down: the thermal bus bar was unable to transfer the heat and the temperatures within it began to rise rapidly. These results corresponded to the calculations carried out before the tests.

A series of experiments has been carried out to determine what happens in the event either of a catastrophic loss of vacuum or of a much smaller vacuum leak. A test facility was designed and assembled in which either of these scenarios could be investigated on a small scale. The test facility contained a 12 litre helium vessel, together with a system of valves, sensors and high-speed data acquisition. To carry out a test, the 12 litre vessel was filled with superfluid helium at 1.8 K. A fast-acting valve was then used to open a large hole in the insulating vacuum to simulate a catastrophic leak, or a very small hole to simulate a small leak. By monitoring the rise of pressure and temperature, and the rate of change of mass of the vessel, the heat flux and venting rate could be calculated. These results can be extrapolated directly to the AMS-02 magnet, and have been used to qualify the cryogenic system for flight on the Space Shuttle.

In addition to these investigations, tests have also been carried out on prototype burst discs. Discs for protecting the vacuum tank have undergone vibration testing followed by controlled bursts. These tests have shown that the discs are not affected by the levels of vibration encountered during a launch. Further tests have been carried out on discs for protecting the helium vessel, which operate at 1.8 K. These discs have been shown to have extremely good leak tightness against superfluid helium.

3. The Silicon Tracker

The AMS-01 Silicon Tracker [9] was the first application in space of the high precision silicon technology developed for position measurements in accelerator experiments [10]. The high modularity, low voltage levels (<100 V), and gas-free operation of the device is well suited to operation in space. The 1998 shuttle test flight demonstrated both the successful adaptation of the technology to the space environment and the feasibility of large area detectors. Silicon micro-strip sensors were originally developed for vertex detectors in colliding-beam experiments in order to provide a few high precision position measurements near the interaction point. The AMS application differs considerably. The tracking information is provided uniquely by the silicon sensors, which implies a large surface area and higher inter-strip capacitances. The major challenges were to maintain the required mechanical precision and low-noise performance in the large scale application, and to do so in outer space.

The silicon tracker is composed of 41.360 x 72.045 x 0.300 mm3 double-sided silicon micro-strip sensors. The sensors have been produced at silicon foundries located in Switzerland [11] and Italy [12] using identical geometries and processing procedures. Over 4000 sensors have been produced to select the ~ 2500 sensors of the very high quality required to assemble the Silicon Tracker units (ladders). Over 2 x 10^7 electrical measurements have been performed using four automatic test stations. This large number of sensors makes the AMS Silicon Tracker the largest precision tracking detector ever built for a space application. Figure 4 shows the principal elements of the silicon ladder and the main components of the readout hybrids.

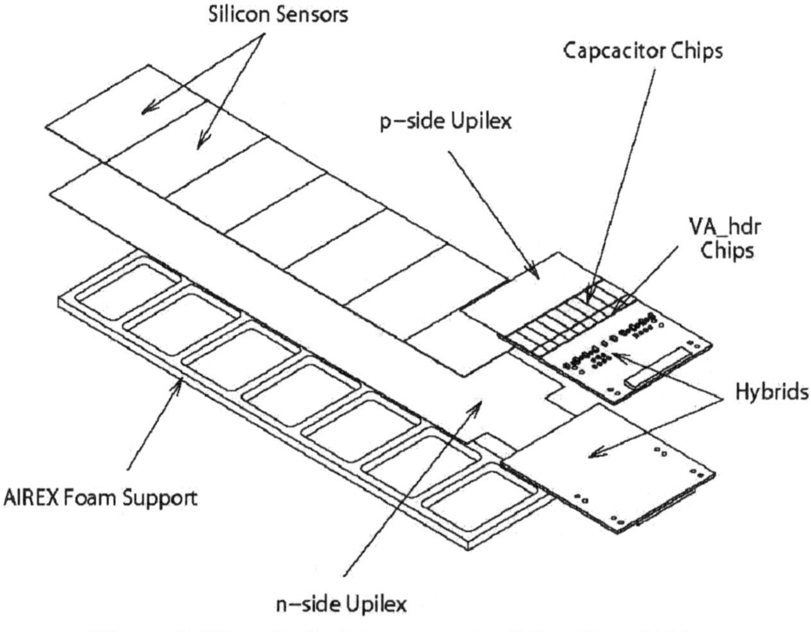

Figure 4: The principal components of the silicon ladder.

The principal goals of the ladder fabrication are to guarantee the required precision for the relative alignment of the silicon sensors (<5μm), and minimize the degradation of the electrical performance due to handling and ultra-sonic bonding. Ladder fabrication was organized between three centers operating with identical procedures derived from the AMS-01. During fabrication the sensor positions on a ladder are recorded with a 3D semiautomatic measuring machine. The results for the sensor alignment for the first 125 (out of 192) AMS-02 ladders is better than 5 μm rms.

The tracker support structure is divided into three sections: a carbon fiber cylindrical shell which supports the planes 2 to 4 located inside the magnet, and two carbon fiber flanges which support the exterior planes 1 and 5 [13]. With respect to the AMS-01 configuration, the number of silicon layers has been increased from 6 to 8 by suppressing one internal plane and equipping both sides of the remaining three internal planes with silicon ladders. When ladders equipping a full tracker plane are produced, they are integrated onto the corresponding support plane. Supervision, quality control and traceability are ensured by a database developed for that purpose. The activities start with acceptance tests of the ladders arriving from all production lines and then proceed in steps toward integration of the complete detector. Installation of the second layer on the other side is more delicate since one has access from only one side for fixation. The assembled plane is finally stored in a light-tight container under dry nitrogen. Figure 5 shows the first fully equipped inner plane.

Figure 5: Plane numbers 2 completely equipped with ladders.

The silicon sensors are grouped together, for readout and biasing, in ladders of different lengths to match the cylindrical geometry of the AMS magnet. The maximum combined strip length in the silicon for a single readout channel is 60 cm. The relatively large input capacitance (30-180 pF), as well as the need for a high dynamic range (<100 MIPs), led to the development of a new front-end readout chip based on the low-noise Viking design, the VA_hdr7 [14].

Each of the 64 channels of the VA_hdr chip consists of a charge sensitive amplifier, a CR-RC semi-Gaussian shaper, and a sample-and-hold stage. An analog multiplexer, shift register and buffer are incorporated in the chip for sequential data output at a maximum clock frequency of 10 MHz. The equivalent noise charge as a function of capacitance load Cdet has been measured to be (350+4Cdet/pF) e^- at a 6 μs peaking time and nominal bias currents. The VA_hdr chips have an average power consumption 0.7 mW per channel. The single channel response of the VA_hdr chip has been measured to be linear up to ~75 MIPs. The strips of the silicon sensors are ac-coupled to the VA_hdr via 700 pF capacitor chips.

The hybrids are mounted on carbon fiber-metal cooling bars, which evacuate the heat generated by the front-end electronics inside the magnet. The presence of the superconducting magnet requires an active cooling system for the tracker. The AMS-02 Tracker Thermal Control System (TTCS) is a two-phase, mechanically pumped loop system. The cooling liquid, CO2 at about 80 bar pressure, is circulated by a pump. Outside of the tracker volume, the fluid passes through a heat exchanger to keep the incoming fluid just at the boiling point while minimizing the pre-heater power required. It is then directed to condensers on the tracker thermal wake and ram radiator panels facing deep space. There, the vapor/liquid mixture is cooled to below the boiling point, and then returns to the pump input, closing the circuit. Ammonia heat pipes embedded in the radiators increase their effectiveness. The relative flow of fluid to the wake and ram radiators is self-adjusting, the fluid will preferentially flow towards the cooler radiator.

Space based particle detection systems have to cope with a far wider range of environmental conditions than those at accelerators. This concerns notably the vibrations during the transport before deployment and the rapid periodic changes in the thermal settings due to solar radiation and cooling while in the shadow of Earth. With the AMS-02 silicon tracker, charged particle tracks are traced at 8 space points in a ~1 m^3 sized B-field to an accuracy of better than 10 μm. The alignment system provides optically generated signals in the 8 layers of the silicon tracker that mimic straight (infinite rigidity) tracks. The AMS-02 tracker is equipped with 2 x 10 pairs of alignment control beams. The beams are narrow (diameter < 0.5 mm) and of small divergence (< 1 mrad). The AMS approach to silicon tracker alignment control using IR laser beams fulfills the requirements of a space borne experiment. It is light weight (3kg), low power (1mW), low dead time (< 1%) and provides a precision exceeding the tracker resolution (8μm) with less than 100 laser shots. The success of the AMS approach in Si tracker alignment control by IR laser beams has lead the team building the largest Si tracker array [15] to develop a similar system for 10 years of operation at the LHC.

An extensive series of tests have been performed to verify the performance of the AMS-02 silicon tracker. For what concerns the spatial resolution the residual distributions of the ladder, described by a Gaussian function and flat background., give widths of the Gaussians of 8.5 and 30 μm respectively for the p- and n-sides. To study the AMS-02 ladder response to light and heavy ions, six ladders were exposed to an ion beam at CERN in October 2003. The tracker Z measurements are compared to those of prototype RICH detector in Figure 6. An excellent correlation to the RICH Z measurement is seen for both side of the silicon sensors, namely the K and the S side, respectively.

If compared to the precursor AMS-01 mission, the performance of the tracker in terms of signal to noise, position resolution and charge resolution has been greatly enhanced. The increase of silicon layers from 6 to 8, together with the more powerful AMS-02 cryomagnet, has significantly increased the physics reach of the AMS-02 detector, making a wide range of physical phenomena accessible during the AMS-02 mission.

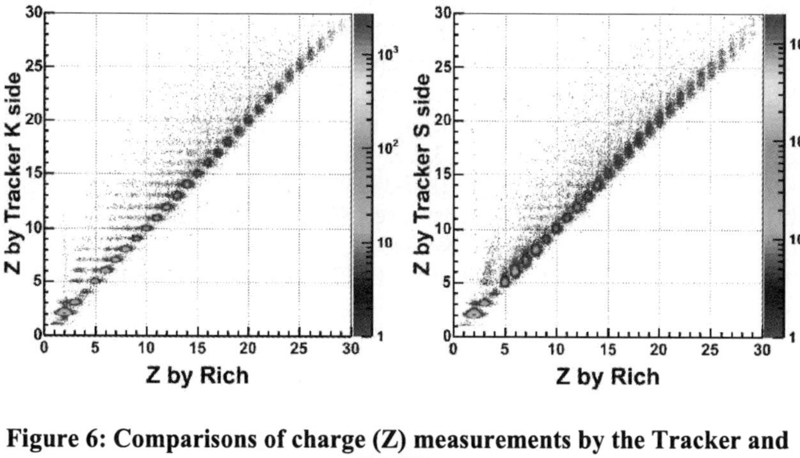

Figure 6: Comparisons of charge (Z) measurements by the Tracker and RICH.

4. Conclusions

AMS-02 is the first precision particle physics experiment on the International Space Station. With its acceptance, resolution and ability to study different particles and nuclei over 3 to 5 years of data taking, it provides an opportunity to accurately explore new regions of physics.

Acknowledgements: the author thanks M. Capell for its support during the preparation of this paper.

References

[1] G. Jungman et al., Phys. Rep. **267** (1996) 195;
J.R. Ellis et al., Nucl. Phys. **B214**, (1998) 3;
E.A. Baltz and J. Edsjo, Phys. Rev. **D59** (1999) 23511;
T. Moroi and L. Randall, Nuc. Phys. **B570** (2000) 455.
[2] P. Salati, et al., Nucl.Phys.Proc.Suppl. **81** (2000) 37.
[3] A.D. Dolgov, Phys. Reports **222** (1992) 309;
[4] J.G. Asbury et al., Phys. Rev. Lett. **18**, 2 (1967) 65;
[5] J.J. Aubert et al., Phys. Rev. Lett. **33**, 23 (1974) 1404.
[6] For example, L3 Collaboration, Nucl. Instr. Meth. A **289** (1990) 35.
[7] B. Blau et al., Grav. and Cosm.5 (2000), Suppl. pp. 1; B. Blau et al., IEEE Trans. on App. Sup. 12 (2002) 349.
[8] S. Harrison, et al., IEEE Trans. on App. Sup. Vol. 13 pp. 1381-1384.
[9] R. Battiston Nucl.Phys.Proc.Suppl. 44 (1995) 274;
J. Alcaraz et al., Nuovo Cim. 112A (1999) 1325;
W.J. Burger, Nucl.Phys.Proc.Suppl. 113 (2002)139.
[10] M. Acciarri et al., Nucl. Instr. and Method. A 351 (1994) 300.
G.F. Dalla Betta, et al., Nucl. Instr. and Method. A 431 (1999) 83.
[11] Colibrys SA , Maladière 83, CH-2007 Neuchâtel.
[12] ITC-irst; Via Sommarive 18, I-38050 Povo.
[13] Oerlikon Contraves; Birchstrasse 155, CH-8050 Zürich.
[14] IDE AS; Veritasveien 9, N-1323 Hovik.
[15] Ostaptchouk et al., 9 May 2001, CMS Note 2001/053, CMS – CERN.

Cosmic Neutrinos and the Energy Budget of Galactic and Extragalactic Cosmic Rays

Francis Halzen
Department of Physics, University of Wisconsin, Madison, WI 53706

Abstract

Although kilometer-scale neutrino detectors such as IceCube are discovery instruments, their conceptual design is very much anchored to the observational fact that Nature produces protons and photons with energies in excess of 10^{20} eV and 10^{13} eV, respectively. The puzzle of where and how Nature accelerates the highest energy cosmic particles is unresolved almost a century after their discovery. We will discuss how the cosmic ray connection sets the scale of the anticipated cosmic neutrino fluxes. In this context, we discuss the first results of the completed AMANDA detector and the science reach of its extension, IceCube.

1 Introduction

Ambitious projects have been launched to extend conventional astronomy beyond wavelengths of 10^{-14} cm, or GeV photon energy. Besides gamma rays, protons (nuclei), neutrinos and gravitational waves will be explored as astronomical messengers probing the extreme Universe. The challenges are considerable:

- Protons are relatively abundant, but their arrival directions have been scrambled by magnetic fields.
- γ-rays do point back to their sources, but are absorbed at TeV-energy and above on cosmic background radiation.
- neutrinos propagate unabsorbed and without deflection throughout the Universe but are difficult to detect.

Therefore, multi-messenger astronomy may not just be an advantage, it may be a necessity for solving some of the outstanding problems of astronomy at the highest energies such as the identification of the sources of the cosmic rays, the mechanism(s) triggering gamma ray bursts and the particle nature of the dark matter.

We here discuss the case for the detection of neutrinos associated with the observed fluxes of high energy cosmic rays and gamma rays; it points, unfortunately, at the necessity of commissioning kilometer-scale neutrino detectors. Though ambitious, the scientific case is compelling because neutrinos will reveal the location of the source(s) and represent the ideal tool to study the black holes powering the cosmic accelerator(s).

Soon after the discovery in the mid-fifties that neutrinos were real particles and not just mathematical constructs of theorists' imagination, the idea emerged that they represent ideal cosmic messengers[1]. Because of their weak interactions, neutrinos reach us unimpeded from the edge of the Universe and from the inner reaches of black holes. The neutrino telescopes now under construction have the capability to detect neutrinos with energies from a threshold of $\sim 10\,\mathrm{GeV}$ to, possibly, $\sim 10^2\,\mathrm{EeV}$, the highest energies observed. Their telescope range spans more than 10 orders of magnitude in wavelengths smaller than $10^{-14}\,\mathrm{cm}$. This is a reach equivalent to that of a hypothetical astronomical telescope sensitive to wavelengths from radio to X-rays. Above $10^5\,\mathrm{TeV}$ the observations are free of muon and neutrino backgrounds produced in cosmic ray interactions with the Earth's atmosphere. Each neutrino is a discovery.[1]

The real challenge of neutrino astronomy is that kilometer-scale neutrino detectors are required to do the science. The first hint of the scale of neutrino telescopes emerged in the nineteen seventies from theoretical studies of the flux of neutrinos produced in the interactions of cosmic rays with microwave photons, the so-called Greissen-Zatsepin-Kuzmin or GZK neutrinos. Since then the case for kilometer-size instruments has been strengthened[2] and the possibility of commissioning such instruments demonstrated[3]. In fact, if the neutrino sky were within reach of smaller instruments, it would by now have been revealed by the first-generation AMANDA telescope. It has been taking data since 2000 with a detector of $0.01 \sim 0.08\,\mathrm{km}^2$ telescope area, depending on the sources[4].

Given the size of the detector required, all efforts have concentrated on transforming large volumes of natural water or ice into Cherenkov detectors. They reveal the secondary muons and electromagnetic and hadronic showers initiated in neutrino interactions inside or near the detector. Because of the long range of the muon, from kilometers in the TeV range to tens of kilometers at the highest energies, neutrino interactions can be identified far outside the instrumented volume. Adding to the technological challenge is the requirement that the detector be shielded from the abundant flux of cosmic ray muons by deployment at a depth of typically several kilometers. After the cancellation of a pioneering attempt[5] to build a neutrino telescope off the coast of Hawaii, successful operation of a smaller instrument in Lake Baikal[6] bodes well for several efforts to commission neutrino telescopes in the Mediterranean[5, 7]. We will here mostly concentrate on the construction and first four years of operation of the AMANDA telescope[4, 8] which has transformed a large volume of natural deep Antarctic ice into a Cherenkov detector. It represents a first-generation telescope as

[1]We will use GeV$= 10^9\,\mathrm{eV}$, TeV$= 10^{12}\,\mathrm{eV}$, PeV$= 10^{15}\,\mathrm{eV}$ and EeV$= 10^{18}\,\mathrm{eV}$ units of energy.

envisaged by the DUMAND collaboration over 20 years ago and a proof of concept for the kilometer-scale IceCube detector, now under construction.

Even though neutrino "telescopes" are designed as discovery instruments covering a large dynamic range, be it for particle physics or astrophysics, their conceptual design is very much anchored to the observational fact that Nature produces protons and photons with energies in excess of 10^{20}eV and 10^{13} eV, respectively. The cosmic ray connection sets the scale of cosmic neutrino fluxes. We will discuss this first.

2 Cosmic Neutrinos Associated with Extragalactic Cosmic Rays

Cosmic accelerators produce particles with energies in excess of 10^8 TeV; we do not know where or how. The flux of cosmic rays observed at Earth is sketched in Fig. 1a,b[9]. The energy spectrum follows a broken power law. The two power laws are separated by a feature dubbed the "knee"; see Fig. 1a. Circumstantial evidence exists that cosmic rays, up to perhaps EeV energy, originate in galactic supernova remnants. Any association with our Galaxy disappears in the vicinity of a second feature in the spectrum referred to as the "ankle". Above the ankle, the gyroradius of a proton in the galactic magnetic field exceeds the size of the Galaxy and it is generally assumed that we are witnessing the onset of an extragalactic component in the spectrum that extends to energies beyond 100 EeV. Experiments indicate that the highest energy cosmic rays are predominantly protons or, possibly, nuclei. Above a threshold of 50 EeV these protons interact with cosmic microwave photons and lose energy to pions before reaching our detectors. This is the GZK cutoff that limits the sources to our local supercluster.

Models for the origin of the highest energy cosmic rays fall into two categories, top-down and bottom-up. In top-down models it is assumed that the cosmic rays are the decay products of cosmological remnants or topological defects associated, for instance, with Grand Unified theories with unification energy $M_{GUT} \sim 10^{24}$ eV. These models predict neutrino fluxes most likely within reach of first-generation telescopes such as AMANDA, and certainly detectable by future kilometer-scale neutrino observatories[10]. They have not been observed.

In bottom-up scenarios it is assumed that cosmic rays originate in cosmic accelerators. Accelerating particles to TeV energy and above requires massive bulk flows of relativistic charged particles. These are likely to originate from the exceptional gravitational forces in the vicinity of black holes. Gravity powers large electric currents that create the opportunity for particle acceleration by shocks, a mechanism familiar from solar flares where particles are accelerated to 10 GeV. It is a fact that black holes accelerate electrons to high energy; astronomers observe them indirectly by their synchrotron radiation. We know that they accelerate protons because we detect them as cosmic rays. Because they are charged, protons are deflected by in-

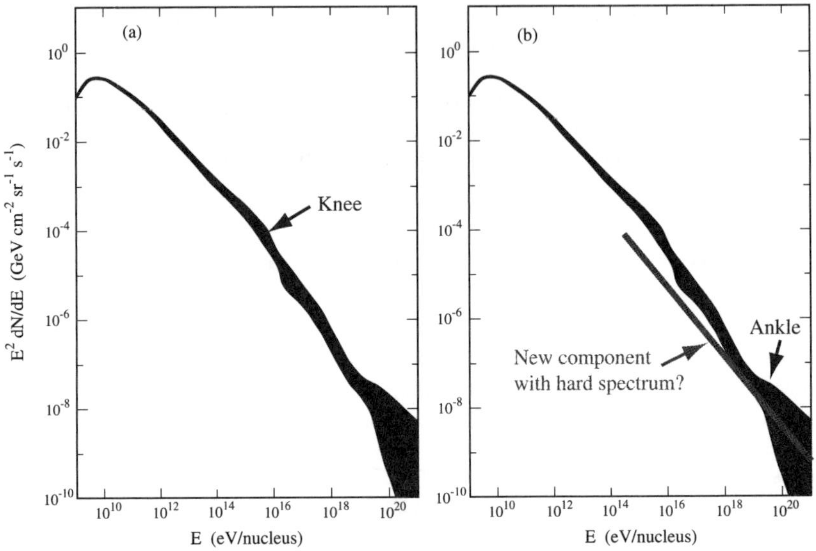

Figure 1: At the energies of interest here, the cosmic ray spectrum consists of a sequence of 3 power laws. The first two are separated by the "knee" (left panel), the second and third by the "ankle". There is evidence that the cosmic rays beyond the ankle are a new population of particles produced in extragalactic sources; see right panel.

terstellar magnetic fields; cosmic rays do not reveal their sources. This is the cosmic ray puzzle.

Examples of candidate black holes include the dense cores of exploding stars, inflows onto supermassive black holes at the centers of active galaxies and annihilating black holes or neutron stars. Before leaving the source, accelerated particles pass through intense radiation fields or dense clouds of gas surrounding the black hole. This results in interactions producing pions decaying into secondary photons and neutrinos that accompany the primary cosmic ray beam as illustrated in Fig. 2. How many neutrinos are produced in association with the cosmic ray beam? The answer to this question provides one rationale for building kilometer-scale neutrino detectors [2]. For orientation, consider a neutrino beam produced at an accelerator laboratory. Here the target and the beam dump absorb all parent protons as well as the secondary electromagnetic and hadronic showers. Only neutrinos exit the dump. If Nature constructed such a "hidden source" in the heavens, conventional astronomy would not reveal it. Cosmic ray sources must be at least partially transparent to protons. Sources transparent only to neutrinos may exist, but they cannot be cosmic-ray

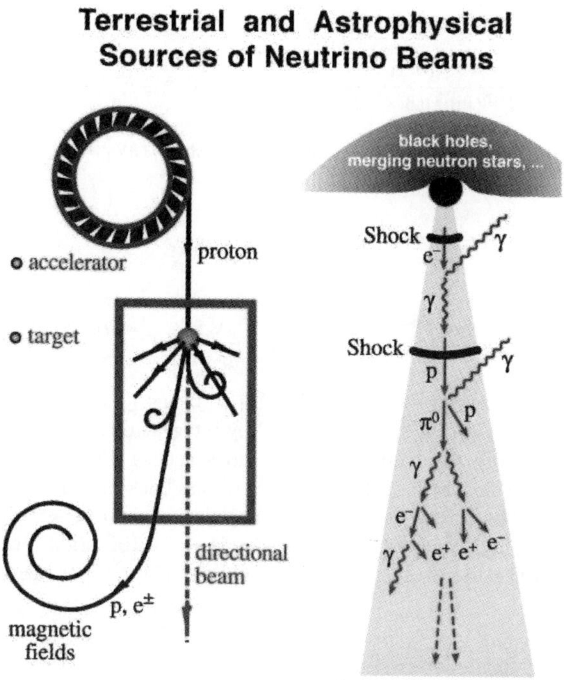

Figure 2: Cosmic beam dump exits: sketch of cosmic ray accelerator producing photons. The charged pions that are inevitably produced along with the neutral pions will decay into neutrinos.

sources.

A generic "transparent" source can be imagined as follows: protons are accelerated in a region of high magnetic fields where they interact with photons and generate neutral and charged pions. The most important process is $p + \gamma \rightarrow \Delta^+ \rightarrow \pi^0 + p$ and $p + \gamma \rightarrow \Delta^+ \rightarrow \pi^+ + n$. While the secondary protons may remain trapped in the acceleration region, roughly equal numbers of neutrons and decay products of neutral and charged pions escape. The energy escaping the source is therefore distributed among cosmic rays, gamma rays and neutrinos produced by the decay of neutrons, neutral pions and charged pions, respectively. The neutrino flux from a generic transparent cosmic ray source is often referred to as the Waxman-Bahcall flux [11]. It is easy to calculate and the derivation is revealing.

Figure 1b shows a fit to the observed spectrum above the "ankle" that can be used to derive the total energy in extragalactic cosmic rays. The flux above the ankle is often summarized as "one 10^{19} eV particle per kilometer square per year per steradian". This can be translated into an energy flux

$$E\left\{E\frac{dN}{dE}\right\} = \frac{10^{19}\,\text{eV}}{(10^{10}\,\text{cm}^2)(3\times 10^7\,\text{sec})\,\text{sr}} = 3\times 10^{-8}\,\text{GeV cm}^{-2}\,\text{s}^{-1}\,\text{sr}^{-1}\,.$$

¿From this we can derive the energy density ρ_E in cosmic rays using the relation that flux = velocity × density, or

$$4\pi\int dE\left\{E\frac{dN}{dE}\right\} = c\rho_E\,.$$

We obtain

$$\rho_E = \frac{4\pi}{c}\int_{E_{\text{min}}}^{E_{\text{max}}}\frac{3\times 10^{-8}}{E}dE\,\frac{\text{GeV}}{\text{cm}^3} \simeq 10^{-19}\,\frac{\text{TeV}}{\text{cm}^3}\,,$$

taking the extreme energies of the accelerator(s) to be $E_{\text{max}}/E_{\text{min}} \simeq 10^3$.

The energy content derived "professionally" by integrating the spectrum in Fig. 2b assuming an E^{-2} energy spectrum, typical of shock acceleration, with a GZK cutoff is $\sim 3\times 10^{-19}\,\text{erg cm}^{-3}$. This is within a factor of our back-of-the-envelope estimate (1 TeV = 1.6 erg). The power required for a population of sources to generate this energy density over the Hubble time of 10^{10} years is $\sim 3\times 10^{37}\,\text{erg s}^{-1}$ per $(\text{Mpc})^3$ or, as often quoted in the literature, $\sim 5\times 10^{44}$ TeV per $(\text{Mpc})^3$ per year. This works out to[12]

- $\sim 3\times 10^{39}\,\text{erg s}^{-1}$ per galaxy,
- $\sim 3\times 10^{42}\,\text{erg s}^{-1}$ per cluster of galaxies,
- $\sim 2\times 10^{44}\,\text{erg s}^{-1}$ per active galaxy, or
- $\sim 2\times 10^{52}$ erg per cosmological gamma ray burst.

The coincidence between these numbers and the observed output in electromagnetic energy of these sources explains why they have emerged as the leading candidates for the cosmic ray accelerators. The coincidence is consistent with the relationship between cosmic rays and photons built into the "transparent" source. In the photoproduction processes roughly equal energy goes into the secondary neutrons, neutral and charged pions whose energy ends up in cosmic rays, gamma rays and neutrinos, respectively.

We therefore conclude that the same energy density of $\rho_E \sim 3\times 10^{-19}\,\text{erg cm}^{-3}$, observed in cosmic rays and electromagnetic energy, ends up in neutrinos with a spectrum $E_\nu dN/dE_\nu \sim E^{-\gamma}\,\text{cm}^{-2}\,\text{s}^{-1}\,\text{sr}^{-1}$ that continues up to a maximum energy E_{max}. The neutrino flux follows from the relation $\int E_\nu dN/dE_\nu = c\rho_E/4\pi$. For $\gamma = 1$ and $E_{\text{max}} = 10^8$ GeV, the generic source of the highest energy cosmic rays produces a flux of ${E_\nu}^2 dN/dE_\nu \sim 5\times 10^{-8}\,\text{GeV cm}^{-2}\,\text{s}^{-1}\,\text{sr}^{-1}$.

There are several ways to sharpen this qualitative prediction:

- The derivation fails to take into account that there are more UHE cosmic rays in the Universe than observed at Earth because of the GZK-effect and it also neglects the evolution of the sources with redshift. This increases the neutrino flux, which we normalized to the observed spectrum only, by a factor $d_H/d_{\rm CMB}$, the ratio of the Hubble radius to the average attenuation length of the cosmic rays propagating in the cosmic microwave background.

- For proton-γ interactions muon neutrinos (and antineutrinos) receive only 1/2 of the energy of the charged pion in the decay chain $\pi^+ \to \mu^+ + \nu_\mu \to e^+ + \nu_e + \bar{\nu}_\mu + \nu_\mu$ assuming that the energy is equally shared between the 4 leptons. Furthermore half the muon neutrinos oscillate into tau neutrinos over cosmic distances. In further calculations we will focus on the muon flux here.

In summary,

$$E_\nu \frac{dN_\nu}{dE_\nu} = \frac{1}{2} \times \frac{1}{2} \times E\frac{dN_{\rm CR}}{dE} \times \frac{d_H}{d_{\rm CMB}} \simeq E\frac{dN_{\rm CR}}{dE} \tag{1}$$

In practice, the corrections approximately cancel. The precise value of the energy where the transition from galactic to extragalactic sources occurs represents another source of uncertainty that has been extensively debated [13]. A transition at a lower energy significantly increases the energy in the extragalactic component and results in an enhancement of the associated neutrino flux.

Waxman and Bahcall referred to their flux as a bound in part because in reality more energy is transferred to the neutron than to the charged pion in the source, in the case of the photoproduction reaction $p + \gamma \to \Delta^+ \to \pi^+ + n$ four times more. Therefore

$$E_\nu \frac{dN_\nu}{dE_\nu} = \frac{1}{4} E\frac{dN_{\rm CR}}{dE}\,. \tag{2}$$

In the end we estimate that the muon-neutrino flux associated with the sources of the highest energy cosmic rays is loosely confined to the range ${E_\nu}^2 dN/dE_\nu = 1 \sim 5 \times 10^{-8}\,{\rm GeV\,cm^{-2}\,s^{-1}\,sr^{-1}}$ depending on the cosmological evolution of the cosmic ray sources.

The anticipated neutrino flux thus obtained has to be compared with the sensitivity of $8.9 \times 10^{-8}\,{\rm GeV\ cm^{-2}\,s^{-1}\,sr^{-1}}$ reached after the first 4 years of operation of the completed AMANDA detector in 2000–2003 [4]. The analysis of the data has not been completed, but a limit of $2 \times 10^{-7}\,{\rm GeV\ cm^{-2}\,s^{-1}\,sr^{-1}}$ has been obtained with a single year of data[14]. On the other hand, after three years of operation IceCube will reach a diffuse flux limit of $E_\nu^2 dN/dE_\nu = 2 \sim 7 \times 10^{-9}\,{\rm GeV\,cm^{-2}\,s^{-1}\,sr^{-1}}$. The exact value of the IceCube sensitivity depends on the magnitude of the dominant high energy neutrino background from the prompt decay of atmospheric charmed particles[3]. The level of this background is difficult to anticipate theoretically and little accelerator data is available in the energy and Feynman-x range of interest[15].

The observed event rate is obtained by folding the cosmic flux predicted with the probability that the neutrino is actually detected in a high energy neutrino telescope;

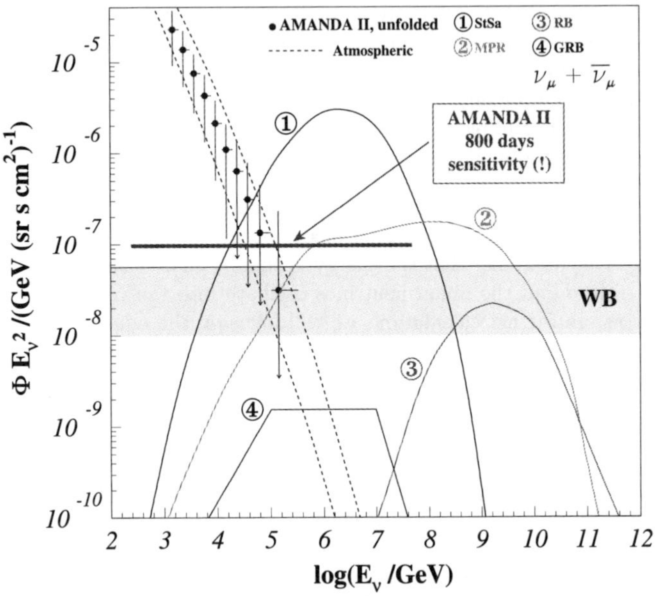

Figure 3: Our estimate of the flux of neutrinos associated with the sources of the highest energy cosmic rays (the shaded range labeled WB) is compared to the sensitivity of the AMANDA experiment reached with 800 days of data. Also shown are fluxes predicted by specific models of cosmic ray accelerators: active galaxies labeled StSa[16] and MPR[17], gamma ray bursts[18] and the diffuse flux produced by cosmic ray producing active galaxies on microwave photons[19] labelled RB. Data for the background atmospheric neutrino flux are from the AMANDA experiment.

only one in a million neutrinos of TeV energy interact and produce a muon that reaches the detector. This probability is given by the ratio of the muon and neutrino interaction lengths in the detector medium, λ_μ/λ_ν[2] and therefore depends on energy. For the flux range estimated above we anticipate $20 \sim 100$ detected muon neutrinos per km^2 per year. Given that its effective area for muon neutrinos exceeds $1\,\text{km}^2$ and that equal fluxes of electron and tau neutrinos are expected, a neutrino signal at the "Waxman-Bahcall" level could result in the observation of as many as one thousand high-energy neutrinos of extraterrestrial origin per year in IceCube [3]. Model calculations assuming that active galaxies or gamma-ray bursts are the actual sources of cosmic rays yield similar event rates than the generic energetics estimate presented.

Gamma ray bursts (GRB), outshining the entire Universe for the duration of the burst, are perhaps the best motivated sources of high-energy neutrinos[20, 21, 22].

The collapse of massive stars to a black hole has emerged as the likely origin of the "long" GRB with durations of tens of seconds. In the collapse a fireball is produced which expands with a highly relativistic velocity powered by radiation pressure. The fireball eventually runs into the stellar material that is still accreting onto the black hole. If it successfully punctures through this stellar envelope the fireball emerges to produce a GRB. While the energy transferred to highly relativistic electrons is thus observed in the form of radiation, it is a matter of speculation how much energy is transferred to protons.

The assumption that GRB are the sources of the highest energy cosmic rays does determine the energy of the fireball baryons. Accommodating the observed cosmic ray spectrum of extragalactic cosmic rays requires roughly equal efficiency for conversion of fireball energy into the kinetic energy of protons and electrons. In this scenario the production of $100 \sim 1000\,\mathrm{TeV}$ neutrinos in the GRB fireball is a robust prediction because neutrinos are inevitably produced in interactions of accelerated protons with fireball photons. Estimates of the flux[18] point again at the necessity of a kilometer-cubed neutrino detector, in agreement with the generic energetics estimates previously presented. Studies of active galaxies as sources of cosmic rays lead to similar conclusions[16].

The case for kilometer-scale detectors also emerges from consideration of "guaranteed" cosmic fluxes. Neutrino fluxes are guaranteed when both the accelerator and the pion producing target material can be identified. We mention three examples. The extragalactic cosmic rays produce ~ 1 event per km^2 year in interactions with cosmic microwave photons[23]. Supernovae producing cosmic rays in the dense star formation regions of starburst galaxies form a hidden source of neutrinos within reach of IceCube[24]. Galactic cosmic rays interact with hydrogen in the disk to generate an observable neutrino flux in a kilometer-scale detector[25].

3 Cosmic Neutrinos Associated with Galactic Cosmic Rays

In the previous section we made an estimate of the neutrino flux from generic accelerators producing the highest energy cosmic rays. We can perform a similar analysis for the galactic cosmic rays by calculating the energy density corresponding to the flux shown in Fig. 1a. The answer is that $\rho_E \sim 10^{-12}\,\mathrm{erg\,cm^{-3}}$. This is also the value of the corresponding energy density $B^2/8\pi$ of the microgauss magnetic field in the galaxy. The power needed to maintain this energy density is $10^{-26}\,\mathrm{erg/cm^3 s}$ given that the average containment time of the cosmic rays in our galaxy is 3×10^6 years. For a nominal volume of the galactic disk of $10^{67}\,\mathrm{cm}^3$ this requires an accelerator delivering $10^{41}\,\mathrm{erg/s}$. This happens to be 10% of the power produced by supernovae releasing 10^{51} erg every 30 years. The coincidence is the basis for the idea that shocks produced by supernovae exploding into the interstellar medium are the origin of the galactic cosmic rays.

With recent observations[26] of the supernova remnant RX J1713.7-3946 the H.E.S.S. array of atmospheric Cherenkov telescopes obtained circumstantial evidence for cosmic ray acceleration and, if confirmed, identified a guaranteed source of cosmic neutrinos[27]. With RX J1713.7-3946, H.E.S.S. may have detected the first site where protons are accelerated to energies typical of the main component of the galactic cosmic rays[28]. Although the resolved image of the source (the first ever at TeV energies!) reveals TeV gamma ray emission from the whole supernova remnant, it shows a clear increase of the flux in the directions of known molecular clouds. This suggests the possibility that protons, shock accelerated in the supernova remnant, interact with the dense clouds to produce neutral pions that are the source of the observed increase of the TeV photon signal. The image shows filaments of high magnetic fields consistent with the requirements for acceleration to the energies observed. Furthermore, the high statistics data for the flux are power-law behaved over a large range of energies without any indication of a cutoff characteristic of synchrotron or inverse-Compton sources. Follow-up observations of the source in radio-waves and X-rays have failed to identify the population of electrons required to generate TeV photons by purely electromagnetic processes; for a detailed discussion see [29]. On the theoretical side, the large B-fields suppress the ratio of photons produced by the inverse Compton relative to the synchrotron. Fitting the data by purely electromagnetic processes is therefore challenging but, apparently, not impossible[29].

Gamma ray telescopes have therefore not succeeded in finding the smoking gun for the supernova origin of the galactic cosmic rays. If the TeV flux of RX J1713.7-3946 is of neutral pion origin, then the accompanying charged pions will produce a guaranteed neutrino flux of roughly 10 muon-type neutrinos per kilometer-squared per year[27] and produce incontrovertible evidence for cosmic ray acceleration. From a variety of such sources we can therefore expect event rates of cosmic neutrinos of galactic origin similar to those estimated for extragalactic neutrinos in the previous section. Supernovae associated with molecular clouds are a common feature of associations of OB stars that exist throughout the galactic plane.

It is important to realize that there is a robust relation between the neutrino and gamma flux emitted by cosmic ray accelerators[27]. The $\nu_\mu + \bar{\nu}_\mu$ neutrino flux (dN_ν/dE_ν) produced by the decay of charged pions in the source can be derived from the observed gamma ray flux by energy conservation:

$$\int_{E_\gamma^{\rm min}}^{E_\gamma^{\rm max}} E_\gamma \frac{dN_\gamma}{dE_\gamma} dE_\gamma = K \int_{E_\nu^{\rm min}}^{E_\nu^{\rm max}} E_\nu \frac{dN_\nu}{dE_\nu} dE_\nu \tag{3}$$

where $E_\gamma^{\rm min}$ ($E_\gamma^{\rm max}$) is the minimum (maximum) energy of the photons that have a hadronic origin. $E_\nu^{\rm min}$ and $E_\nu^{\rm max}$ are the corresponding minimum and maximum energy of the neutrinos. The factor K depends on whether the π^0's are of pp or $p\gamma$ origin. Its value can be obtained from routine particle physics. In pp interactions 1/3 of the proton energy goes into each pion flavor. In the pion-to-muon-to-electron decay chain

2 muon-neutrinos are produced with energy $E_\pi/4$ for every photon with energy $E_\pi/2$. Therefore the energy in neutrinos matches the energy in photons and $K = 1$. The flux has to be reduced by a factor 2 because of oscillations. For $p\gamma$ interactions $K = 1/4$. The estimate should be considered a lower limit because the observed photon flux to which the calculation is normalized may have been attenuated by absorption in the source or in the interstellar medium.

In the case of supernova remnants the calculation of the neutrino flux can be performed on the back-of-the-envelope. Let's specialize to TeV photons. From a source such as RX J1713.7-3946 a flux of 10^{-11} photons per cm^2 second is detected. This is consistent with theoretical expectations. As previously pointed out, a few supernovae per century transferring a fraction ~ 0.1 of their energy, or about $\mathrm{W}_{CR} = 10^{50}$ erg, into the acceleration of cosmic rays can accommodate the observed flux up to the "knee" in the spectrum. The acceleration takes place in the high magnetic fields created in the shock expanding into the interstellar medium. In the interaction of the shocked protons with the interstellar proton density of $n \sim 1\,\mathrm{cm}^{-3}$, neutral pions are produced decaying into TeV photons close to the observed rate, or[28]

$$E\frac{dN_\gamma}{dE}(> E) = 10^{-11}\,(\frac{photons}{\mathrm{cm}^2\mathrm{s}})(\frac{W_{CR}}{10^{50}\mathrm{erg}})(\frac{n}{1\mathrm{cm}^3})(\frac{d}{1\mathrm{kpc}})^{-2} \tag{4}$$

Here d is roughly the distance appropriate for RX J1713.7-3946 and we therefore obtain a flux consistent with the H.E.S.S. observation. We can finesse the more formal derivation by simply assuming that each TeV gamma ray is accompanied by a neutrino from a charged pion to obtain an event rate of 10 detected neutrinos per decade of energy per km^2 year, a result readily obtained from the relation

$$\frac{dN_{events}}{d(lnE)}(> E) = 10^{-11}\,(\frac{photons}{\mathrm{cm}^2\,\mathrm{s}})\,area\;time\,(\frac{\lambda_\mu}{\lambda_\nu}), \tag{5}$$

where the last factor represents, as before, the probability that the neutrino is detected. It is approximately 10^{-6} for the TeV energy considered here. From several such sources IceCube will detect a flux of neutrinos similar to the one associated with extragalactic sources.

In summary, the energetics of galactic as well as extragalactic cosmic rays points at the necessity to build kilometer-scale detectors to observe the associated neutrino fluxes that will reveal the sources. The case for doing neutrino astronomy with kilometer-scale instruments can also be made in other ways[2] and, as is usually the case, the estimates of the neutrino fluxes pointing at the necessity of such detectors are likely to be optimistic.

Acknowledgments

I thank my IceCube collaborators as well as Concha Gonzalez-Garcia and Tom Gaisser for discussions. This research was supported in part by the National Science Foundation under Grant No. OPP-0236449, in part by the U.S. Department of Energy under

Grant No. DE-FG02-95ER40896, and in part by the University of Wisconsin Research Committee with funds granted by the Wisconsin Alumni Research Foundation.

References

[1] K. Greisen, Ann. Rev. Nucl. Part. Sci. **10**, 63 (1960), M.A.Markov in Proceedings of the 1960 International Conference on High Energy Physics, E.C.G. Sudarshan, J.H.Tinlot and A.C.Melissinos Editors, 578 (1960).

[2] T. K. Gaisser, F. Halzen, and T. Stanev, Phys. Rept. **258**, 173 (1995) [Erratum **271**, 355 (1995)], hep-ph/9410384; J.G. Learned and K. Mannheim, Ann. Rev. Nucl. Part. Science **50**, 679 (2000); F. Halzen and D. Hooper, Rept. Prog. Phys. **65**, 1025 (2002), arXiv:astro-ph/0204527.

[3] J. Ahrens et al. (IceCube Collaboration), Astropart. Phys. **20**, 507 (2004), astro-ph/0305196 and http://icecube.wisc.edu.

[4] A. Achterberg et al., Proceedings of the 29th International Cosmic Ray Conference, Pune, India, 2005, arXiv:astro-ph/0509330; G. Hill, invited talk at the same conference.

[5] J. Babson *et al.*, DUMAND Collaboration, Phys. Rev. D **42**, 3613 (1990).

[6] V. A. Balkanov et al. (Baikal Collaboration), Nucl. Phys. Proc. Suppl. **118**, 363 (2003).

[7] E. Migneco *et al.*, Nucl. Phys. Proc. Suppl. **136**, 61 (2004).

[8] A. Karle for the AMANDA collaboration, *Observation of Atmospheric Neutrino Events with AMANDA*, Proceedings of the 26th International Cosmic Ray Conference, Salt Lake City, Utah (1999); E. Andres et al. (AMANDA Collaboration), Nature **410**, 441 (2001); Phys.Rev.D **66**, 012005 (2002), arXiv:astro-ph/0205109.

[9] T. K. Gaisser, Proceedings of the 31st International Conference on High Energy Physics, Amsterdam, The Netherlands, July 2002.

[10] D. V. Semikoz and G. Sigl, JCAP **0404**, 003 (2004), arXiv:hep-ph/0309328.

[11] J. N. Bahcall and E. Waxman, Phys. Rev. D **64**, 023002 (2001).

[12] T. K. Gaisser, OECD Megascience Forum, Taormina, Italy, 1997, arXiv:astro-ph/9707283.

[13] Markus Ahlers, Luis A. Anchordoqui, Haim Goldberg, Francis Halzen, Andreas Ringwald, Thomas J. Weiler, arXiv:astro-ph/0503229.

[14] J. Ahrens et al. (AMANDA Collaboration), Phys. Rev. Lett. **90**, 251101 (2003), arXiv:astro-ph/0309585.

[15] P. Gondolo, G. Ingelman and M. Thunman, Nucl. Phys. Proc. Suppl. **48**, 472 (1996), arXiv:hep-ph/9602402.

[16] F. W. Stecker and M. H. Salamon, Astrophys. J. **512**, 521 (1992), arXiv:astro-ph/9808110; A. Atoyan and C. D. Dermer, Phys. Rev. Lett. **87**, 221102 (2001), arXiv:astro-ph/0108053 and references therein. F. W. Stecker, Phys. Rev. D. **72**, 107301 (2005), arXiv:astro-ph/0510537 for a recent update.

[17] K. Mannheim, R. J. Protheroe and J. P. Rachen, Phys. Rev. D **63**, 023003 (2001), arXiv:astro-ph/9812398.

[18] D. Guetta et al., Astropart. Phys. **20**, 429 (2004), arXiv:astro-ph/0302524

[19] J. P. Rachen and P. L. Biermann, Astron. & Astrophys. **272**, 161 (1993), arXiv:astro-ph/9301010.

[20] E. Waxman and J. N. Bahcall, Phys. Rev. Lett. **78**, 2292 (1997), arXiv:astro-ph/9701231.

[21] M. Vietri, Phys. Rev. Lett. **80**, 3690 (1998), arXiv:astro-ph/9802241.

[22] M. Bottcher and C. D. Dermer, M. Bottcher and C. D. Dermer, arXiv:astro-ph/9801027, Astrophys. J. **574**, 65 (2002), arXiv:astro-ph/0005440.

[23] R. Engel, D. Seckel and T. Stanev, Phys. Rev. D **64**, 093010 (2001), astro-ph/0101216, and references therein.

[24] A. Loeb and E. Waxman, Jan. 2006, submitted to Phys Rev. Lett.

[25] V. S. Berezinsky and V. A. Kudryavtsev, *Sov. Astron. Lett.* **14** 873 (1998)

[26] Talks at Gamma 2004, Heidelberg, Germany, 2004; H. J. Volk, E.G. Berezhko, Leonid T. Ksenofontov, Submitted to Astron. & Astrophys, arXiv:astro-ph/0409453.

[27] J. Alvarez-Muniz and F. Halzen, Ap. J. **576**, L33 (2002).

[28] F. Aharonian, L. O'C. Drury, & H.J. and Völk, *A&A*, **285**, 645 (1994)

[29] D. Berge et al. (HESS collaboration), 3rd International Symposium on High-Energy Gamma-ray Astronomy, Heidelberg, Germany.

THEORETICAL ASPECTS OF HIGH ENERGY NEUTRINOS AND GRB

P. MÉSZÁROS and S. RAZZAQUE

Department of Astronomy and Astrophysics, Department of Physics, Pennsylvania State University, University Park, PA 16802, USA
E-mail: nnp@astro.psu.edu, soeb@astro.psu.edu

Abstract: Neutrinos at energies ranging from sub-TeV to EeV from astrophysical sources can yield interesting physical information about fundamental interactions, about cosmic rays and about the nature of the sources and their environment. Gamma-ray bursts are a leading candidate source, and their expected neutrino emission can address a number of current questions, which may be answered with forthcoming experiments such as IceCube, Auger, ANITA and KM3NeT.

1. Introduction

The origin of the observed ultrahigh-energy (UHE) cosmic-rays (CRs) above the "ankle", roughly at EeV ($= 10^{18}$ eV) energy, of the CR energy spectrum is most probably extra-galactic. Any galactic origin at this energy, due to small magnetic deflections, would result in an anisotropic distribution of their arrival direction contrary to the observed data. The requirement that they are not attenuated by the cosmic microwave background through photo-meson ($p\gamma$) interactions constrains them to have originated within a radius of 50-100 Mpc, the so-called "GZK" volume.[1,2] Two broad classes of models suggested are the "top-down" scenarios, which attribute UHE-CRs to the decay of fossil Grand Unification defects, and the "bottom-up" scenarios, which assume UHECRs are accelerated in astrophysical sources.

The observed UHECR energy injection rate into the universe is $\sim 3\times10^{44}$ erg Mpc^{-3} yr^{-1} above the ankle. This is similar to the $\sim$ 0.1-1 MeV γ-ray energy injection rate by the local gamma-ray bursts (GRBs) which led to postulating that GRBs are the sources of UHECRs [3,4]. This coincidence has been corroborated using new data and further considerations[5,6,7], making GRBs promising candidates for UHECRs. Other candidates, in the bottom-up scenario, are active galactic nuclei (AGNs), and cluster accretion shocks. An unavoidable by-product of UHECR acceleration is the production of UHE neutrinos, via $p\gamma$ and pp, pn interactions. We limit our discussion to UHE neutrinos from GRBs here.

2. Nature of the High Energy Emission from GRBs

In the most widely accepted GRB model, the *fireball shock model*, the prompt γ-rays are produced by shocks in the plasma material ejected in a jet moving relativistically (with a bulk Lorentz factor $\Gamma \gtrsim 100$), usually taken to be (*internal shocks*), or in other versions an *external shock* (see, e.g. [8]). Such jets can arise from the core collapse of massive stars, convincingly shown to be the progenitor of long GRBs, or from mergers of compact binary systems (neutron star-neutron star, black hole-neutron star), which may be implicated in producing short GRBs. Late time collision of the jet material with an external medium (external shocks) produce a long lasting x-ray, UV and optical radiation, collectively known as the GRB afterglow.

The highly relativistic nature of the outflows is inferred from and constrained by the observations of GeV photons which avoid attenuation by $\gamma\gamma \to e^{\pm}$ production *in situ.* The probable mechanism(s) responsible for the observed photons is/are synchrotron radiation or/and inverse Compton (IC) scattering by high energy electrons. These electrons are accelerated by the relativistic shocks via the Fermi mechanism in the tangled magnetic field, resulting in a power-law energy distribution. The high bulk Lorentz factors result in synchrotron spectra which in the observer frame extend beyond 100 MeV, and IC scattering of such synchrotron photons leads to the expectation of GeV and TeV spectral components.[9] While $\lesssim 18$ GeV photons have been observed,[10] TeV photons are likely to be degraded to lower energies by $\gamma\gamma$ pair production, either in the source itself,[11] or (unless the GRB is at very low redshifts) in the intervening intergalactic medium.[12,13]

GRBs are likely to be more luminous in neutrinos, gravitational waves and cosmic rays compared to sub-GeV electromagnetic channels which comprise a small fraction of the burst kinetic energy. A significant amount of baryons (neutrons and protons) are expected to be present in the GRB jet along with leptons, each with $\Gamma m_p c^2 \gtrsim 100$ GeV bulk kinetic energy in the observer frame. Protons are also expected to co-accelerate with electrons in the internal and external shocks by the same Fermi mechanism. Using the shock parameters inferred from broad-band photon spectral fits, one infers that protons can be accelerated to Lorentz factors up to $\lesssim 10^{11}$ in the observer frame, i.e. to the GZK energy of $E_p \sim 10^{20}$ eV.

3. High Energy Neutrinos

High energy neutrinos, detectable by the neutrino telescopes such as IceCube in the $\sim$ 100 GeV-EeV range, are produced in the GRBs in a way similar to the beam-dump experiments in particle accelerators. Shock-accelerated protons interacting with ambient radiation and/or plasma material by photonuclear ($p\gamma$) and/or inelastic nuclear (pp/pn) collisions produce charged pions ($\pi^{\pm}$) and neutral pions (π^0). Neutrinos are produced from $\pi^{\pm}$ decays along with muons and electrons.

Such neutrinos may serve as diagnostics of the presence of relativistic shocks, and as probes of the acceleration mechanism and the magnetic field strength. The flux and spectrum of EeV neutrinos depends on the density of the surrounding gas, while the TeV-PeV neutrinos depend on the fireball Lorentz factor. Hence, the detection of very high energy neutrinos would provide crucial constraints on the fireball parameters and GRB environment. Lower energy ($\lesssim$TeV) neutrinos originating from sub-stellar shocks, on the other hand, may provide useful information on the GRB progenitor.

3.1. *Neutrinos contemporaneous with the gamma-rays*

With an initial $\Gamma_i = 300\ \Gamma_{300}$ and a variability time scale $\delta t = 10^{-3}\delta t_{-3}$ s, internal shocks in the GRB jet take place at a radius $r_i \sim 2\Gamma_i^2 c\delta t \sim 5 \times 10^{12}\delta t_{-3}\Gamma_{300}^2$ cm. The fireball becomes optically thin at a radius $\lesssim r_i$ allowing observed γ-ray emission. Shock accelerated protons interact dominantly with observed synchrotron photons with $\sim$MeV peak energy in the fireball to produce a Δ^+ resonance as $p\gamma \to \Delta^+$. The threshold condition to produce a Δ^+ is $E_p E_\gamma = 0.2\Gamma_i^2$ GeV^2 in the observer frame, which corresponds to a proton energy of $E_p = 1.8 \times 10^7 E_{\gamma,\mathrm{MeV}}^{-1}\Gamma_{300}^2$ GeV. The short-lived Δ^+ decays either to $p\pi^0$ or to $n\pi^+ \to n\mu^+\nu_\mu \to ne^+\nu_e\bar{\nu}_\mu\nu_\mu$ with roughly equal probability. It is the latter process that produces high energy neutrinos in the GRB fireball, contemporaneous with the γ-rays.[14] The secondary π^+ receive $\sim$ 20% of the proton energy in such an $p\gamma$ interaction and each secondary lepton roughly shares 1/4 of the pion energy. Thus each flavor (ν_e, $\bar{\nu}_\mu$ and ν_μ) of neutrino is emitted with $\sim$ 5% of the proton energy, dominantly in the PeV ($= 10^{15}$ eV) range, with equal ratios.

The diffuse muon neutrino flux from GRB internal shocks due to proton acceleration and subsequent $p\gamma$ interactions is shown as the short dashed line in Fig. 1. The flux is compared to the Waxman-Bahcall limit of cosmic neutrinos from optically thin sources, which is derived from the observed cosmic ray flux.[15] The fluxes of all three neutrino flavors (ν_e, ν_μ and ν_τ) are expected to be equal after oscillation in vacuum over astrophysical distances.

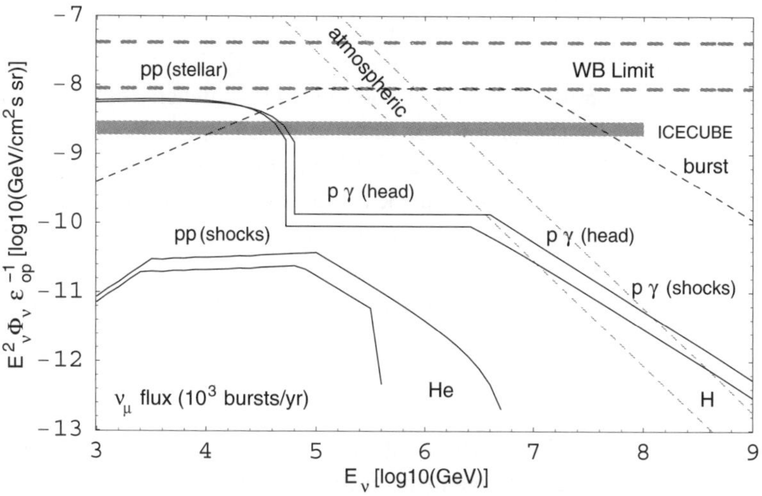

Figure 1. Diffuse ν_μ flux arriving simultaneously with the γ-rays from shocks outside the stellar surface in observed GRB (dark short-dashed curve), compared to the Waxman-Bahcall (WB) diffuse cosmic ray bound (light long-dashed curves) and the atmospheric neutrino flux (light short-dashed curves). Also shown is the diffuse muon neutrino precursor flux (solid lines) from sub-stellar jet shocks in two GRB progenitor models, with stellar radii $r_{12.5}$ (H) and r_{11} (He). These neutrinos arrive 10-100 s before the γ-rays from electromagnetically detected bursts (with similar curves for ν_μ, ν_e and ν_τ).

3.2. *Neutrinos from the GRB afterglow*

The GRB afterglow arises when relativistic jetted plasma material starts being slowed down by the external medium (e.g. the interstellar medium, or ISM), driving a blast wave ahead of the jet. This produces an external forward shock or blast wave, and a reverse shock in the jet. The external shock takes place at a radius $r_e \sim 4\Gamma_e^2 c\Delta t \sim 2 \times 10^{17}\Gamma_{250}^2 \Delta t_{30}$ cm which is well beyond the internal shock radius.[15] Here $\Gamma_e \approx 250\Gamma_{250}$ is the bulk Lorentz factor of the ejecta after the partial energy loss from emitting γ-rays in the internal shocks, and $\Delta t = 30\Delta t_{30}$ s is the duration of the GRB jet. Neutrinos are produced in the external reverse shock due to $p\gamma$ interactions of shock accelerated protons predominantly with synchrotron soft x-ray photons produced by electrons. The energy of the neutrinos from the afterglow would be in the EeV range as more energetic protons interact with these soft photons to produce Δ^+. The efficiency of proton to pion conversion by $p\gamma$ interactions in the external shocks (afterglow) is typically smaller than in the internal shocks because $r_e \gg r_i$, implying lower photon density.

In the case of a massive star progenitor the GRB jet may be expanding into a stellar wind much denser than the typical ISM density of $n \simeq 1$ cm^{-3}, which is emitted by the progenitor prior to its collapse. For a wind with mass loss rate of $\sim 10^{-5} M_\odot$ yr^{-1} and velocity of $v_w \sim 10^3$ km/s, the wind density at the typical external shock radius would be $\simeq 10^4$ cm^{-3}. The higher density implies a lower Γ_e, and hence a larger fraction of proton energy lost to pion production. Protons of energy $E_p \gtrsim 10^{18}$ eV lose all their energy to pion production in this scenario producing EeV neutrinos.[16]

3.3. *Precursor neutrinos*

In the long duration GRBs, the relativistic jet is expected to be launched near the central black hole resulting from the collapse of the stellar core, hence the jet is initially buried deep inside the star. As the jet burrows through the stellar material, it may or may not break through the stellar envelope.[17] Internal shocks in the jet, while it is burrowing through the stellar interior, can produce high energy neutrinos due to accelerated protons, dominantly below ~ 10 TeV, through pp and $p\gamma$ interactions.[18] The jets which successfully penetrate through the stellar envelope result in GRBs (γ-ray bright bursts), while the jets which choke inside the stars do not produce GRBs (γ-ray dark bursts). However, in both cases high energy neutrinos can be produced in the internal shocks, which slice through the stellar envelope since they interact very weakly with matter.

These neutrinos from the relativistic buried jets are emitted as precursors ($\sim$ 10-100 s prior) to the neutrinos emitted from the GRB fireball in case of an electromagnetically observed burst. In the the case of a choked burst (electromagnetically undetectable) no direct detection of neutrinos from individual sources is possible. However the diffuse neutrino signal is boosted up in both scenarios. The diffuse neutrino flux from two progenitor star models are shown in Fig. 1, one for a blue super-giant (labeled H) of radius $R_* = 3 \times 10^{12}$ cm and the other a Wolf-Rayet type (labeled He) of radius $R_* = 10^{11}$ cm. The Waxman-Bahcall diffuse cosmic ray bound,[19] the atmospheric flux and the IceCube sensitivity to diffuse flux are also plotted for comparison. The neutrino component which is contemporaneous with the gamma-ray emission (i.e. which arrives after the precursor) is shown as the dark dashed curve, and is plotted assuming that protons lose all their energy to pions in $p\gamma$ interactions in internal shocks.

3.4. *Early np decoupling non-thermal neutrinos*

Neutrons are expected to be present in considerable numbers in the GRB jet ($n_n \simeq n_p$) because of a *neutronized* core similar to that in supernovae in the case of long GRB, and from neutron star material in the case of a short GRB. In the long GRB, the core collapse neutronization leads to copious thermal ($\sim$ 10 MeV) neutrinos, but due to their low energy, their cross section is too small for detection at cosmological distances. However, in both long and short GRB outflows, neutrons are present, and are initially coupled to the protons by elastic nuclear scattering. If the initial acceleration of the fireball is very high, the neutrons can eventually decouple from the fireball, when the comoving expansion time falls below the nuclear scattering time. Protons, on the other hand, continue accelerating and expanding with the fireball as they are coupled to the electrons by Coulomb scattering. The relative velocity between the protons and neutrons, in such a case, can get high enough for inelastic interactions (np) above the pion production threshold of $\sim$ 140 MeV, leading to $\sim$ 10 GeV neutrinos in the observer's frame.[20,21,22] Highly neutron-enriched ($n_n \sim 10n_p$) jet in case of short GRBs may lead to $\sim$ 50 GeV neutrinos, as the relative velocity between the protons and neutrons increases substantially, which are detectable from a nearby burst.[23]

4. GRB-Supernova Connection

A fraction of long GRBs have recently been shown to be associated with supernovae of type Ib/c.[24] A GRB jet loaded with baryons would then leave long-lasting UHE CR, neutrino and photon signatures in those supernova remnants which were associated with a GRB at the time of their explosion. One example may be the SN remnant W49B which is probably a GRB remnant. A signature of a neutron component in the relativistic jet outflow would be a TeV γ-ray signature due to inverse Compton interactions following neutron decay.[25]. Another example may be some of the HESS unidentified sources.[26] Neutron decay would also give rise to TeV neutrinos. The imaging of the surrounding emission could provide new constraints on the jet structure of the GRB.

Cosmic-rays accelerated in the GRB remnant, similar to SN remnants which are observed as TeV γ-ray sources such as RX J1713.7-3946, [27] would also be expected to produce UHE neutrinos. Expected neutrino and γ-ray energy, commonly originating from $p\gamma$ and/or pp/pn interactions, would be higher in case of GRB remnants because of the higher expansion velocity.

5. Neutrino Flavor Astrophysics

High energy neutrinos from astrophysical optically thin sources are expected to be produced dominantly via $p\gamma$ interactions. Subsequent decay of π^+ and neutrino flavor oscillations in vacuum lead to an observed anti-electron to total neutrino flux ratio of $\Phi_{\bar{\nu}_e} : \Phi_\nu \simeq 1 : 15$.[28] At high energy this ratio may be lower even,[29] since the muons suffer significant electromagnetic energy loss prior to decay.[30] In the case of pp/pn interactions, typically attributed to optically thick sources, $\pi^\pm$ are produced in pairs and the corresponding expected flux ratio on Earth is $\Phi_{\bar{\nu}_e} : \Phi_\nu \simeq 1 : 6$. However even in the optically thin sources the nominal $\Phi_{\bar{\nu}_e} : \Phi_\nu$ ratio may be enhanced above $1 : 15$ by $\gamma\gamma \to \mu^\pm$ interactions and subsequent $\mu^\pm$ decays.[31] The targets are usual synchrotron photons and UHE incident photons are provided by the $p\gamma \to p\pi^0 \to p\gamma\gamma$ channel itself. This mechanism yields an enhancement ratio $\Phi_{\bar{\nu}_e} : \Phi_\nu \simeq 1 : 5$ solely from $\mu^\pm$ decays.

Measurement of the $\bar{\nu}_e$ to ν flux ratios may be possible by IceCube at the Glashow resonant interaction $\bar{\nu}_e e \to W^- \to$ anything at $E_\nu \simeq 6.4$ PeV.[32] Any enhancement over the $1 : 15$ ratio, e.g., from a single nearby GRB would then suggest a $\gamma\gamma$ origin. However, the flux of $\gamma\gamma$ neutrinos depend on the source model such as magnetization, radius etc. We have plotted the $\bar{\nu}_e$ to ν flux ratio in Fig. 2, which includes the contribution from $p\gamma$ and $\gamma\gamma$ channels, from a GRB internal shocks with different model parameters. The solid, dashed, dotted-dash and dotted lines correspond to the magnetization parameter $\varepsilon_B = 10^{-1}$, 10^{-2}, 10^{-3} and 10^{-4} respectively. The shocks take place at the photosphere ($r_{\rm ph}$) and at a radius $10r_{\rm ph}$.[31] Note that the ratio is enhanced from the $p\gamma$ value of $1/15$ in the small energy range where $\gamma\gamma$ interactions contribute significantly. This result then may be used to learn about the GRB model parameters.

6. Conclusions

Although fireball shock model is the leading GRB scenario, there is no strong direct proof so far for the internal shock or the reverse shock origin of the observed radiation. High energy neutrino emission from GRBs would

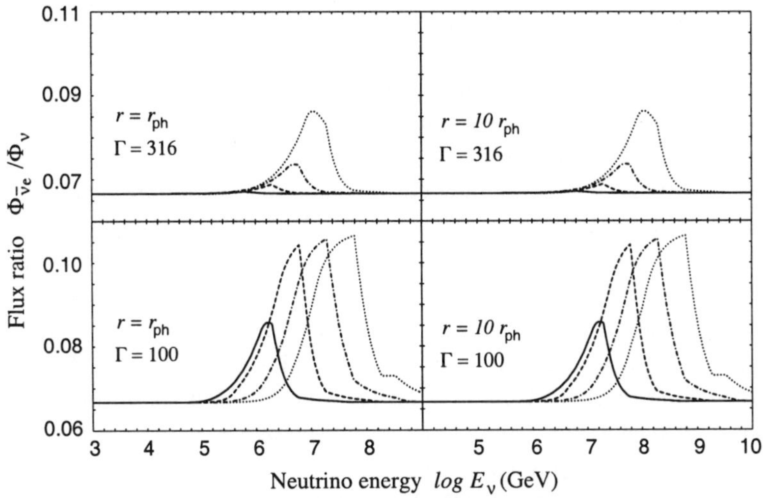

Figure 2. Expected anti-electron to total neutrino flux ratio: Φ_{ν_e}/Φ_ν on Earth from a GRB after vacuum oscillations as function of neutrino energy. The fluxes are both from the $p\gamma \rightarrow n\pi^+$ and $\gamma\gamma \rightarrow \mu^+\mu^-$ interactions. Depending on the GRB model parameters such as the internal shock radius (at the photosphere $r_{\rm ph}$ and at $10r_{\rm ph}$), bulk Lorentz factor (Γ) and magnetization ($\varepsilon_B = 10^{-1}$, 10^{-2}, 10^{-3} and 10^{-4} denoted by solid, dashed, dotted-dash and dotted lines respectively), the flux ratio may be enhanced from the nominal 1/15 value in certain energy ranges.

serve as a direct test for this, as well as for "baryonic" jet models, where the bulk of the energy is carried by baryons. On the other hand, an alternative Poynting flux dominated GRB jet model would have to rely on magnetic dissipation and reconnection, accelerating electrons and hence also accelerating protons– but there would be much fewer protons to accelerate and probably to much lower energy.

The *Pierre Auger Observatory*, a CR detector currently under construction, will have very large (~ 3000 km^2 each for its two location in the Southern and Northern hemisphere) area.[33] It will help to disentangle the two scenarios (top-down or bottom-up) and will reveal whether a GZK feature indeed exists by greatly improving the UHECR count statistics. Within the bottom-up scenario, the directional information may either prove or significantly constrain the alternative AGN scenario, and may eventually shed light on whether GRBs are indeed the sources of UHECRs.

Upcoming experiments such as IceCube,[34] ANITA,[35] KM3NeT,[36] and Auger[33] are currently being built to detect high energy astrophysical neutrinos. They can provide very useful information on the particle acceleration, radiation mechanism and magnetic fields, as well as about the sources and their progenitors. Direct confirmation of a GRB origin of UHECRs is difficult but the highest energy neutrinos may indirectly serve that purpose pointing directly back to their sources. Most GRBs are located at cosmological distances (with redshift $z \sim 1$) and individual detection of them by km scale neutrino telescopes may not be possible. The diffuse neutrino flux is then dominated by a few nearby bursts. The likeliest prospect for UHE ν detection is from these nearby GRBs in correlation with electromagnetic detection.

The prospect for high energy neutrino astrophysics is very exciting, with AMANDA already providing useful limits on the diffuse flux from GRBs[37,38] and with IceCube[39,40] on its way. The detection of TeV and higher energy neutrinos from GRBs would be of great importance for understanding the astrophysics of these sources such as the hadronic vs. the magnetohydrodynamic composition of the jets, as well as the CR acceleration mechanisms involved. High energy neutrinos from GRBs may also serve as probes of the highest redshift generation of star formation in the Universe, since they can travel un-attenuated, compared to the conventional electromagnetic astronomical probes.

Acknowledgments

Work supported by NSF grant AST0307376 and NASA grant NAG5-13286.

References

1. K. Greisen, *Phys. Rev. Lett.* **16**, 748 (1966).
2. G. T. Zatsepin and V. A. Kuzmin, *JETP Lett.* **4**, 78 (1966) [*Pisma Zh. Eksp. Teor. Fiz.* **4**, 114 (1966)].
3. M. Vietri, *Astrophys. J.* **453**, 883 (1995).
4. E. Waxman, *Phys. Rev. Lett.* **75**, 386 (1995).
5. E. Waxman *Astrophys. J.*, **606**, 988 (2004).
6. M. Vietri, D. de Marco and D. Guetta, *Astrophys. J.* **592**, 378 (2003).
7. S. Wick, C. Dermer, A. Atoyan, *Astropar. Phys.*, **21**, 125 (2004)
8. P. Mészáros, *Annu. Rev. Astron. Ap.*, **40**, 137 (2002).
9. P. Mészáros, M. J. Rees and H. Papathanassiou, *Astrophys. J.* **432**, 181 (1994).
10. K. Hurley *et al.*, *Nature* **372**, 652 (1994).
11. S. Razzaque, P. Mészáros and B. Zhang, *Astrophys. J.* **613** 1072 (2004).
12. P. Coppi and F. Aharonian, 1997, *Astrophys. J.* **487**, L9 (1997).
13. O. C. de Jager and F. W. Stecker, *Astrophys. J.* **566**, 738 (2002).
14. E. Waxman and J. Bahcall, *Phys. Rev. Lett.* **78**, 2292 (1997).

15. E. Waxman and J. N. Bahcall, *Astrophys. J.* **541**, 707 (2000).
16. Z. G. Dai and T. Lu, *Astrophys. J.* **551**, 249 (2001).
17. P. Mészáros and E. Waxman, *Phys. Rev. Lett.* **87**, 171102 (2001).
18. S. Razzaque, P. Mészáros and E. Waxman, *Phys. Rev.* **D68**, 083001 (2003).
19. E. Waxman and J. Bahcall, *Phys. Rev.* **D59**, 023002 (1999).
20. E. V. Derishev, V. V. Kocharovsky and Vl. V. Kocharovsky, *Astrophys. J.* **521**, 640 (1999).
21. J. N. Bahcall and P. Mészáros, *Phys. Rev. Lett.* **85**, 1362 (2000).
22. P. Mészáros and M. J. Rees, *Astrophys. J.* **541**, L5 (2000).
23. S. Razzaque and P. Mészáros, *Astrophys. J.* (submitted), astro-ph/0601652.
24. M. Della Valle, astro-ph/0504517.
25. K. Ioka, S. Kobayashi and P. Mészáros, *Astrophys. J.* **613**, L171 (2004).
26. A. Atoyan, J. Buckley and H. Krawczynski, *Astrophys. J*, **642**, L43 (2006)
27. J. Alvarez-Muniz and F. Halzen, *Astrophys. J.* **576**, L33 (2002).
28. J. G. Learned and S. Pakvasa, *Astropart. Phys.* **3**, 267 (1995).
29. T. Kashti and E. Waxman, *Phys. Rev. Lett.* **95**, 181101 (2005).
30. J. P. Rachen and P. Mészáros, *Phys. Rev.* **D58**, 123005 (1998).
31. S. Razzaque, P. Mészáros and E. Waxman, *Phys. Rev.* **D** (in press), astro-ph/0509186.
32. L. A. Anchordoqui, H. Goldberg, F. Halzen and T. J. Weiler, *Phys. Lett.* **B621**, 18 (2005).
33. http://www.auger.org/
34. http://icecube.wisc.edu/
35. http://www.ps.uci.edu/ anita/
36. http://km3net.org/
37. M. Stamatikos *et al.*, *AIP Conf. Proc.* **727**, 146 (2004).
38. J. Becker *et al.*, *Astropart. Phys.* **25**, 118 (2006).
39. J. Ahrens *et al.*, *New Astron. Rev.* **48**, 519 (2004).
40. P. O. Hulth, in *NO-VE 2006, Neutrino Oscillations in Venice*, Italy (astro-ph/0604374).

PRECISE MEASUREMENT OF LOW ENERGY (<TEV) COSMIC-RAY FLUX WITH BESS*

AKIRA YAMAMOTO
FOR THE BESS COLLABORATION†
High Energy Accelerator Research Organization (KEK), Tsukuba, Ibaraki, 305-0801, Japan,

The BESS experiment precisely measured cosmic-ray flux in the energy range below 1 TeV with a high-precision, magnetic-rigidity spectrometer. It has provided fundamental cosmic-ray data, and has extended search for low energy antiparticle of novel cosmic origins. The BESS-Polar program is being carried out with long duration flight in Antarctica to realize unprecedented sensitivity in low energy cosmic-ray observation. This report describes recent results from BESS and the progress in the BESS-Polar program.

1. Introduciton

The Balloon-borne Experiment with a Superconducting Spectrometer (BESS) has been carried out since 1993 for precise measurement of cosmic-ray fluxes and sensitive search for cosmic-ray antiparticle of cosmic origin [1-5]. The absolute flux and spectral shape of primary cosmic rays are the basis to discuss the origin and the propagation history of the cosmic rays in the Galaxy. It is very important to minimize uncertainties of the primary cosmic-ray spectra in study of atmospheric neutrino oscillation, and the primary spectra are fundamental as input to calculate spectra of cosmic-ray secondary antiproton, background to the one of cosmic origin, produced by cosmic-ray interactions with the interstellar gas. Table 1 summarizes various measurements carried out with a series of the BESS experiment continuously upgraded. We report here the representing fundamental information: cosmic-ray spectra in the energy range below TeV and low energy cosmic-ray antiproton spectra annually measured with aiming at search for cosmic-ray antiparticle of cosmic origin. The status of the long duration flight in Antarctica, BESS-Polar, is also reported.

---

*This work is supported with "Kakenhi" by MEXT and JSPS and partly by RESCEU, University of Tokyo in Japan, and supported by NASA with grants in the US.

* The BESS Collaboration formed by KEK, University of Tokyo, Kobe University, ISAS/JAXA, in Japan, and by NASA/GSFC, Univ. of Maryland, and Dever Univ., in the U.S.A.

Table 1. Major observation/search with the BESS experiment.

Atm depth. [g/cm2]	Particles	E. range [GeV, GeV/n]	Uncertainty [%]	Year	Ref
5	P, He	1-120, 1-54	5, 10	1998	[6]
	P. He	1- 540, 1-250	15, 20	2002	[7]
	P, He	0.18 – 20		1993, 1997-02	[8,9]
	p-bar	0.18/0.1-4.2	10 @ 2 GeV	1093 – 2004	[10-19]
	D-bar	0.2-1	Upper Limit	1997-2000	[20]
	He-bar	1-14	Upper Limit	1993-2004	[21-24]
5 - 28	P, He, mu	0.5-10		2001	[25]
– (26)	P-bar	0.2 – 3.4		2001	[26]
742	mu	0.6 – 100		1999	[27]
	P, p-bar	0.25 – 3.3		1999	[28]
940~994,	Mu	0.6 - 21	3 (system.)	1995, 97, 99	[27, 29]
994	p-bar	0.2 – 3.4		1997	[26]

2. Magnetic Rigidity Spectrometer, BESS

The BESS instrument features a uniform, solenoidal magnetic field, a large geometrical acceptance with a horizontally cylindrical spectrometer configuration [3, 30 31]. Figure 1 shows the cross sectional views of the BESS spectrometer, in 2000, the upgraded BESS-TeV, 2002, and BESS-Polar, 2004. The thin solenoid magnet design allows incoming cosmic rays particles to penetrate the solenoidal magnetic field with minimum interactions, and may maximize the use of the magnetic field (1 T) in a compact design. In the volume of the solenoid, the rigidity of particles is measured by using drift chambers (JET/IDCs). Since the magnetic field is highly uniform inside the solenoid, the high resolution has been realized. These features provide high statistics and high resolution in momentum analysis. The particle identification is made by using Time-of-Flight (TOF) counters, the JET chamber, and by silica-aerogel Cherenkov counters. A shower counter was installed to identify electrons and muons in specific flights. The spectrometer was developed with a maximum detectable rigidity (MDR) of 200 GV in the BESS original design [30], and was improved to be 1400 GV in the BESS-TeV program [31]. Retaining the basic features of the BESS detector, JET/IDCs were upgraded with twice the measurement points and a better position resolution. The arc-shaped drift chambers called outer drift chambers (ODCs) were installed both on the top and bottom of the spectrometer and twice the long track as the previous BESS was obtained by four measurement points of each ODC. Calibration of the spectrometer system was carried out by using the accelerator beam at KEK [32]. In BESS-Polar program, an entirely new spectrometer has been developed with much reduced material in the spectrometer to focus on the lowest energy region in the cosmic-ray measurement [4-6]. A thin superconducting solenoid magnet provides a magnetic field of 0.8 T with a wall material density of 2.3 g/cm2, including the cryostat wall [33]. The recent development of high strength aluminum stabilized superconductor has enabled the coil design to be much thinner and much more transparent [34]. The warm bore of the cryostat acts as a pressure vessel for a central tracking detector (JET). No outer pressure-vessel is provided, and time-of-flight (TOF) counters and silica-aerogel Cherenkov counters (ACC) are placed in vacuum. A set of middle-TOF counters is additionally installed under the JET chamber for triggering the very low energy particles before stopping in the lower detector components. In this approach, the total material in the upper half of the detector is reduced to be 4.5 g/cm2, corresponding to 1/4 of the previous BESS spectrometer material to make the event trigger. A large solar power system is prepared for the detector operation for 20 days with a power supplying capacity of 900 W.

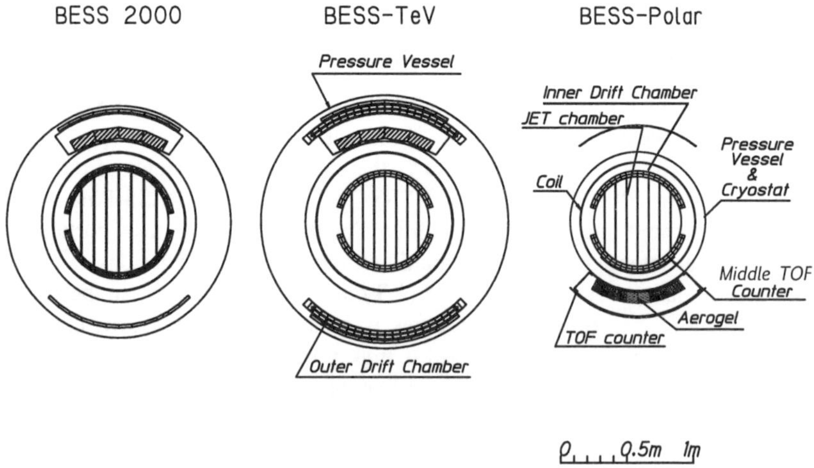

Fig. 1. Cross sections of BESS spectrometer and its upgrades to BESS-TeV and BESS-Polar.

3. Proton and Helium spectra below TeV

The proton and helium spectra were precisely measured by using BESS, and BESS–TeV spectrometer [6,7]. The first measurement with BESS, 1998, was made in the energy range up to 120 GeV for protons and 54 GeV/n for helium nuclei with the overall uncertainty less than 5 % for proton and 10 % for helium. The second measurement was carried out with BESS-TeV, 2002, and the energy range was extended up to 540 GeV for protons and 250 GeV/*n* for helium nuclei. The overall uncertainties were less than 15% for protons, 20% for helium nuclei. The measured results are shown in Fig. 2 in comparison with other experiments [7]. Although a small systematic shift of 2–3% was found in absolute proton flux between the results of BESS-TeV and BESS-98 from 30 to 100 GeV, the results are well consistent within the overall uncertainty of 5% [7]. Our resultant spectral shape of protons and helium nuclei is well consistent to that measured by AMS I [35]. Above 30 GeV, the absolute proton flux measure by BESS-TeV and AMS shows good agreement within 5%. Discrepancies in the spectra below 10 GeV for protons and 5 GeV/*n* for helium nuclei come from the difference in solar activity (around minimum in 1998 and near maximum in 2002).

At high energies, the spectrum *F* may be parameterized by a power law in kinetic energy, E_k, as [7]:

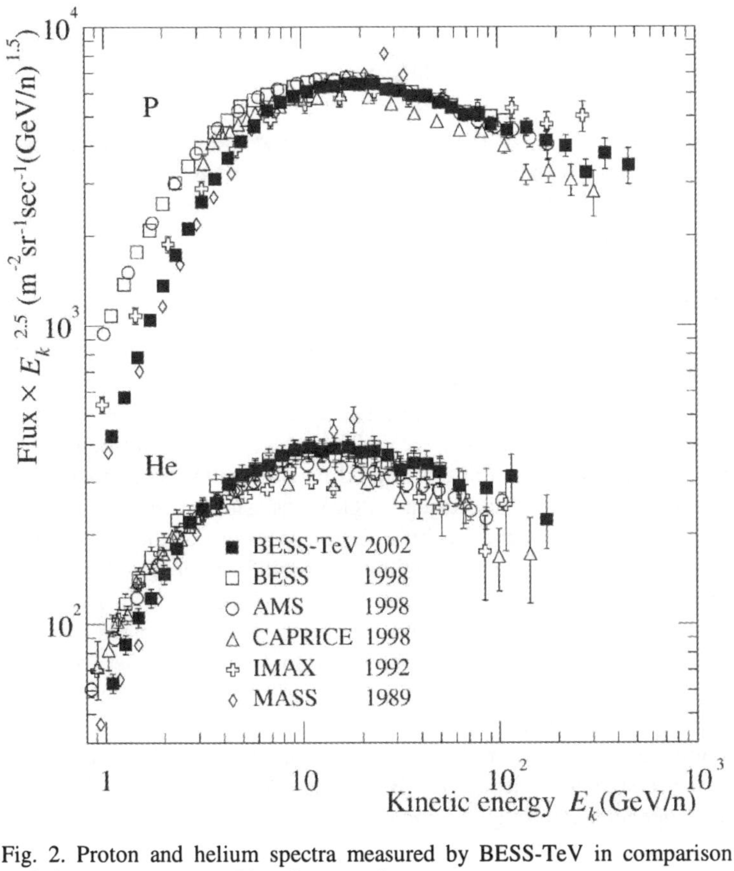

Fig. 2. Proton and helium spectra measured by BESS-TeV in comparison with the spectra measured in other experiments.

$$F = \Phi E^{-\gamma_k}.$$

The fitting range was chosen to be 30~540 GeV for protons and 20~250 GeV/*n* for helium nuclei so that the solar modulation effect was negligible. The best fit values and uncertainties for protons (Φ_p and γ_p) and helium nuclei (Φ_{He} and γ_{He}) were obtained as:

$$\Phi_p = (1.37\pm0.06(\text{sta.})\pm 0.11(\text{sys.})) \times 10^4\ (\text{m}^2\ \text{sr sGeV})^{-1},$$

$$\gamma_p = 2.732\pm 0.011(\text{sta.})\pm 0.019(\text{sys.})$$

and

$$\Phi_{He} = (7.06\pm 0.94(\text{sta.})\pm 1.17(\text{sys.})) \times 10^3\ (\text{m}^2\ \text{sr s (GeV}/n))^{-1},$$

$$\gamma_{He} = 2.699\pm 0.040(\text{sta.})\pm 0.044(\text{sys.}),$$

respectively [7]. The BESS-TeV (-2002) measured result with the above best-fit parameters are may be smoothly linked to the results from ATIC-I [*] and from RUNJOB [**]. The BESS-TeV result contributed to the precise calculation of the atmospheric neutrino flux.

4. Measurement of low-energy antiproton

"Mass-identified" low energy antiprotons below 1 GeV were first reported in BESS-1993 [10, 11] and more than thousand low energy antiprotons have been observed in the subsequent flights [12-19]. The energy spectrum was measured in an energy range 0.18 to 4.2 GeV. The cosmic-ray antiprotons of mostly secondary origin has been well understood by observing the characteristic peak of the energy spectrum at around 2GeV as shown in Fig. 3 (a). The measurements show, however, a little flatter spectrum at the very low energy region below 1 GeV in the solar minimum period of 1995 – 1997 [15]. It is therefore expected to extend the observation through the next solar minimum before drawing any conclusions on possible mixing of antiprotons from primary origins [36-38]. It is also very important to study the affect of the solar modulation on the spectrum, because the low energy cosmic rays are much influenced by solar conditions.

The antiproton to proton ratio is an ideal probe to study the solar modulation and its charge-sign dependence. The BESS experiment has been uniquely able to track this dependence through a solar activity variation during a period of 1997 ~

2004 in concluding the field polarity transition in 2000 [15-19]. Fig. 3 (b) shows temporal variation of the p-bar/p ratio according to the change of the solar modulation [32, 39]. We have observed stable p-ba/p ratio in the positive polarity phase through 1999 and a sudden increase in 2000 following the solar field reversal. The time dependent antiproton/proton ratio has been discussed with the spherically symmetric model [39] and the drift model [40]. Our measurement better agrees with the drift model, in the low energy region, although the spherical model may be consistent in higher energy region [41].

A search for antideuterons has been carried out using the data obtained from four balloon flights from 1997 through 2000, and no candidate was found [20].

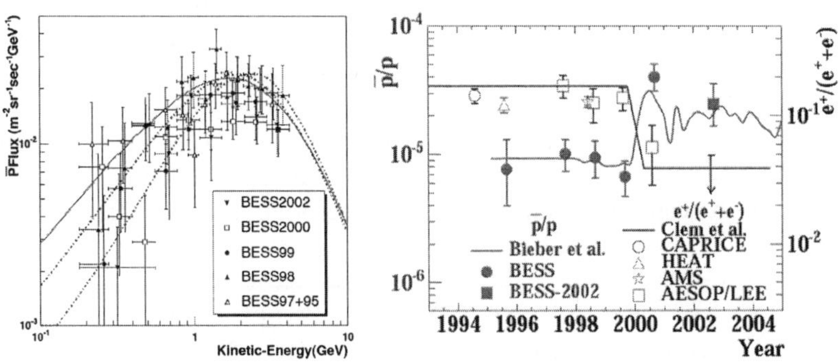

(a) (b)

Fig.3. (a) Energy spectra of antiprotons and (b) p-bar/p ratio during a period of the solar magnetic field polarity reversal at 2000. As shown in comparison, e+/e- ratio measured by Clem et al. [42] is consistently reversed to the p-bar/p ratio.

We have obtained, for the first time, a 95 % confidence level upper limit of 1.9 x 10^{-4} $(m^2 s.sr.\ GeV/n)^{-1}$ to the differential flux of the cosmic ray antideuterons at the top of atmosphere in an energy region between 0.17 and 1.15 GeV/n, including a systematic error of 10 %. This leads to an upper limit on the evaporation rate of local PBHs to be 2.3 x 10^0 pc^{-3} yr^{-1} (95 % C.L.).

The search for antihelium has been continuously carried out since 1993 [20-24]. No antihelium candidate has been detected in 6.6 x 10^6 observed helium events in a rigidity range of 1 – 14 GV accumulated in seven flights, 1993 – 2000. The resultant upper limit of the antihelium/helium flux ratio at the top of the atmosphere has been decreased down to < 6.8 x 10^{-7} with a 95 % confidence level under an assumption that antihelium and helium would have the same spectrum. Further accumulated data in 2004 is still in analysis.

5. Progress in BESS-Polar Project

A long duration balloon (LDB) flight over Antarctica (around the south pole) can provide an excellent opportunity for a high-statistical, low-energy antiproton measurement. It flies at the low geomagnetic cut-off region about 10 days for a single circumnavigation, so we can expect much higher statistics by one LDB Antarctic flight. Thus the BESS flight around the South-Pole, BESS-Polar, has been prepared since 2001 [3-5] and the first flight has been realized in December,2004. 900 million cosmic-ray events were recorded on the hard disk drives without event filtering, and the recorded data were securely recovered. The data analysis in progress, and four times larger statistics in low energy antiproton measurement has been obtained, with 8 days observation and with effectivly 2/3 of geometrical acceptance [19]. The second flight with the BESS-Polar spectrometer is to be realized in the end of solar minimum period, in 2007. As shown in Fig.4, the cosmic-ray flux is to be highest in the end of solar minimum, and we intend to realize the next BESS-Polar flight to be realized in Dec. 2007, with the detector upgrade based on the experience obtained in the BESS-Polar I, 2004. Thus unprecedented statistics of the cosmic-ray observation is expected to search for antiparticle of cosmic origin such as evaporation of PBH

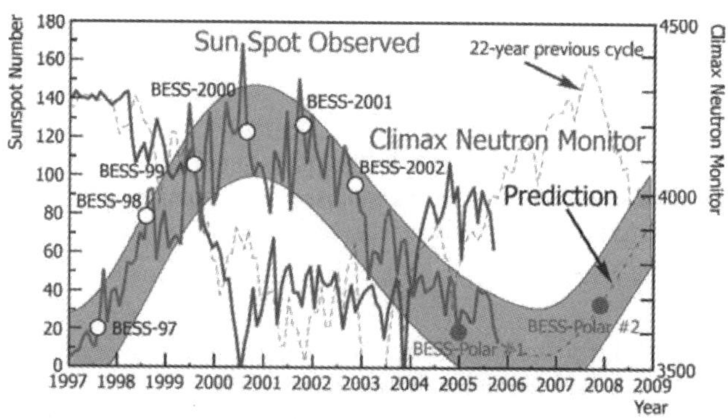

Fig. 4. Solar activity and the BESS-Flight record and plan.

6. Conclusion

The BESS experiment precisely measured comic-ray flux in the energy range below 1 TeV. It provided fundamental data of cosmic rays with various observations, and searched for low energy antiparticle with novel cosmic origins. The BESS-Polar program is being carried out with long duration flight in Antarctica to realize unprecedented statistics in cosmic-ray observation in solar minimum, to study fundamental cosmic-ray physics and to search for the antiparticle of cosmic-origin.

References

1. S. Orito, *KEK Report* 87-19,111, (Proc. of the ASTROMAG Workshop, (1987).
2. A. Yamaomoto et al, *Adv. Space Res.* **14**, (2) 75 (1994).
3. A. Yamamoto et al., *Adv. Space Res.,* **30** (5) 125 (2002).
4. J. Mitchell et al., *Nucl. Phys B (Proc Suppl.)* **134**, 31 (2004).
5. T. Yoshida et al., *Adv. Space Res.,* **33** (10) 755. (2004).
6. T. Sanuki et al., *Astrophys. J.*, **545**, 1135 (2000).
7. S. Haino et al., *Phys. Lett.* **B 594**, 35 (2004).
8. J. Z. Wang et al., *Astrphys. J.*, **564**, 244 (2002).
9. Y. Shikaze to be published.
10. K. Yoshimura et al., *Phys. Rev. Lett*, **75**, 3792 (1995).
11. A. Moiseev et al., *Astrophys. J.* , **474**, 479 (1997).
12. K. Yoshimura et al., *Phys. Rev. Lett*, **75**, 3792 (1995).
13. A. Moiseev et al., *Astrophys. J.* , **474**, 479 (1997).
14. H. Matsunaga et al., *Phys. Rev. Lett.,*.**81**, 4052 (1998).
15. S. Orito et al., *Phys. Rev. Lett*, **84**, 1078 (2000).
16. T. Maeno et al., *Astroparticle Phys*, **16**, 121 (2001).
17. Y. Asaoka et al., *Phys.Rev. Lett.* **88**, 051101 (2002).
18. S. Haino et al., *Proc. ICRC-29*, Pune (2005).
19. S. Matsuda, PhD thesis to be sumitted.
20. H. Fuke, *Phys. Rev. Lett.*, **95,** 081101 (2005).
21. J. Ormes et al., *Astrophys. J.*, **482**, 187 (1997).
22. T. Saeki et al., *Phys. Lett.*, **B 422,** 319 (1998).
23. M. Sasaki et al., *Nucl. Phys.* **B** (Proc. Suppl.) **113**, 202 (2002).
24. M. Sasaki et al., to be presented in COSPAR-06, Beijing (2006).
25. K. Abe et al., *Phys. Lett.*, **B 564**, 8 (2003).
26. K. Yamato et al., *Phys. Letter* **B 632** , 475 (2004).
27. T. Sanuki et al., *Phys. Lett.* **B 541**, 234 (2002).
28. T. Sanuki et al., *Phys. Lett.* **B 577,** 10 (2003).
29. M. Motoki et al., *Astroparticle Physics*, **19** , 113 (2003).
30. Y. Ajima et al., *Nucl. Instr. & Method.*, **A 443**, 71 (2000).
31. S. Haino et al., *Nucl. Instr. & Methods,* **A 594,** 35 (2004).
32. Y. Asaoka et al, *Nucl. Inst. & Methods,* **A 489**, 170 (2002).
33. Y. Makida et al., *IEEE Trans. Appl. Superc.*, **15** 1248 (2005).
34. A. Yamamoto et al., *Nucl. Phys* **B** (Proc. Suppl.) **78,** 56 (1999).
35. J. Alcaraz et al., *Phys Lett.* **B 490,** 27 (2000).
36. K. Maki et al., *Phys Rev. Lett.*, **76** , 3474 (1996),
37. T. Mmitsui et al., *Phys. Lett.* **B 389** , 169 (1996).
38. K. Yoshimura et al., *Adv. Space Res.* **27** (4) 693 (2001)..
39. S. Haino et al., *Proc. ICRC-29*, Pune, (2005).
40. L.A. Fisk et al.,J. Geophys. Res., 76 (1997) 221.
41. Bieber et al., *Phys. Rev. Lett.*, **83** , 674 (1999).
42. J. M. Clem et al., *Proc. ICRC-28*, Tsukuba (2003).

THE PAMELA COSMIC RAY TELESCOPE ON BOARD RESURS-DK1 SATELLITE: AN OVERVIEW OF HELIOSPHERIC OBSERVATION CAPABILITIES

M. CASOLINO

INFN and University of Rome Tor Vergata
Dept. of Physics
Via Della Ricerca Scientifica 1
00133 Roma, Italy
E-mail: casolino@roma2.infn.it

PAMELA is a satellite-borne experiment with the main purpose to measure the antiparticle component of cosmic rays over an extended energy range (80 MeV – 190 GeV for $\overline{p}$ and 50 MeV – 270 GeV for e^+) and with unprecedented accuracy. Other physics objectives are the measurement of the galactic, heliospheric and trapped components of cosmic rays (80 MeV – 700 GeV for p, 50 MeV – 400 GeV for e^-). The apparatus consists of a permanent magnetic spectrometer equipped with a double-sided silicon microstrip tracking system and surrounded by a scintillator anticoincidence system. A silicon-tungsten imaging calorimeter, complemented by a scintillator shower tail catcher performs the particle identification task. Fast scintillators are used for Time-of-Flight measurements and to provide the primary trigger. A neutron detector improves the particle identification, extends the range of particle measurements to the TeV region and allows to study solar and trapped neutron component. In this work we will focus on the description of the observational characteristics of the detector in the field of solar, heliospheric and trapped cosmic rays.

1. Detector Description

The PAMELA experiment is devoted to the study of cosmic rays, primarily its antiparticle component. The core of the instrument is a permanent magnet spectrometer[1] equipped with a double-sided, microstrip silicon tracker. Under the spectrometer lies a sampling electromagnetic calorimeter[5], composed of tungsten absorber plates and single-sided, strip silicon detector planes. A Time-of-Flight (ToF) system, made of six layers of plastic scintillator strips arranged in three planes is employed for particle identification at low energies and albedo rejection[3]. At the bottom of the instrument is a neutron detector[14], composed of ^{3}He counters enveloped in polyethylene moderator. Plastic scintillator counters are used for anticoincidence counting[18]; another scintillator between the calorimeter and the neutron detector is used to register the tail of particle showers. Data acquisition and reduction is performed by an on board CPU[6] which also stores data on a 2 Gbyte memory before transmission to the satellite and to the ground. After downlink, data are processed[7] immediately to assess in real time the conditions of the detectors. The instrument is housed in a pressurized container located on the side of the Russian Resurs-DK1 satellite, devoted primarily to Earth observations. The launch is scheduled in Summer 2006 with a Soyuz Frigat vector from the cosmodrome of Baikonur. The satellite will fly on a quasi-polar (inclination 70°), elliptical (altitude 350–600 km) orbit with an expected mission length of 3 years.

The orbit of the satellite, its long observational lifetime, and the structure of the detectors allow PAMELA to address several items of cosmic-ray physics, ranging from the study of the antiparticle component, with a statistic and over an energy range unreached by previous balloon-borne experiments[a], to the study of solar and trapped cosmic rays. The results will increase our knowledge of cosmic ray origin and propagation, as well as shed some light on some cosmological questions. In this work we will focus on the observational capabilities of PAMELA in respect to heliospheric cosmic rays.

2. Jovian electrons

Since the discovery made by Pioneer 10 of Jovian electrons with an energy between 1 and 25 MeV, at about 1 AU (Astronomical Unit) from Jupiter [23,11], several interplanetary missions have measured this component of cosmic rays. Currently we know that Jupiter is the strongest electron source below 25 MeV in the heliosphere within a radius of 11 AUs. Its spectrum has a power law with spectral index $\gamma = 1.65$, increasing above 25 MeV, where the galactic component becomes dominant. At 1 AU from the Sun IMP-8 has detected Jovian electrons in the range between 0.6 and 16 MeV and measure their modulation by the passage of Coronal Interaction Regions (CIR) with 27 days periodicity[11,10]. There are also long term modulation effects related to the Earth-Jupiter synodic year (13 month).

[a] A contemporary measurement with BESS[24] in the course of its next polar flight and with AMS[2] on the ISS will be fundamental in the reduction of the systematics of the instruments and improvement of observations. In case of common measurements with BESS it will be particularly important to compare data acquired close to the South Pole.

This is due to the fact that Jovian electrons are bound to move spiraling along the interplanetary magnetic field lines and thus when the two planets are on the same solar wind line the e^- transit from Jupiter to the Earth is eased and the flux increases. Vice versa when the two planets lie on different spiral lines the electron flux decreases.

With PAMELA it will be possible to study for the first time the high energy Jovian electron component and test the hypothesis of reacceleration at the solar wind Termination Shock (TS). It is known that cosmic rays originating outside the heliosphere can be accelerated at the solar wind TS. This applies also to Jovian electrons, which are transported outward by the solar wind, reach the TS and increase their energy through the process of shock acceleration. Some of these reaccelerated electrons are scattered back in the heliosphere[19] reaching also Earth. Jovian electrons are dominant in the 50–70 MeV range (where they represent the primary non-reaccelerated Jovian component) but at higher energies represent a fraction (about 1%) of the galactic flux. A precise measurement of their spectrum would provide information of the acceleration processes in Jupiter's magnetosphere; furthermore their long and short term modulation would give information on propagation phenomena in the inner heliosphere. In this range their detection is reduced by geomagnetic cutoff effects, reducing observations only in the regions close to the poles. At higher energy (from 70 up to $\simeq 2GeV$) the galactic and Jovian components will be separable by studying the temporal profile due to synodic modulation. The large energy range allows to gather a large number of events of reaccelerated electrons of Jovian origin in an energy range where they have never been observed.

Furthermore it is possible that the reacceleration of electrons at the solar wind TS is modulated by the solar cycle. With three years of observations toward the solar minimum it will be possible to detect also this effect. In addition to these phenomena, charge dependent modulation effects can be studied by comparing the temporal dependence of electron and positron spectra (the latter of galactic nature).

3. Solar Energetic Particles

The launch of PAMELA is expected at the minimum of solar activity, with observations in the phase going toward the next solar maximum. Considering events with an energy above the 80 MeV trigger rate[b], it is possible to

[b]Lower energy events could be studied using the single rate and studying the interaction of the solar particles with the geomagnetic field, for instance the lowering of the

estimate[22] about 10 significant solar events during the experiment lifetime. The rate of background particles hitting the top trigger scintillator could be very high for intense solar events, hence a different trigger configuration has to be set in these cases. The usual trigger involves a coincidence of at least an element of each of the three scintillators (one on top, one before and one after the magnetic spectrometer). During solar particle events a devoted trigger mask (e.g. excluding the top scintillator) can be programmed from ground[c]. The observation of solar energetic particle (SEP) events with a magnetic spectrometer will allow several aspects of solar and heliospheric cosmic ray physics to be addressed for the first time:

3.1. *Protons*

PAMELA will be able to measure the spectrum of cosmic-ray protons from 80 MeV up to almost 1 TeV and therefore will be able to measure the solar component over a very wide energy range (where the upper limit will be limited by statistics and thus dependent on the spectral shape). These measurements will be correlated with other instruments in different points of the Earth's magnetosphere to give information on the acceleration and propagation mechanisms of SEP events. Up to now there has been no direct measurement[16] of the high energy (>1 GeV) proton component of SEPs. The importance of a direct measurement of this spectrum is related to the fact[21] that there are many solar events where the energy of protons is above the highest ($\simeq$100 MeV) detectable energy range of current spacecrafts, but is below the detection threshold of ground Neutron Monitors[4]. However, over the PAMELA energy range, it will be possible to examine the turnover of the spectrum, where we find the limit of acceleration processes at the Sun. Our instrument has a maximum trigger rate of about 60 Hz and a geometrical factor of 20.5 cm^2 sr. This implies that we will be able to read all events with an integral flux (above 80 MeV) up to $\simeq$ 4 particles / cm^2 s sr. For such events we expect[8] about 2×10^6 particles/day (assuming a spectral index of $\gamma = 3$ we have 2×10^3 events / day above 1 GeV). It is necessary to note that larger solar particle events will saturate the trigger and the mass memory reducing the study of the event temporal distribution.

---

geomagnetic cutoff[17,15].

[c]This can be selected on the basis of information coming from the satellite monitoring system (e.g. SOHO, ACE, GOES). In this way observation and memory filling would therefore vary according to the event type (impulsive, gradual) and intensity.

3.2. *Positrons*

Positrons are produced mainly in the decay of π^+ coming from nuclear reactions occurring at the flare site. Up to now, they have only been measured indirectly by remote sensing of the gamma ray annihilation line at 511 keV. Using the magnetic spectrometer of PAMELA it will be possible to separately analyze the high energy tail of the electron and positron spectra at 1 AU from the Sun obtaining information both on particle production and charge dependent propagation in the heliosphere.

3.3. *Nuclei*

PAMELA can identify light nuclei up to Oxygen and isotopes of Hydrogen and Helium. Thus we can investigate into the light nuclear component related to SEP events over a wide energy range. Applying the same estimates as above, we can expect $\simeq 10^4$ ^{4}He and $\simeq 10^2$ ^{3}He nuclei for gradual events, and more for impulsive ones. This statistics will allow us to examine in detail the amount of the ^{3}He and deuterium up to 3 GeV/c. These measurements will help us to better understand the selective acceleration processes in the higher energy impulsive[20] events.

3.4. *Neutrons*

Neutrons are produced in nuclear reactions at the flare site and can reach the Earth before decaying. Although there is no devoted trigger for neutrons in PAMELA, the background counting of the neutron detector will measure in great detail the temporal profile and distribution of solar neutrons. The background counting system keeps track of the the number of neutrons which hit the neutron detector in the time elapsed since last trigger. The counter is reset each time it is read allowing for a precise measurement of background neutron conditions during the mission. On the occurrence of solar events, neutrons are expected to reach Earth before protons as they have no charge. They are not deflected by any magnetic field and will be directly recorded by PAMELA (if it is not in Earth's shadow). In addition, the number of observable solar "neutron" events is higher than the protons ones since neutrons are not bound by the interplanetary field lines and thus can reach Earth independently from the point of production on the Sun. Neutrons can also be indirectly observed by observing the increase of protons produced from neutron decay. These protons can reach the Earth propagating on interplanetary field lines not originally connected to the production site of the Solar surface[12].

4. Conclusions

In this work we have briefly described some of the observational possibilities of PAMELA in relation to solar and heliospheric physics. This will be the first time a magnetic spectrometer telescope in low Earth orbit will be operational for long duration observation. It will thus be possible to perform direct measurements in an energy range and with a precision up to now never reached in direct observations.

References

1. O. Adriani et al., *Nucl. Instr. and Meth. in Phys. Res. A* **511**, 72, (2003).
2. J. Alcaraz et al., *Phys. Lett. B*, **472**, 215, (2000).
3. G.C. Barbarino et al., *Nucl. Phys. (Proc. Suppl.) B* **125**, 298, (2003).
4. G.A. Bazilevskaya and A.K. Svirzhevskaya, *Sp. Sci. Rev.* **85**, 431, (1998).
5. M. Boezio et al., *Nucl. Instr. and Meth. in Phys. Res. A* **487**, 407, (2002).
6. M. Casolino et al., "The central processing unit of Pamela experiment", in press on *Adv. Sp. Res.*
7. M. Casolino et al., "YODA++, On ground mission control and data handling system for Pamela experiment", in press on *Adv. Sp. Res.*
8. M. Casolino et al., Proc. 28th ICRC, OG 1.5, 3477 Tsukuba, (2003).
9. M. Casolino et al., "Cosmic-ray observations of the heliosphere with the PAMELA experiment", in press on *Adv. Sp. Res.*
10. D.L. Chenette, *Jour. Geophys. Res.* **85**, 2243, (1980).
11. J.H. Eraker, Astrophys. Jour. **257**, 862, (1982).
12. P. Evenson, P. Meyer and K. R. Pyle, *Astrophys. Jour.* **274**, 875, (1983).
13. H. Fichtner, M. S. Potgieter, S. E. S. Ferreira, B. Heber, and R. A. Burger, Proc. 27th ICRC, SH 3666, Hamburg, (2001).
14. A.M. Galper et al., Proc. 27th ICRC, Hamburg,OG1.06, 2219 (2001).
15. R.A. Leske et al., *Jour. Geophys. Res.* A 12, 30011, (2001).
16. L. Miroshnichenko, *Solar Cosmic Rays*, Kluwer, (2001).
17. R.C. Ogliore et al., Proc. 27th, ICRC, SH3.6, 4221 Hamburg, (2001).
18. M. Pearce et al., Proc. 28th ICRC, OG 1.5, 2125, Tsukuba, (2003).
19. M.S. Poitgieter and S.E.S. Ferreira, *Jour. Geophys. Res.* A 7, SSH 1, (2002).
20. D.V. Reames, *Sp. Sci. Rev.* **90**, 413, (1999).
21. J.M. Ryan, *Sp. Sci. Rev.* **93**, 581, (2000).
22. M.A. Shea and D.F. Smart, Proc. 27th ICRC, SH1.07, 3401, Hamburg, (2001).
23. J.A. Simpson et al., *Science* **4122**, 306, (1974).
24. T. Yoshida et al, *Adv. Sp. Res.* **33**, 1755, (2004).

PARTICLE ACCELERATION IN KINETIC PLASMA PROCESSES

M. HOSHINO, S. ZENITANI, K. NAGATA, AND Y. TAKAGI

The University of Tokyo,
7-3-1 Hongo, Bunkyo, Tokyo 113-0033, JAPAN
E-mail: hoshino@eps.s.u-tokyo.ac.jp

The acceleration of high energy particles in astrophysical sources has attracted our attention for a long time, and their origins remain a central problem in astroplasma physics. We review recent progress that was made to understand both the dynamic structure of the source region and the connection with particle acceleration, and specifically discuss that not only the stochastic acceleration such as the first order Fermi acceleration, but also the direct acceleration seen in the shock front or in the reconnection region can contribute to the origins of the astrophysical nonthermal particles.

1. Introduction

Kinetic plasma processes are thought to be responsible for the generation of high energy particles in astrophysics, and shock waves and magnetic reconnection are postulated to a central "engine" of the production of energetic particles. The pulsar winds, gamma-ray bursts and blazers are the examples of the high energy particle acceleration sites emitting the non-thermal radiation. Among many acceleration mechanisms in plasma, they are basically classified into two kinds of particle acceleration mechanisms: one is a stochastic/Fermi acceleration under turbulent magnetic fields, and the other is a direct acceleration of charged particles where the particle gains successively energy from an electric field. So far both mechanisms of the stochastic/Fermi acceleration and the direct acceleration, more or less, have advantage and disadvantage for explaining the observed high energy particles.

In the Fermi acceleration, the particle can be accelerated by a multiple interaction of particle with turbulent fields. The first order Fermi acceleration at shock particularly met with great success, because it predicts a power-law energy spectrum with its power-law index of 2 that depends weakly on the shock compression ratio and because the index is believed to be close to that of the original energy spectrum in many astrophysical sources. However, the Fermi process at shock does not appear to produce spectra with its power-law index $s < 2$, and it is known that some high energy sources seen in the pulsar winds, gamma-ray bursts and blazers show harder spectra with $s < 2$, especially at the lower energy band[1,2,3]. Another open issue is the so-called injection problem, namely, how a certain part of particles are extracted from the thermal population and are injected into the Fermi acceleration process. This injection problem remains a subject of ongoing discussion in shock physics.

In the direct acceleration, it is known that several mechanisms can produce a power-law energy spectrum with a harder spectrum[4], but the generation of the characteristic power-law energy spectrum, in general, is not obvious. One of the advantages of the direct acceleration is the fact that the acceleration time scale is fast. The acceleration time scale in the direct acceleration is of the order of $(c/v)\Omega_c^{-1}$, while that in the first order Fermi acceleration is of the order of $(c/v)^2\Omega_c^{-1}$, where Ω_c, v and c are the cyclotron frequency, the shock speed and the speed of light, respectively. Therefore, the high energy particle in the course of the direct acceleration process can survive against radiation loss processes, and we could expect the higher maximum energy in the direct acceleration than that in the first order Fermi acceleration.

Recently the evidence accumulated from the observations on those astrophysical objects indicates that the energy spectra are relatively soft at the higher energy, while they are hard at the lower energy[5]. Therefore, the direct acceleration as well as the stochastic acceleration is believed to play an important contribution to the high energy particle sources seen in the astrophysical objects. The combination of the direct and the stochastic acceleration may be an alternative model to generate the high energy particles. In this paper, we study the kinetic plasma processes on relativistic shocks and reconnection in details, and discuss that the direct accelerations can serve the generation of the hard energy spectrum.

2. Relativistic Shock Waves

Relativistic shocks and their associated energetic particles have received considerable attention in recent years, because many sources of high energy radiation in the universe appear to involve highly relativistic plasma outflow from the central object. Recently the computer simulation by using particle-in-cell (PIC) simulation is encouraging the study of the particle acceleration. The basic principles of the PIC simulation solve the full set of Maxwell's equations self-consistently, and the particles are advanced in time using the relativistic Lorentz force equation. We review several simulation studies below.

2.1. *Perpendicular Shocks with Surfing Acceleration*

Many plasma instabilities as the energy dissipation for maintaining the shock structure are known to operate in the shock front region. In addition, it is also expected that the high energy, non-thermal particles are generated. In fact, non-thermal distribution is known to be often generated in association with some plasma instabilities in collisionless plasmas. In the direct acceleration, however, the formation of the non-thermal, power-law spectrum remains an elusive problem.

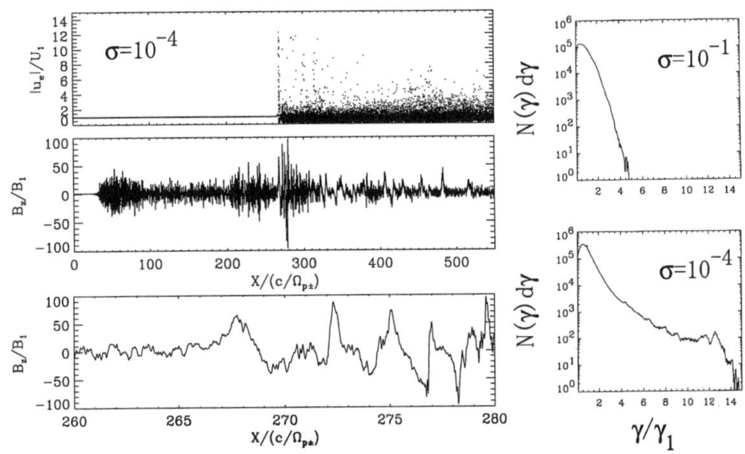

Figure 1. Shock structure for a perpendicular shock with $\sigma = 10^{-4}$ (left) and energy spectra for $\sigma = 10^{-1}$ and 10^{-4} (right).

Let us first discuss about the structure and its dynamics of a relativistic, perpendicular magnetosonic shock wave in electron-positron plasmas, where the magnetic field lines are perpendicular to the shock normal direction. A key parameter to control the shock dynamics is the so-called σ value, which is defined by $\sigma = B^2/(8\pi N m \gamma c^2)$, where B, N, m and γ are the upstream magnetic field, the density, the mass of particle, and the Lorentz factor of the upstream bulk flow speed, respectively. Based on the particle-in-cell (PIC) simulation studies, Langdon et al.[6] and Gallant et al.[7] discuss that the relativistic perpendicular shock is not necessarily good candidate of cosmic ray accelerator for a moderate σ, while Hoshino[8] argued that the non-thermal particles can be generated around the shock front for a small σ due to the shock surfing acceleration.

Figure 1 shows a relativistic shock structure with $\sigma = 10^{-4}$ and the energy spectra for two different σ parameters of $\sigma = 10^{-1}$ and 10^{-4}. The left-hand panels show, from top, the phase space of total momentum $|u_e|$, the magnetic field B_z, and the expanded view of B_z around the shock front. We can find a coherent, large-amplitude wave of B_z and its associated strong acceleration at the shock front $X \sim 268$. This kind of structure cannot be found for a moderate σ parameter of $\sigma > 10^{-1}$ (not shown here). The right-hand panels show the comparison of the particle acceleration for two different σ. For $\sigma = 10^{-1}$, the downstream energy spectrum is approximated by a relativistic Maxwellian, while for $\sigma = 10^{-4}$ nonthermal energy spectrum can be obtained.

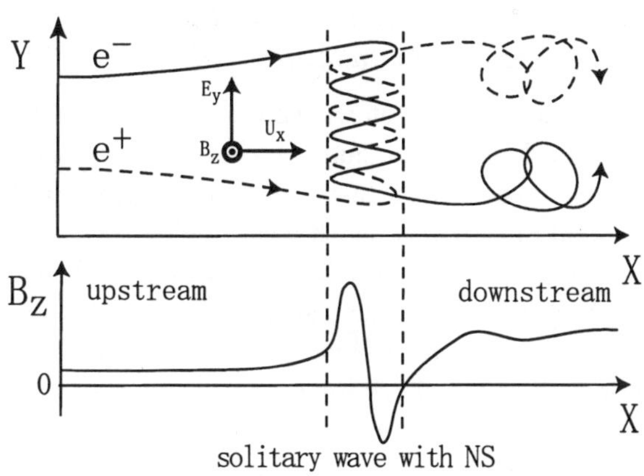

Figure 2. Schematic diagram of a shock surfing mechanism at a relativistic shock front.

Let us quickly mention the acceleration mechanism of the shock surfing. The shock surfing acceleration is usually discussed for the case of the ion-electron shock waves[9,10,11,12]. Due to the inertia difference between ions and electrons, the ion can penetrate deeper into the shock downstream, while the electron are quickly decelerated at the shock front. In order to maintain the charge neutrality at the shock front, the polarization electric field is produced. Then a part of the incoming ions, which energy is less than the potential energy of the polarization electric field, can be reflected from the shock front, and will be returned back by the Lorentz force of the upstream magnetic field, and then may recoil again. During this process, ions results in a drift along the surface of the shock front, which is parallel to the motional electric field. As a result, the particles are accelerated.

In electron-positron plasmas, an electrostatic potential structure is not induced at the shock front, while the current sheet formed at the shock front plays an alternative role of the particle trapping at the shock front region. The magnetic field polarity of the coherent, large-amplitude waves seen at the shock front in Figure 1 changes its direction, and the current sheet is formed. It is known that the structure is almost stable during the nonlinear evolution of a shock wave, and both positrons and electrons can be effectively trapped inside the current sheet. Figure 2 summarizes the idea of the magnetosonic surfing acceleration.

2.2. *Turbulence at Shock Front: Weibel Instability and Synchrotron Maser Instability*

The shock dissipation process at the shock front is required to maintain the collisionless shock structure, and for a relativistic magnetosonic shock with a finite σ the synchrotron instability can provide the mechanism of the energy dissipation[7]. For a shock with $\sigma \ll 1$, i.e., a weak magnetic field case, however, the synchrotron instability may not be effective. In the limit of $\sigma = 0$, no electromagnetic instability could be realized except for secondary electromagnetic instability. A mechanism of the self-generation of the magnetic field at the shock front seems to be a solution to maintain a relatively stable shock structure. Recently, the Weibel instability is proposed to be promising mechanism as the energy dissipation as well as the generation of magnetic field, because the temperature anisotropy can be easily generated at the shock front region. For example, Hededal et al.[13] studied the generation of the magnetic field and the plasma heating around the shock front by using a PIC simulation, and discussed that the magnetic fields implied by the modeling of GRB afterglows can be explained by the instability. The magnetic fields generated by such a plasma instability may be responsible to synchrotron and/or jitter radiation.

The plasma composition is another key agent of a collisionless shock wave. If heavy ions are contaminated in the pair plasmas, the electromagnetic waves emitted by the ions play an important role on energy dissipation as well as particle acceleration. Hoshino et al.[4] discussed that the upstream bulk flow energy carried by ions can be quickly transferred into the pair plasmas in downstream through the electromagnetic waves excited by a synchrotron maser instability of ions, and the downstream pair plasma develops into a power-law energy spectrum. They discussed that this mechanism can be operative for a relatively wide parameter range of σ. The synchrotron waves emitted by ions have the same polarization as the gyro-motions of positrons, and the energy transfer from ions to positrons can happen. If the number density of ions is small compared with the pair plasma density, the polarization of the synchrotron wave becomes linear, and as a result both electrons and positron can be accelerated.

2.3. *First Order Fermi Acceleration vs Surfing Acceleration*

The study of PIC simulations on the shock structure and dynamics is encouraging our understanding of the particle acceleration mechanism, and several direct acceleration mechanisms can be confirmed to be the non-thermal particle accelerators in astrophysical settings. However, it has so far not been possible to demonstrate the first order Fermi process in a self-consistent PIC simulation, and nor is the competing processes between the direct acceleration and the stochastic acceleration understood, because the time scale of the Fermi acceleration is longer than the typical calculation time of simulation, and because most of PIC simulation focus on the super-luminal, quasi-perpendicular shocks where the Fermi acceleration is not necessary effective. However, the shock wave with turbulent fields is believed to posses both the direct and the stochastic acceleration mechanisms.

By assuming the prescribed electromagnetic turbulence in the background plasma, Takagi and Hoshino[14] studied the particle acceleration in a system where both the shock surfing and the stochastic/Fermi acceleration simultaneously take place by means of a test particle orbit simulation[15]. The motions of a large number of particles are calculated under the Lorentz equation. The polarization electric field at the shock front is also included in order to make the shock surfing process effective.

Figure 3 shows the energy spectrum obtained by the above test particle

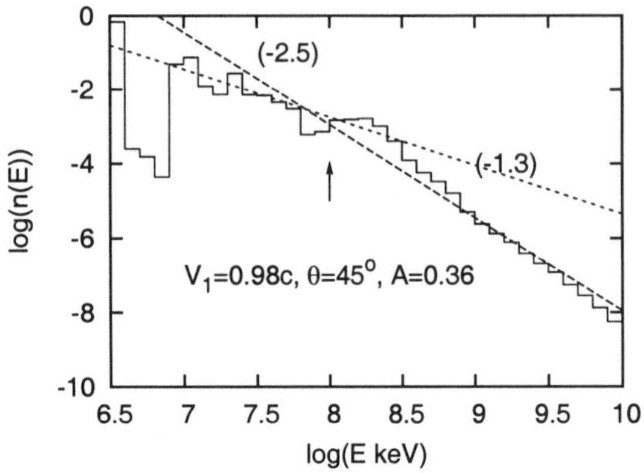

Figure 3. Energy spectrum for a relativistic shock wave where both surfing acceleration and Fermi acceleration processes are included.

orbit calculation for a relativistic shock with the upstream flow speed of $v_1 = 0.98c$ and the shock angle of $\theta_{Bn} = 45°$. The level of turbulence is $\delta B/B_0 = 1$, and the shock potential is assumed to be $e\phi/(\gamma_1 - 1)m_i c^2 = 0.36$. One can see a hard energy spectrum at lower energy with the spectral index of $s \sim 1.3$, while a soft spectrum at high energy with the index of $s \sim 2.5$. This result shows that the shock surfing acceleration can occurs even if the turbulent waves are excited around the shock front, and the surfing mechanism can inject the upstream cold plasmas into the supra-thermal regime in which the first ordered Fermi acceleration becomes effective. The surfing acceleration may give a solution of the so-called injection problem.

2.4. *Collision between Shock Wave and Current Sheets*

Most of the shock acceleration studies assume that the shock upstream region is a uniform medium except for the existence of turbulent waves, namely, a low-entropy, uniform, supersonic plasma is assumed to be transported toward the shock. However, the upstream region may have discontinuities and/or non-uniform structure. For example, the pulsar wind is believed to contain many current sheets, because a striped wind should be formed if the rotation axis of the neutron star is not aligned with the magnetic moment axis[16]. The interaction between the shock and other discontinuities/shock is a fundamental theoretical issue[17].

Shown in Figure 4 is a result about the interaction with the shock and the current sheet[18], which could mimic the situation of the pulsar wind

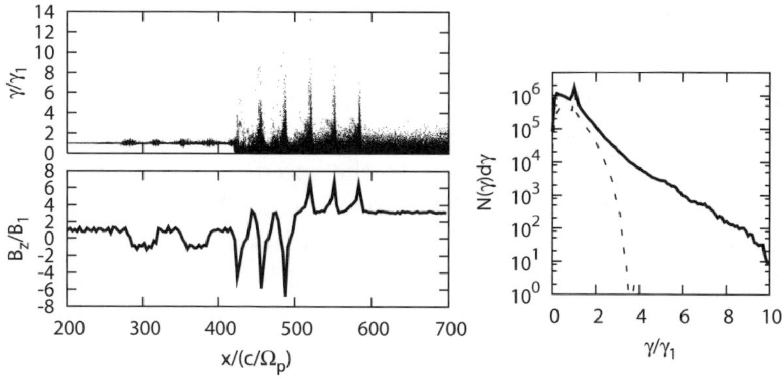

Figure 4. Interaction between the relativistic shock and current sheets. The shock structure (left) and the energy spectra (right).

region. Except for the current sheet injection in upstream, the shock upstream structure is same as the relativistic perpendicular shock of $\sigma = 10^{-1}$ shown in Figure 1. The current sheet structure in upstream is described by the so-called Harris solution, which is known as Vlasov equilibrium of the current sheet with the aniti-parallel magnetic field configuration. The left-hand panels show the shock structure, and solid line in the right-hand is the energy spectrum in downstream. For reference, the energy spectrum without current sheet in upstream is shown by dashed line. Nothing happens before the interaction between the current sheet and the shock front in a one-dimensional system, but after the interaction the current sheet can be separated into another tangential discontinuity and the emission of magnetosonic wave in downstream. We can find high energy particles around the large amplitude magnetosonic waves in downstream.

3. Relativistic Magnetic Reconnection

Magnetic reconnection is known to be one of the powerful mechanisms of the energy conversion from the magnetic field to the kinetic energy, and it has been extensively studied in the solar corona and in the terrestrial magnetosphere[19]. Reconnection is also believed to be important in radio galaxies and quasars[20]. In contrast with the shock acceleration, magnetic reconnection could operate in a much larger fraction in the universe, because many plasma states have magnetic fields whose structure contains the neutral sheet where the magnetic field polarity changes its direction.

3.1. *Current Sheet Acceleration*

The dynamics of magnetic reconnection has been investigated in the non-relativistic plasma state so far, but it is also postulated that reconnection play a key role for the case of the relativistic plasmas, where the Alfven speed $v_A = c\sqrt{\sigma/(1+\sigma)}$ is close to the speed of light. For a large σ regime, the magnetic field for the outside plasma sheet is strong, and the plasma temperature in the neutral sheet becomes a relativistic hot. In the relativistic regime, the electric field generated by the reconnection flow is of the same order of the magnetic field, and the acceleration efficiency is strongly enhanced.

Zenitani and Hoshino[21] first investigated nonlinear evolution of the current sheet and its particle acceleration by using the PIC simulation, and demonstrated that the reconnection is one of the important direct acceleration mechanisms which can generate the power-law energy spectrum. They discussed that the relativistic Speiser/meandering motion around the X-type neutral line is a main acceleration process of the power-law energy spectrum.

The acceleration rate around the X-type neutral region can be given by $d\varepsilon/dt = eEc$, where ε is the particle energy, E is the reconnection electric field in association with the global vortex plasma flow, and c is the speed of light. The particle acceleration lasts until its ejection from the X-type region, and the escape from the acceleration region can be provided by the Lorentz force of the reconnecting magnetic field. The typical ejection time τ can be estimated by the gyro-period of the reconnecting magnetic field, i.e., $\tau = \varepsilon/eBc$. Therefore, the loss rate of the accelerated particle can be given by $-dN/(Ndt) = 1/\tau$, where N is the particle number. From the above equations, we can easily find $N \propto \varepsilon^{-s}$, where $s = E/B \sim 1$ for the relativistic reconnection. The power-law index with $s \sim 1$ was confirmed around the X-type region by a large-scale PIC simulation by Jaroschek et al.[22] It is also demonstrated that the energy spectrum becomes soft with the index $1 < s < 3$, if the spectrum is integrated in a whole reconnection domain with the multiple magnetic islands.

3.2. *Guide Field Effect on Reconnection*

It should be noted that the current sheet is unstable not only for the magnetic reconnection mode but also for the drift-kink mode. The drift-kink mode is excited along the electric current direction, and the current sheet is deformed into a meandering structure, while the reconnection mode

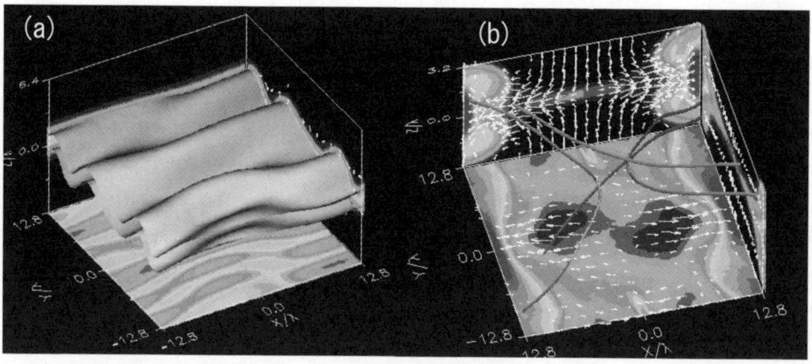

Figure 5. Relaltivistic drift-kind mode withough guide field (left) and relativistic magnetic reconnection with guide field (right) in a three-dimensional system.

happens in the plane perpendicular to the electric current. In the non-relativistic regime, it is known that the growth of the reconnection mode is faster than that of the drift-kink mode, but for the case of the relativistic hot plasma sheet the drift-kind mode dominates for an anti-parallel magnetic field geometry. However, if the guide magnetic field parallel to the electric current is added into the slab geometry with anti-parallel magnetic field, the drift-kink mode is strongly stabilized by the tension force of magnetic field line. This suppression mechanism is similar to the case for Kelvin-Helmholtz instability. Having these three-dimensional effects in mind, Zenitani and Hoshino[23] investigated the non-linear evolution of relativistic reconnection and drift-kink mode, which are shown in Figure 5. The left-hand panel shows the nonlinear evolution of the Harris current sheet without a guide magnetic field, while the right-hand panel is the case with a guide magnetic field. They found that the high energy, nonthermal particles can be effectively produced for the current sheet with a guide magnetic field.

4. Discussion

We studied the connection between the plasma dynamics and the particle acceleration both in shock waves and in magnetic reconnection, and discussed that several direct acceleration mechanisms such as the shock surfing and the current sheet acceleration are capable of producing particles with a power-law spectrum. It is found that a hard energy spectrum with $s < 2$ can be generated in those direct acceleration processes, and the power-law indices of the energy spectra depend on the properties of the dynamical plasma structures. In a relativistic reconnection, the spectrum around the X-type region can be approximated by $s \sim 1$, and the shock surfing acceleration can provide the spectrum with $s \sim 1.3$.

So far an understanding of the direct acceleration mechanism is not simple compared with the most widely investigated Fermi acceleration model, but recent progress of the direct acceleration seems to provide the wealth of information on particle acceleration, which can serve the improvements in radio, X-ray and gamma ray observations of nonthermal phenomena in astrophysical systems.

References

1. A. Konopelko et al., *Astrophys. J.*, **597**, 851 (2003).
2. A. Mastichiadis et al., *Astrophys. J. Lett.*, **618**, 75 (2005).
3. L. Moran et al., *Astron. Astrophys.*, **432**, 467 (2005).
4. M. Hoshino et al., *Astrophys. J.*, **390**, 454 (1992).
5. J. G. Kirk and P. Duffy, *J. Phys. Nucl. Part. Phys.*, **25**, 163 (1999).
6. A. B. Langdon, J. Arons, and C. E. Max, *Phys. Rev. Lett.* **61**, 779 (1984).
7. Y. Gallant et al., *Astrophys. J.* **391**, 73 (1992).
8. M. Hoshino, *Prog. Theor. Phys. Suppl*, **143**, 149 (2001).
9. R. Z. Sagdeev and V. D. Shapiro, *JETP Lett. Engl. Transl.*, **101**, 279 (1973).
10. R. Sugihara and Y. Mizuno, *J. Phys. Soc. Jpn*, **47**, 1290 (1979).
11. T. Katsouleas and J. M. Dawson, *Phys. Rev. Lett.*, **51**, 392 (1983).
12. M. Hoshino and N. Shimada, *Astrophys. J.*, **572**, 880 (2002).
13. C. B. Hededal et al., *Astrophys. J.* **617**, L107 (2004).
14. Y. Takagi and M. Hoshino, in preparation
15. J. Niemiec and M. Ostrowski, *Astrophys. J.* **610**, 851 (2004).
16. F. V. Coroniti, *Astrophys. J*, **349**, 538 (1990).
17. J. G. Kirk, *Phys. Rev. Lett.*, **92**, 181101 (2004).
18. K. Nagata and M. Hoshino, in preparation
19. e.g., M. Hoshino, R. L. Stenzel, and K. Shibata, Eds., *Earth, Planets, and Space*, **53**, 409 (2001).
20. M. C. Begelman, R.D. Blandford, and M. J. Rees, *Rev. Mod. Phys.*, **56**, 255 (1984).
21. S. Zenitani and M. Hoshino, *Astrophys. J.* **562**, L63 (2001).
22. C. H. Jaroschek et al., *Plasma Phys.* **11**, 1151 (2004).
23. S. Zenitani and M. Hoshino, *Phys. Rev. Lett.* **95**, 095001 (2005).

THE SWIFT GAMMA-RAY BURST MISSION: FIRST RESULTS

N. GEHRELS[1], on behalf of the Swift Team

[1]*NASA/GSFC, Greenbelt, MD 20771, USA*

Abstract. Since its launch on 20 November 2004, the Swift mission is detecting ~100 new gamma-ray bursts (GRBs) each year, and immediately (within two minutes) starting simultaneous X-ray and UV/optical observations of the afterglow. It has already collected am impressive database of bursts, including prompt emission to higher sensitivity than BATSE, uniform monitoring of afterglows, and rapid follow-up by other observatories notified through the GCN.

Keywords: Gamma Ray Bursts, Astrophysics.
PACS: 98.70Rz.

Introduction

Despite impressive advances over the roughly three decades since GRBs were first discovered (Klebesadel et al. 1973), the study of bursts remains highly dependent on the capabilities of the observatories which carried out the measurements. The era of the Compton Gamma Ray Observatory (CGRO) led to the discovery of more than 2600 bursts in just 9 years. Analyses of these data led to the conclusion that GRBs are isotropic on the sky and occur at a frequency of roughly two per day all sky (Briggs 1996).

The BeppoSAX mission made the critical discovery of X-ray afterglows (Costa et al. 1997). With the accompanying discoveries by ground-based telescopes of optical (van Paradijs et al. 1997) and radio (Frail et al. 1997) afterglows, GRBs could start to be studied within the astrophysical context of identifiable objects in a range of wavelength regimes. Successful prediction of the light curves of these afterglows across the electromagnetic spectrum has given confidence that GRBs are the signal from extremely powerful explosions at cosmological distances, which have been produced by extremely relativistic expansion (Wijers, Rees & Mesaros 1997).

The Swift mission selected by NASA in 1997 combines the sensitivity to discover new GRBs with the ability to point high sensitivity X-ray and optical telescopes at the location of the new GRB as soon as possible. From this capability Swift has the goal to answer the following questions:

1. What causes GRBs?

2. What physics can be learned about black hole formation and ultra-relativistic out- flows?

3. What is the nature of subclasses of GRBs?

4. What can GRBs tell us about the early Universe?

The general operations of the Swift observatory are as follows. The wide-field Burst Alert Telescope (BAT) detects the bursts in the 15-350 keV band and determines the position to a few arcminute accuracy. The position is provided to the spacecraft, built by Spectrum Astro General Dynamics, which repoints to it in less than 2 minutes. Two narrow-field instruments then observe the afterglow: the X-Ray Telescope (XRT) and UV-Optical Telescope (UVOT). Alert data from all three instruments is sent to the ground via NASA's TDRSS relay satellite. The full data set is stored and dumped to the Italian Space Agency's Malindi Ground Station.

Swift was built by an international team from the US, UK, and Italy. After five years of development it was launched from Kennedy Space Center on 20 November 2004. The spacecraft and instruments were carefully brought into operational status over an eight-week period, followed by a period of calibration and operation verification, which ended with the start of normal operations on 5 April 2004. A complete description of the Swift mission can be found in Gehrels et al. (2004).

As of 1 April 2006, the Swift achievements include: discovery of 126 new GRBs by the Swift Burst Alert Telescope (BAT) instrument (with a typical error region of less than 2 arcmin radius); observation of 103 X-ray afterglows by the Swift X-Ray Telescope (XRT) instrument (with a typical error region of less than 3 arcsec radius); and observations of 30 afterglows by the Swift UV-Optical Telescope (UVOT) instrument (with a typical error region of less than 1 arcsec). More than half of the afterglow observations start within two minutes of the BAT GRB trigger with a record of only 54 seconds. Afterglow data have been obtained from non-Swift discovered bursts with typical response times of 3 hours.

Swift Highlights

2.1 BAT Detected GRBs

The BAT (Barthelmy, et al. 2005a) on Swift has detected 126 GRBs between when it was turned on in mid-December 2004 and 1 April 2006. Thus in 466 days of operation, the BAT has detected GRBs at a rate of about 99 bursts per year, close to the rate of 100 bursts estimated prior to launch.

Spectral analysis of the BAT bursts shows them to be consistent with the population of GRBs seen by CGRO, both in the ratio of the fluxes in the 25-50 keV and 50-100 keV energy bands, and in flux and duration distributions.

2.2 XRT Detected GRBs

The XRT (Burrows, et al. 2005) is performing the first rapid-response observations of the X-ray afterglow of GRBs. In the first ~100 cases, all but 5 of the BAT GRB triggers resulted in detection of an X-ray counterpart for the BAT source. In 3 cases the XRT observations started while the BAT was still detecting hard X-ray prompt emission from the GRB.

The Swift afterglow observations are rapid, with more than half of the observations started in less than 300 seconds after the burst. When XRT arrives this quickly it is very common to see a fast X-ray decline within the first 300 seconds.

In addition to the BAT detected events, Swift can also observe GRBs discovered by other satellites. Swift has discovered X-ray afterglow emission in 5 cases for HETE-2 and 5 cases INTEGRAL. In a particularly impressive case, Swift was able to respond to the ground control commands and start observations of the HETE-2 GRB050408 within 40 minutes of the GRB.

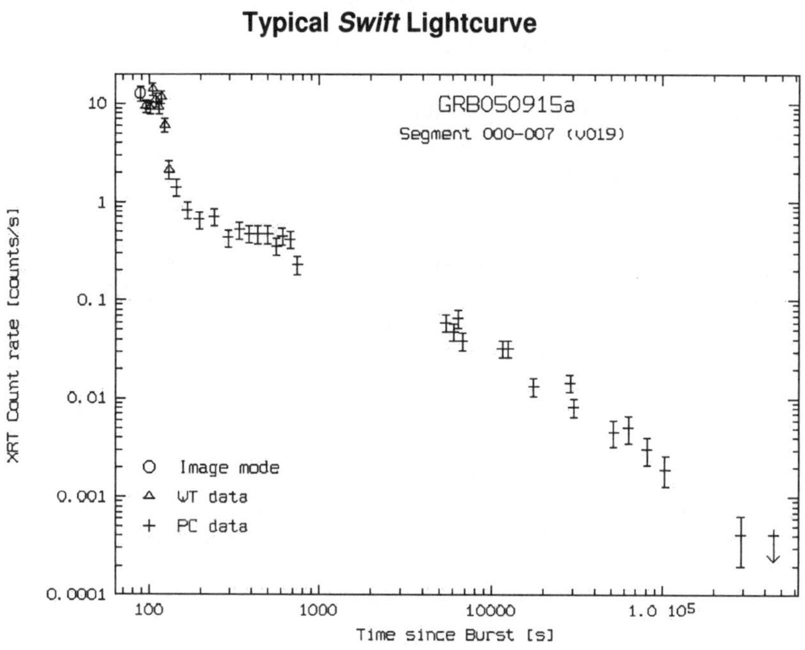

Fig. 1. Typical X-ray afterglow light curve.

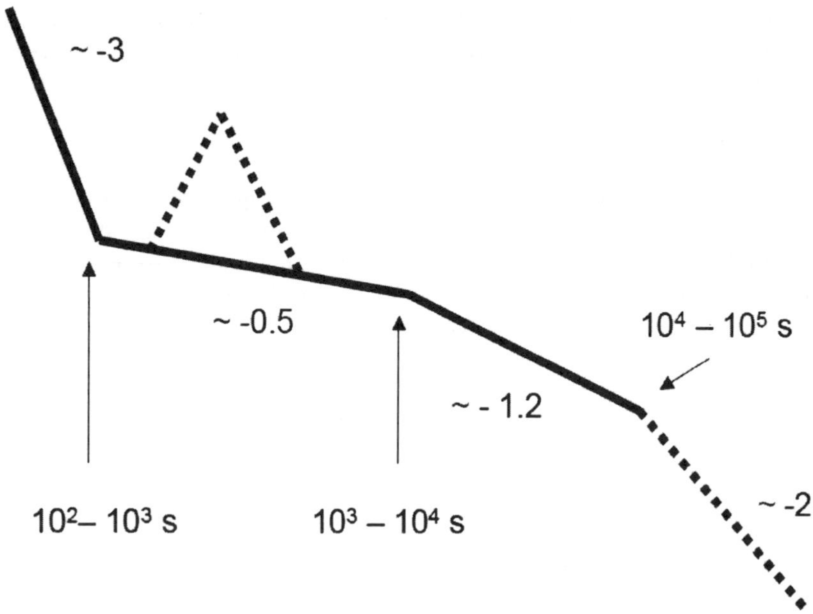

Fig 2. Generic lightcurve observed by XRT. Numbers shown next to the segment are the power-law index of the time decay. From Zhang et al. (2006).

2.3 UVOT Detected GRBs

The UVOT (Roming, et al. 2005a) is co-aligned with the XRT and so observes the GRB afterglows just as promptly as the XRT. Despite these prompt observations the UVOT has detected far fewer UV/optical counterparts than the XRT.

Of the first 95 GRBs observed by the UVOT, only 30 had detected emissions. The UVOT has generated important upper limits for these early times, which are lower than those for bursts studied by previous missions. Combining UVOT plus ground-based observations ~64 optical afterglows have been detected.

Reasons for this low number include the possibility that the Swift bursts are more distant than previous bursts; that a substantial number of GRBs have intrinsic dust extinction which suppresses the optical/UV emission compared to the I and R bands typically reported for earlier afterglows; or the possibility that some afterglows come from high magnetic field regions in the outflow which suppresses the optical and UV emission. These possibilities are discussed in Roming et al. (2005b).

Although not every GRB produces detectable UV or optical flux, several bursts have produced early time light curves, including GRB050318 (Still et al. 2005), GRB050319 (Mason et al. 2006) and GRB050525A (Blustin et al. 2006).

2.4 XRT Early Light Curve Behavior

Swift has opened up a new regime for GRB afterglow studies. Never before has it been possible to study the X-ray behavior on timescales of minutes after the GRB happens. Swift has frequently started observations within a few minutes of the detection of GRBs by the BAT (with a record of only 52 seconds).

These extremely prompt observations have given rise to new findings. In roughly 40% of the cases, the X-ray afterglows can be characterized by a three-part light curve (see Figure 1 and 2). First comes an extremely rapid decay of a very bright source. At these early times the decay can be fit by a power law of index in the range of 2.5 or greater. After a few minutes the decay rate flattens, and we can fit it with an index approximately equal to 1 (plus or minus ~0.5). Finally after a delay ranging from hours to days, the decay rate steepens again, sometimes resulting in a behavior interpreted as a jet break (see Zhang et al. (2005) and Nousek et al. (2005) for summary papers). Tagliaferri et al (20005) and Barthelmy et al. (2005b) each consider two early XRT afterglows. They show that the X-ray emission during the prompt phase (estimated from

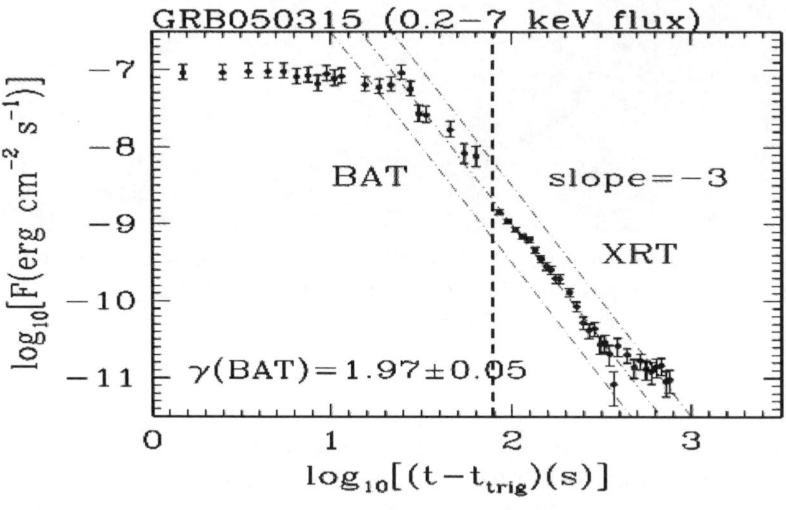

Fig. 3. The BAT spectrum is extrapolated to the 0.2 – 7 keV band. The early XRT lightcurve connects smoothly to the prompt emission. From Barthelmy et al. (2005b).

extrapolation of the BAT spectrum) connects to the bright early XRT afterglow (see Figure 3). This suggests that the bright early afterglow is an extension of the prompt phase.

Swift also detects strong X-ray flares in afterglows at early times. In one case (GRB050502b) the X-ray flux increased by a factor of roughly 100. The dramatic flaring events seem to be superposed on a background, which follows the multipart behavior mentioned above. Burrows et al. (2005b) discuss the flaring behavior seen in GRB050502b and GRB050408. Flares are seen in 20%-30% of the observed afterglow.

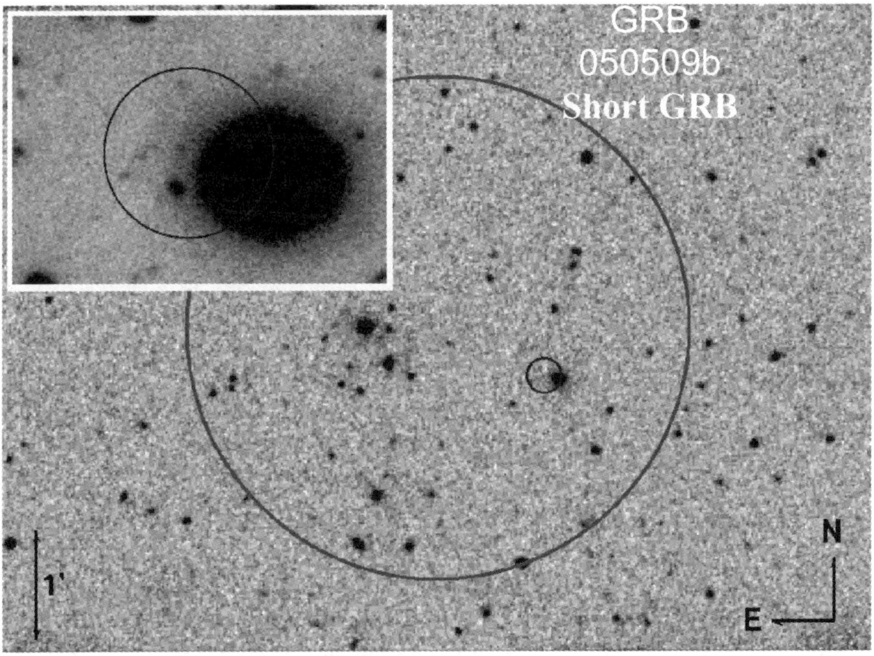

Fig. 4. Localization of short GRB 050509b. Large circle is BAT position; small circle is XRT position. The inset shows a bright elliptical galaxy in the XRT circle. From Gehrels et al. (2005).

2.5 Short GRBs

As of late April 2006, the BAT has detected 10 GRBs in the short-hard class. The first one of them (GRB 050202) had no prompt slew and no counterparts. From the next 2 events we have learned a great deal. GRB 050509b (Gehrels et al. 2005) had the first detection of an X-ray afterglow which gave an error circle with a bright elliptical galaxy (cD galaxy in a cluster) in it (Figure 4). GRB 050724 (Barthelmy et al. 2005c) had an XRT afterglow, plus Chandra, optical and radio detections. The sub-arcsecond positions located it once again in the outer regions of a bright elliptical. The fact that these ellipticals have very low star formation rates argues strongly against a collapsar origin like that for long bursts. Also, the redshifts for the two are in the z = 0.2 to 0.3 range, a factor of ~3 closer than typical long GRBs. The evidence to date is consistent with an origin of short burst in merging binary neutron stars. HETE-2 short burst GRB050709 had a redshift of z=0.16. (Villasenor et.al. 2005)

The picture for the last 7 short GRBs seen by Swift is still not well understood. GRB050813 appears to have a faint host at z=1.8. GRBs 050906 and 051105A did not have XRT detections despite rapid slews. GRB 050925 was in the galactic plane and had a soft spectrum; it may be a new galactic SGR.

2.6 GRB Redshifts

As of late April 2006, redshifts have been determined for 32 Swift GRBs. The average redshift (excluding short GRBs) is z=2.7. This is significantly higher than the pre-Swift average of z=1.2. The sensitivity of the Swift instruments is leading to a sampling of more distance GRBs.

On September 4, 2005, Swift detected a long, smooth GRB (Cusumano et al. 2006). The redshift was found to be the very large value of z =6.29, one of the highest redshift objects ever seen. The light curve for this GRB is shown in Figure 5.

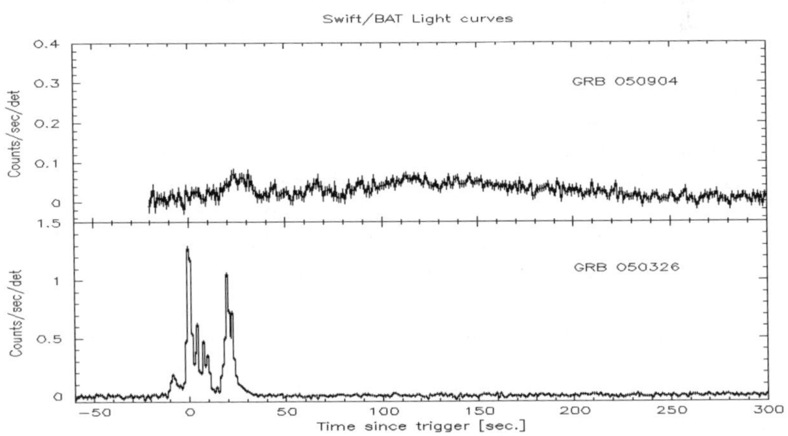

Fig. 5. Lightcurve of high-redshift GRB 050904 compared to a typical GRB. The long smooth nature of the lightcurve is due to cosmologic time dilation as the photons propagated to us from z=6.29.

2.7 Giant Flare from SGR 1806-20

On 27 December 2004, the Solar System was struck by the brightest gamma-ray transient ever observed. Every orbiting gamma-ray observatory detected the flash produced by the soft gamma-ray repeater SGR 1806-20. Although Swift was not pointed toward the target, the flux was so high that the BAT detector was swamped by more than a billion gamma-rays per cm2 passing through the structure of the spacecraft.

Palmer et al. (2005) present the Swift data on this dramatic event. Although the emitting system is located more than 10 kpc from the Earth, the energy flux was brighter than the full Moon for the 0.2 seconds. This giant flare was more than 100 times more luminous than the two previous flares seen in 1998 from SGR1900+14 and in 1979 from SGR0526-66.

Such events may be the cause of at least some short GRBs, in that the rapid, extremely bright flash of gamma rays had a similar duration and energy profile to a short GRB. Such an event in an external galaxy would be detectable out to 60 Mpc.

2.8 UV/Optical & X-ray Observations of SN2005am

Type Ia supernovae are critical to our understanding of the fundamental fabric of our Universe. They are the standard candles used to measure distances over the range in which cosmological effects become significant. Observations of nearby supernovae in the ultraviolet can be quite important for understanding and calibrating their light curves and luminosities.

Missions with UV capability such as the International Ultraviolet Explorer (IUE) and the Hubble Space Telescope began these studies, but they are limited in the intrinsically slower operational response time than offered by Swift. Thus Swift has been an ideal observatory for early observations of nearby bright supernovae, of which SN2005am is a prime example.

Brown et al. (2005) present ultraviolet and optical light curves for SN2005am, starting four days prior to maximum light, and extending to 69 days after peak. In addition, when the target was bright enough, Swift was able to carry out moderate-resolution grism UV/optical measurements. These data for SN2005am are the best sampled in time, and cover the widest range of any Type Ia supernova follow-up to date.

Conclusions

The Swift observatory is performing excellent scientific observations at high efficiency and with important progress toward its mission objectives.

The BAT is working flawlessly. The positional agreement with the XRT and ground-based detections suggests that the typical on-board positional accuracy for GRBs is roughly 65 arcsec, exceeding the pre-launch predictions.

The UVOT has demonstrated excellent UV and optical performance on GRBs and other sources. Source positions are acurate to 0.3 arcsec.

The XRT has demonstrated excellent X-ray sensitivity and rapid responsiveness. The average accuracy for the XRT positions confirmed with XMM or ground-based optical detection is 2.6 arcsec. XRT is observing afterglows at a level of 100 to 1000 times fainter than Beppo-SAX. This rapid acquisition with sensitive X-ray detection is revealing new lightcurve behaviors.

As Swift observations become more numerous we are building up a substantial database of prompt gamma-ray emission and early (to late) X-ray and UV/optical light curves. From these, new insights into GRB formation and GRB environments are being gleaned.

REFERENCES

Barthelmy, S., et al. 2005a, Sp Sci Rev. 120, 143.
Barthelmy, S., et al. 2005b, ApJ, 635, L133.
Barthelmy, S., et al. 2005c, Nature, 438, 994.
Blustin, A., et al. 2006, ApJ, 637, 901.
Briggs, M. S. 1996, ApJ, 459, 40.
Brown, P. J., et al. 2005, ApJ, 635, 1192.
Burrows, D.N., et al. 2005a, Sp. Sci. Rev. 120, 165.
Burrows, D. N., et al. 2005b, Sci., 309, 1833.
Costa, E., et al. 1997, Nature, 387, 783.
Cusumano, G. et al. 2006 Nature, 440, 164.
Frail, D. A., et al. 1997, Nature, 389, 261.
Gehrels, N., et al. 2004, ApJ, 661, 1005.
Gehrels, N., et al. 2005, Nature, 437, 851, 2005.
Klebesadel, R.W., Strong, I.B., & Olson, R.A. 1973, ApJ, 182, L85.
Mason, K., et al. 2006, ApJ, 639, 311.
Nousek, J., et al., 2005, ApJ, accepted (astro-ph 0508332).
Palmer, D., et al. 2005, Nature, 434, 1107.
Roming, P. et al., 2005a, Sp. Sci. Rev., 120, 95.
Roming, P., et al. 2005b, ApJ, (astro-ph 0511751).
Still, M., et al. 2005, ApJ, 635, 1187.
Tagliaferri, G., et al. 2005, Nature, 436, 985.
Van Paradijs, J., et al. 1997, Nature, 386, 686.
Villasenor, J.S. et al. 2005, Nature, 437, 855.
Wijers, R. A. M. J., Rees, M. J., & Meszaros, P. 1997, MNRAS, 288, L51.
Zhang, B., et al., 2005, ApJ, (astro-ph 0508321).

GAMMA-RAY BURST: PROBLEMS DELINEATED BY HETE-2 AND OTHER OBSERVATIONS

N. KAWAI

Department of Physics, Tokyo Institute of Technology,
2-12-1 Ookayama, Meguro-ku, Tokyo 152-8551, Japan
E-mail: nkawai@phys.titech.ac.jp

Highlights of the recent observations of gamma-ray bursts are presented. In the optical afterglow of GRB030329 detected by HETE-2, strong evidence for the association of long GRBs with the core-collapse supernovae was found. With the wide energy coverage, the nature of the X-ray rich GRBs and X-ray flashes have been studied systematically with HETE-2, and they are found to have many properties in common with the classical GRBs, suggesting that they are a single phenomenon. In July 2005, HETE-2 localized a short gamma-ray burst GRB 050709, which led to the discovery of the first optical afterglow of a short GRB in a low-redshift galaxy, providing evidence that the origins of most short GRBs are different from those of long GRBs. The extremely intense flare of the soft gamma repeater SGR 1806−20 was observed in December 2004. The possible connection with the short population GRBs are discussed.

1. Introduction

Gamma-ray bursts (GRBs) are short episodes of explosive gamma-ray emission from random positions in the sky, which typically last for tens of seconds or less. Since their unexpected discovery in 1960's by nuclear test surveillance satellites,[1] their origin had been a mystery for 30 years. Even their distance was unknown.

The discovery of their afterglows in X-ray and optical bands in 1997,[2,3] at last, convincingly showed that GRBs are the most violent explosion in the Universe that the mankind knows (except for the Big Bang) at cosmological distances with a typical redshift of > 0.5. The only plausible candidates capable of releasing such large energies may be the core-collapse of a star with tens of solar masses, and merger of relativistic compact stars (i.e. neutron stars or black holes), either of which results in a release of huge gravitational binding energy.

We note, however, that afterglows then had been found only for the "long soft" GRBs, one of the two populations of GRBs characterized by their durations. The clue for "short hard" GRBs was first discoverd in 2005, which we present later in this paper.

For the study of GRBs, detection of the afterglows and the study of their prompt burst emission are both important. The High Energy Transient Explorer 2 (HETE-2) is the first satellite designed specifically for the study of GRBs. The primary goals of the HETE-2 mission are the broadband (2 keV to over 400 keV) observation of the burst emission of GRBs, and the prompt distribution of precise GRB coordinates to the astronomical community for immediate follow-up observations.[4,5,6] It was launched in October 2000, and has localized $\sim$ 80 GRBs in 2001–2005, among which, 20 GRBs have known redshifts.

In November 2004, *Swift*, a larger satellite dedicated for the GRB study[7] was successfully launched. It was fully commissioned in April 2005, and is now localizing GRBs at a rate of ~100 per year. *Swift* and HETE-2 together are contributing to make a significant progress in understanding GRBs.

2. Long population GRBs

In a widely accepted scenario, accretion of stellar remnants to the newly born black hole causes collimated, repeated ejection of relativistic shells ("jets") with a Lorentz factor of hundreds. The collision of the shells forms shocks. Synchrotron radiation from the shocks is relativistically beamed and observed as a gamma-ray burst. The relativistic shells propagate into the interstellar medium to form external shocks, where the accelerated particles emit afterglows in a broad band (radio to X-rays) that decay with time following a power-law with indices of 1–2.[8,9]

2.1. *Supernova association*

There had been increasing circumstantial and tantalizing evidence that GRBs are associated with core collapse supernovae (SNe). The detection and localization[10] of GRB 030329 by HETE-2 led to a dramatic confirmation of the GRB-SN connection. GRB 030329 was among the brightest 1% of GRBs ever seen (see Figure 1). Given that GRBs typically occur at $z \sim$1–2, the probability that the source of an observed burst should be as close as GRB 030329 (z=0.167) is one in several thousand.

About ten days after the burst, the spectral signature of an energetic Type Ic supernova emerged[11] on top of the usual GRB afterglow continuum of non-thermal synchrotron emission. The underlying supernova has been

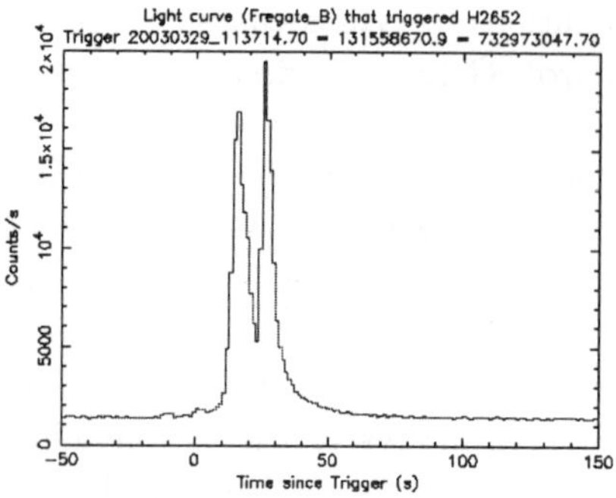

Figure 1. GRB 030329 time history observed by HETE-2 FREGATE.

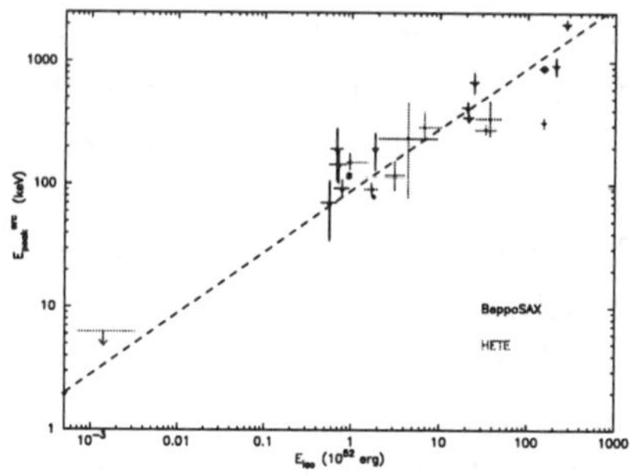

Figure 2. Distribution of HETE-2 and *Beppo*SAX bursts in the ($E_{\rm iso}$,$E_{\rm peak}$)-plane.

designated SN 2003dh. The spectrum of SN 2003dh is strikingly similar to that of the Type Ic supernova SN 1998bw, which was putatively associated with GRB980425.[12] The clear detection of SN 2003dh in the afterglow of GRB 030329 confirmed decisively the connection between long GRBs and core collapse supernovae.

2.2. *X-ray Flashes, Spectral Energy Peak, and Radiated Energy*

HETE-2 is detecting X-ray flashes (XRFs), which are similar to ordinary "classical" GRBs in many ways except that XRFs have larger fluence in the X-ray band (2–30 keV) than in the gamma-ray band (30–400 keV).[13] The good sensitivity of HETE-2 down to 2 keV makes it ideal for detecting and studying XRFs. Using twelve BeppoSAX GRBs with measured redshifts, Amati et al.[14] showed that the spectral peak energy at the source frame $E^{\rm src}_{\rm peak}$ and the isotropic-equivalent radiated energy $E_{\rm iso}$ are tightly correlated, and follows a relation $E^{\rm src}_{\rm peak} \propto E^{1/2}_{\rm iso}$. With the 10 HETE GRBs/XRFs with measured redshifts (Fig. 2) we have confirmed this relation. Furthermore, we extended this relation by three orders of magnitude in $E_{\rm iso}$.[15,16] These results provide strong evidence that XRFs and regular long GRBs form a continuum, and are a single phenomenon. The identification of a star forming galaxy as the host of the XRF 020903,[17] and the detection of the supernova component in its afterglow,[18] further confirm that XRFs are essentially a subclass of long GRBs. The extended Amati et al relation ($E^{\rm src}_{\rm peak} \propto E^{1/2}_{\rm iso}$) suggest that the $E^{\rm src}_{\rm peak}$ and $E_{\rm iso}$ are controlled by some single parameter, which differentiate XRFs and GRBs. Understanding this key parameter should certainly lead to the understanding of the energetics and radiation mechanism of GRBs. The correlation between the radiated energy and the spectral peak energy has been further refined with the correction for the jet opening angle.[19] It has been suggested that this relation can be used to constrain the cosmological parameters (Ω_M and Ω_Λ) at higher redshifts than those possible with Type Ia supernovae.[20]

2.3. *Long population GRBs: what are known and what are not*

Here is a very simplified summary of the current understanding on long population GRBs (LGRBs).

- Origin of LGRBs in the core collapse of massive stars (The "collapsar" scenario) is widely accepted. It is directly supported with a few solid cases of association of LGRBs with type Ic supernovae. There are also circumstantial evidence such as the locations of LGRBs in star forming galaxies.
- Fireball and synchrotron shock model is also widely accepted as the radiation mechanism of GRBs. The "external" shock formed by the collision of the relativistic jet material with the interstellar medium has been the favored explanation for the afterglow emission, while the prompt burst emission has been explained in terms of "internal shocks" formed by mutual collisions of ultra-relativistic shells. However, recent Swift observations of early X-ray afterglows (within $\sim 10^4$ s of the triggers) seem to indicate that they have properties difficult to explain with the simple external shock model.
- Little is understood about the physics of the relativistic jets that power the GRB emission: formation, acceleration and collimation (radiatively driven or magnetically), and prolonged activity of the central engine which may last even for one day or longer in some cases.
- Only a tiny fraction of massive stars produce GRBs at their deaths. It is not yet known what is special about them. There are theoretical speculations and observational inference suggesting low metallicities and high angular momentum of GRB progenitors, but they need to be confirmed by direct observations.

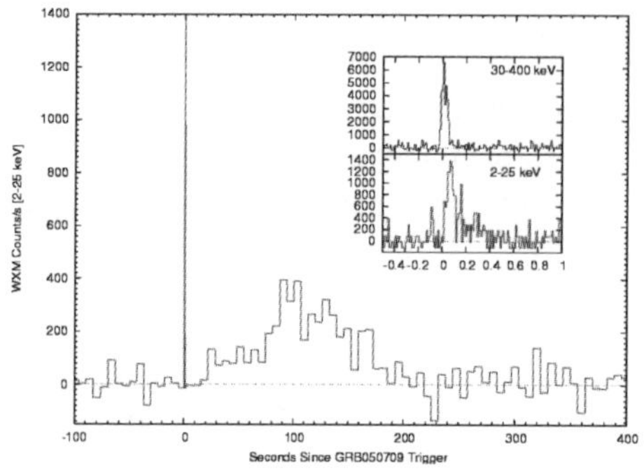

Figure 3. The X-ray light curve of GRB 050709, a short gamma-ray burst detected by HETE-2, showing the short spike followed by a long soft emission lasting for >100 s. The inset shows the initial spike in the 30–400 and 2– 25 keV bands with a high time resolution.

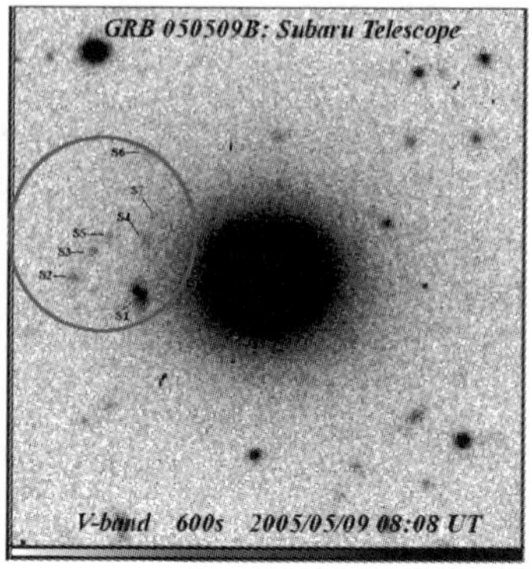

Figure 4. The optical image of the field containing the *Swift* XRT error circle of GRB 050509B obtained with Subaru Suprime Cam. (Kosugi et al. 2005)

3. Short population GRBs

About 20% of the GRBs in the BATSE catalog form a distinct class called short hard GRBs, with durations shorter than 2 seconds and more energies in higher energy gamma-rays.[21] Without detection of optical counterparts or arcsec localizations, their origin has remained a deep mystery until recently.

3.1. *Afterglow detections*

A substantial progress in understanding of short GRBs has been made in May-July 2005. X-ray afterglows were detected for three short GRBs: GRB 050509B and GRB 050724 localized by *Swift*, and GRB 050709 localized by HETE-2.[22,23,24]

In each of the arcsecond positions of the short GRBs, galaxies at low redshifts ($z \sim 0.2$) were found, though the association of GRB 050509B is not so certain as no optical or NIR afterglow was detected for this GRB, and the *Swift* XRT error circle includes distant faint galaxies in addition to the outskirt of a elliptical galaxy (see Figure 4; Kosugi et al.[25]) at a relatively low redshift z =0.226.[26]

HETE-2 detected and localized GRB 050709 (Figure 3), which showed a short hard spike (duration <0.2 s) followed by a long soft X-ray lasting for > 100 s in 2–10 keV band. In addition to the X-ray afterglow, a fading object was detected in the optical and near infrared bands at the edge of a star-forming galaxy at $z = 0.16$.[27] An X-ray, optical and radio afterglow emission was detected for the *Swift*-detected GRB 050724 at the position of an elliptical galaxy at z =0.26.[28]

The implied output energies of GRB 050709 and GRB 050724 based on the association with the low-z galaxies were orders of magnitude lower than those of long GRBs.

The probable association of two short GRBs with elliptical galaxies, in which one would not expect core collapse events to occur, seem to support the neutron star or black hole merger scenario. However, since GRB 050709 is associated with a dwarf star-forming galaxy, the other scenario involving young stellar population cannot be ruled out. The giant flares from soft gamma repeaters (magnetar flares) may also contribute some fraction of the observed short GRBs, though GRB 050709 is not likely to be the case because its radiated energy was orders of magnitude larger than that of the giant flare or SGR 1806−20 in December 2004.[29] In reality, short GRBs may consist of events of heterogeneous classes. Clearly more observations are necessary to convincingly understand their origin.

3.2. *Giant flares of soft gamma repeaters*

On 27 December 2004, a giant flare from the soft gamma repeater SGR 1806−20 was detected by almost all the radiation detectors deployed in the space. It is the third of this kind, following the flare on March 5, 1979 from SGR 0520−66 in LMC and one on August 27, 1998 from SGR 1900+14. These sources are called "soft gamma repeaters", as named for the their activitiy of repeatedly producing short bursts in soft gamma-ray bands. They are all rotating neutron stars with 5–8 s periods. Their large period derivative can be interpreted as strong surface magnetic field in excess of 10^{14} gauss, and thus they are called "magnetars". Two of them are located on the Galactic plane, and one is in LMC. These three flares are produced by three different sources, but their light curves are remarkably similar: an extremely intense and short spike with durations shorter than a fraction of a second followed by oscillating tails modulated with their spin periods (7.56 s in the case of SGR 1806−20, which lasted for several minutes. In the December 2004 event, the first 0.1-second spike was so intense that it saturated almost all the instruments designed for detecting gamma-rays, such as those on INTEGRAL,[30] RHESSI,[29] and Swift.[31] Its X-ray flux was hundreds of times brighter than those of the largest solar flares recored near the earth. It was most likely the highest cosmic hard X-ray flux ever recorded by human-built radiation detectors. The LEP (Low Energy Particle) detector on GEOTAIL, a Japanese magnetospheric satellite, however, had a sufficiently small effective area for hard X-rays, so that it succeeded in measuring the unsaturated peak profile of the intense first spike.[32] In the first 200 ms of the flare, while most of the other instruments were saturated, GEOTAIL measured multi-peaked structure on time scale of 10's of seconds. This structure probably indicate that there were repeated energy injection. At the peak, the photon flux was $\sim 10^7\,\mathrm{photons\,s^{-1}\,cm^{-2}}$, equivalent to $\sim 20\,\mathrm{erg\,s^{-1}\,cm^{-2}}$, and the fluence (integrated energy flux) in the first 300 ms amounted to $\sim 2\,\mathrm{erg\,cm^{-2}}$. This peak flux is 8–9 orders of magnitude larger than those of typical GRBs, and the fluence is 6–7 orders of magnitudes larger. Since the distance to SGR1806−20 is estimated to be about 10 kpc , the total radiated energy is estimated to be several times 10^{46} erg. Here, we argue, that the beaming (or collimation) factor of the SGR flare should small ($\sim$ a few) unlike that of long GRBs (typically ~ 100). The oscillating tail should be radiated to most of the solid angle of the sky. If the beaming factor of the initial spike is ~ 100, like that of long GRBs, then we should be observing 100 times more numerous "tail-only" flares, whose initial spikes miss the earth because of its small sold angle. However, no such "tail-only" events has been observed by satellites, while the oscillating tail should be bright enough to trigger small gamma-ray instruments on satellites. The lack of tail-only events indicate that the initial spike is not strongly beamed either. Accordingly, the correction factor to derive the true radiated energy should not be much different from 1, and the true radiated energy should not be much smaller than the isotropic value of 3×10^{47} erg.

The frequency ($\approx$ 1 in 30 years per galaxy) and the luminosity ($\approx 10^{47}\,\mathrm{erg\,s^{-1}\,cm^{-2}}$) of the SGR giant flares have been compared to those of short population GRBs, and it was found that SGR flares can account for only a minor fraction of the observed short GRBs.[29] On the other hand, if the observed rate of SGR flares are universally applicable to nearby galaxies, small, but a significant number of extragalactic SGR flares must have been recorded among the BATSE short GRBs, and naturally Swift is expected to detect some, possibly with the oscillating tails, which should be an unambiguous signature of its origin.

Acknowledgments

The HETE mission is supported in the US by NASA; in Japan, in part by the Ministry of Education, Culture, Sports, Science, and Technology; and in France, by CNES.

References

1. Klebesadel, R. et al. 1972 Astrophys. J. 182, L85.
2. Costa, E. et al. 1997, Nature 387, 783.
3. van Paradijs, J. et al. 1997, Nature 386, 686.
4. Ricker, G.R. et al. 2002, Gamma-Ray Bursts and Astronomy, eds. G. R. Ricker and R. Vanderspek (AIP, New York, 2002), 3.
5. Shirasaki, Y. et al. 2003, Publ. Astron. Soc. Japan 55, 1033.
6. Atteia, J.-L. et al. 2002, Gamma-Ray Bursts and Astronomy, eds. G. R. Ricker and R. Vanderspek (AIP, New York, 2002), 17.
7. Gehrels, N. et al. 2004, Astrophys. J. 611, 1005. p
8. Meszaros, P., Rees, M.J. 1992, MNRAS 258, 41p.
9. Piran, T. 1999 Phys. Rep. 314, 575.
10. Vanderspek, R. et al. 2004, Astrophys. J. 617, 1251.
11. Stanek, K. A. et al. 2003, Astrophys. J. 591, L17.
12. Galama, T. et al., 1998, Nature 395, 670.
13. Heise, J. 2000, Proc. 2nd Rome Workshop: Gamma-Ray Bursts in the Afterglow Era, eds. E. Costa, F. Frontera, J. Hjorth (Springer-Verlag, 2000), 16.
14. Amati, L. et al. 2002, Astron. Astrophys. 390, 81.
15. Sakamoto, T. et al. 2004, Astrophys. J. 602, 875.
16. Sakamoto, T. et al. 2005, Astrophys. J. 629, 311.
17. Soderberg, A. et al. 2004, Astrophys. J. 606, 994.
18. Soderberg, A. et al. 2005, Astrophys. J. 627, 877.
19. Ghirlanda, G. et al. 2004, Astrophys. J. 616, 331.
20. Ghirlanda, G. et al. 2004, Astrophys. J. 613, L13.
21. Kouveliotou, C. et al. 2003, Astrophys.J. 413, L101.
22. Gehrels, N. et al. 2005, Nature 437, 851.
23. Villasenor, J. S. et al. 2005, Nature 437, 855.
24. Barthelmy S. D. et al. 2005, Nature 438, 994.
25. Kosugi, G. et al. 2005, GCN Circ. 3422.
26. Prochaska, J. X. et al. 2005, GCN Circ. 3390.
27. Fox, D. et al. 2005, Nature 437, 845.
28. Berger, E. et al. 2005, Nature 438, 988.
29. Hurley, K. et al. 2005, Nature 434, 1098.
30. Mereghetti, S. et al. 2005, ApJ 624, L105.
31. Palmer, D. 2005, Nature 434, 1107.
32. Terasawa, T. 2005, Nature 434, 1110.

The non-thermal high energy emission from GRBs - theoretical predictions

Ehud Nakar

Theoretical Astrophysics, Caltech, Pasadena, CA 91125 USA
udini@tapir.caltech.edu

Gamma-Ray bursts (GRBs) are considered to be one of the most promising sources of ultra-high energy cosmic rays (UHECRs), detectable neutrino flux and bright GeV photons. Here I briefly review the theoretical predictions for the production of these types of emission in GRBs.

1. Introduction

Gamma-ray bursts (GRBs) are the most violent explosions in the universe. Releasing a fraction of solar rest-mass energy in the form of soft γ-rays, a GRB outshine the entire observed γ-ray universe combined for a few seconds. Thus, unsurprisingly, GRBs are among the most promising candidate sources for high-energy particles and radiation. GRBs are expected to be the main transient sources of GeV photons that will soon be probed by GLAST[a] and Agile[b]. They are also suggested as the acceleration sites of the observed extra-galactic ultra-high energy cosmic rays ($> 10^{18}$ eV; UHECR) as well as a promising source of $> 10^{14}$ eV neutrinos that might be detected by upcoming neutrino detectors like IceCube[c] and ANTARES[d].

Here I briefly review the theoretically predicted contribution of GRBs to the cosmic budget of non-thermal high energy emission. I begin by describing the observations (mainly $\lesssim MeV$ photons) and the generally accepted fireball model that explains these observations (§2). Then I review the predicted emission of cosmic-rays (§3), neutrinos (§4) and GeV photons (§5). I summarize in §6.

[a]http://glast.gsfc.nasa.gov/
[b]http://people.roma2.infn.it/ agile/
[c]http://icecube.wisc.edu/
[d]http://antares.in2p3.fr/

2. Observations and the fireball model

An extensive observational effort in the last two decades has provided an extensive set of GRB observations ranging from radio to gamma-rays. The observational progress led to a theoretical picture in which the source of the GRB radiation is energy dissipated by highly relativistic outflow (see [1,2] for recent comprehensive reviews).

GRBs are detected as a short flash of non-thermal 10keV-1MeV photons, usually called called the "prompt" gamma-ray emission. The prompt emission is followed by an "afterglow" - X-ray, UV-optical and radio emission that is observed for days, weeks and even years after the burst. The energy output during the prompt emission and the afterglow is comparable. The properties of the prompt emission suggested that GRBs are divided into two groups [3] - short/hard and long/soft. Recently the different physical origin of the two phenomena was confirmed [4–6] although it seems that similar physical processes are involved in both types of events. The observed energy output of long and short GRBs (both of a single burst and the sum of the whole population) is listed in table 1. Note that the rates are the observed ones and the energy output per burst is the isotropic equivalent. Long GRBs are beamed to an average angle of [7] ≈ 0.15 rad while short GRBs are likely to be beamed but the average beaming is yet unknown. As a result the energy output per burst is smaller by the beaming factor while the rate is larger by the same factor. The total energy output remains unchanged.

Table 1. Energy output in 10keV-1MeV photons

	Long GRBs	Short GRB
Observed all sky rate [8,9]	$\approx 500\ \mathrm{yr}^{-1}$	$\approx 170\mathrm{yr}^{-1}$
Observed local rate density [7,9–11]	$\sim 0.5\ \mathrm{Gpc}^{-3}\ \mathrm{yr}^{-1}$	$\sim 20\ \mathrm{Gpc}^{-3}\ \mathrm{yr}^{-1}$
$E^{\mathrm{a}}_{\gamma,iso}$	$10^{52} - 10^{54}$ erg †	$10^{49} - 10^{51}$ erg
$L^{\mathrm{a}}_{\gamma,iso}$	$10^{50} - 10^{52}$ erg/s	$10^{49} - 10^{52}$ erg/s
Average observed flux [b]	$\sim 10^{-8}\ \mathrm{GeV/cm^2/s/sr}$	$\sim 5 \cdot 10^{-10}\ \mathrm{GeV/cm^2/s/sr}$
Local energy output rate [c]	$\sim 10^{53}\ \mathrm{erg\ Gpc^{-3}\ yr^{-1}}$	$\sim 10^{51}\ \mathrm{erg\ Gpc^{-3}\ yr^{-1}}$
Total energy output [d]	$\sim 10^{66}$ erg	$\sim 10^{63}$ erg

Note: † Recent observations suggest that there is a subclass of faint ($E_{\gamma,iso} \sim 10^{49}$ erg) long GRBs with a significantly higher local rate density[12,13]. The luminous long GRBs, however, still dominate the energy production rate.

[a] Isotropic equivalent quantities per burst

[b] The total observed soft γ-ray flux from GRBs, averaged over long time ($\gg 1$ day).

[c] The energy output density rate in soft γ-rays of the entire local GRB population.

[d] The energy output in soft γ-rays of the entire GRB population in the observed universe.

The enormous non-thermal luminosity of the prompt emission indicates that the emitting source is moving ultra-relativistically (Lorentz factor $\Gamma > 100$) toward the observer[14]. The emitting source is most likely a relativistic wind that is driven by accretion onto a newly born black hole. The composition of this wind is still a major open questions. The leading candidates are baryon-dominated flow, a Poynting-flux-dominated flow or a combination of the two (mildly magnetized plasma). The generally accepted model explaining these observations is known as the internal-external fireball model. In this model the prompt emission results from internal dissipation of the energy within the wind. In the baryonic flow model by internal shocks[15,16] while in the a Poynting-flux-dominated flow by reconnection of the magnetic field as a result of current instabilities[17,18]. The afterglow is produced when the relativistic wind interacts with the ambient medium. In a baryonic wind two shocks are generated, a short lived reverse shock that propagates back into the wind and a long lived forward shock that runs into the external medium. Once the reverse shock dies away all the wind energy is dissipated into the external medium and a single blast-wave propagates into the ambient medium. This blast wave is the source of the long-lasting afterglow. The evolution of a Poynting-flux-flow differ during the dissipation of the flow energy to the external medium. There is no reverse shock, instead the interaction with the external medium reflects electromagnetic waves back into the magnetized wind. Once the electromagnetic energy is dissipated to the ambient medium the evolution of the blast-wave is expected to be similar to the baryonic wind case.

Below, I discuss various models for high energy emission within the context of this internal-external fireball model (for different view see [19]). The fireball model was developed mostly to explain the long burst observations which are far more detailed than these of short GRBs. However, the recent discovery of short burst afterglows and the similarity in many properties of the prompt emission of short and long bursts[20,21] suggests that similar mechanisms operate in both phenomena. In any case, since the long GRBs are much better understood and as they govern the energy output of GRBs, I will consider hereafter only long GRBs.

3. Ultra-high energy cosmic rays

GRBs are a promising sources of the observed extra-galactic cosmic rays above 10^{18} eV (UHECRs) through diffusive shock accleration[22–25](DSA; for reviews of the process see [26,27]). The observed GRB energy flux in soft γ-rays (see table 1) is comparable to the observed flux of UHECRs. Therefore GRBs can be the source of UHECRs provided that they accelerate high energy cosmic rays and produce γ-rays with similar efficiencies [e]. The observed γ-ray spectrum is also consistent with this idea, suggesting that electrons are accelerated in the source to high energy power-law distribution with an index $p \approx 2.2$, which is similar to the UHECR spectrum. This spectrum is also that expected as a result of DSA in relativistic shocks[28–30]. Thus the suggested picture is one in which both electrons and protons are efficiently accelerated in shocks that take place within GRBs.

However, global energy and spectral considerations are not enough. The conditions in the shock should be such that protons can be accelerated to $\sim 10^{20}$ eV. The three major factors that limit the acceleration are time, confinement and cooling. Assuming that the coherence length of the magnetic field in the upstream (unshocked wind) is larger than R/Γ, the criterion that the acceleration is faster than the adiabatic cooling is[22] $R_L < R/\Gamma$ where R_L is the Larmor radius of the accelerated particles in the upstream field and R [Γ] is the radius [Lorentz factor] of the relativistic wind. This condition also guaranties that the accelerated particle is confined and that it has sufficient time to be accelerated. Writing this condition using the parameters of the relativistic wind one obtains[22]:

$$\varepsilon_B > 0.02 E_{p,20}^2 \Gamma_{2.5}^2 L_{w,52}^{-1}, \qquad (1)$$

where ε_B is the fraction of magnetic field energy density in the unshocked wind[f] out of the total wind energy density (including rest mass energy) as measured in the wind rest frame. E_p is the energy of the accelerated proton, L_w is the wind luminosity as measured in the observer frame and Q_x denotes $Q/10^x$ in cgs units. The requirement that the acceleration is faster than synchrotron cooling is satisfied for[22] $R > 10^{12} cm E_{p,20}^3 \Gamma_{2.5}^{-2}$. In a baryonic wind both conditions can be satisfied during the internal shocks ($R \sim 10^{12} - 10^{15}$ cm) and the reverse shock ($R \sim 10^{16} - 10^{17}$ cm) given that $100 \lesssim \Gamma \lesssim 1000$ and that the wind is mildly magnetized. The constraint on Γ is conveniently similar to the constraint obtained from opacity considerations while the magnetization of the wind is currently poorly constrained.

---

[e]Wick et. al.,[25] suggest that GRBs are the source of all cosmic rays down to the knee ($\sim 10^{15}$ eV) in which case the GRB energy in cosmic-rays is ~ 100 time larger than the one observed in soft γ-rays

[f]ε_B here is not to be confused with the magnetic field density in the shocked downstream which is the site of the synchrotron emission. While the observed emission indicates a significant downstream field there are currently only indirect evidence suggesting that the wind is magnetized[31,32].

The pre-existing magnetization in the ambient medium, which is the upstream of the external shocks, is expected to be much smaller than the one required by Eq. 1. In principal the upstream magnetic field might be significantly amplified by back reaction of the accelerated particles[33,34]. However the accelerated particle precursor cannot run further than a distance of R/Γ^2 ahead of the shock and therefore the maximal coherence length of an amplified field is limited to R/Γ^2. It was shown recently[34] that even if the upstream magnetic field is amplified to ($\varepsilon_B \sim 1$) with the maximal coherence length (R/Γ^2), UHECRs cannot be accelerated. Therefore UHECRS are not expected to be produced in the external shock.

To conclude, acceleration of UHECRs by DSA in GRBs is plausible if the relativistic wind carries with it from the source a magnetic field that is close to equipartition with the baryons. If the magnetic field is too low UHECRs cannot be confined while if the magnetic field is too strong (Poynting-flux-dominated flow) no internal or reverse shocks take place. In Poynting-flux-dominated flow the electric potential drop can be large enough to accelerate UHECRs[17]. However the electric potential is a strict upper limit and it is hard to prevent energy loses from preventing UHECR acceleration. Currently there is no calculation that account for energy loses that show that UHECRs can be accelerated in Poynting flux dominated flow or that estimate the energy output in cosmic-rays in this model.

4. Neutrinos

If GRB outflows are baryonic then we expect a dense environment of high energy photons, protons and neutrons, to exist during some of the GRB phases. As a result, GRBs are predicted to be sources of non-thermal neutrinos, via photo-meson and nucleon-nucleon interactions, over a wide range of energies (in addition to the thermal MeV neutrinos that are produced in supernovae that accompany GRBs).

Internal shocks are predicted to produce $\sim 10^{14}$ eV neutrinos [35–38]. These shocks produce photons with typical energy $\epsilon_\gamma \sim 1$ MeV (the prompt emission). Pion production is expected if protons are accelerated in these shocks to high energies. Pions are efficiently produced when the photon energy, as measured in the rest frame of the accelerated protons, is comparable to the pion rest mass, implying typical proton energy $E_p \sim 10^{16}\epsilon_{\gamma,MeV}^{-1}\Gamma_{2.5}^2$ eV. Since the neutrinos produced from the pion decay cascade carry about 5% of the original proton energy, the typical neutrino energy is $\epsilon_\nu \sim 5 \cdot 10^{14}$ eV. The fraction of the energy that is carried by neutrinos depends on the photon density in the source and thus may vary significantly from burst to burst. Waxman & Bahcall[35] estimate that in a bright ($L_\gamma \sim 10^{52}$ erg/s) and variable (variability time scale of ~ 10 ms) burst neutrinos can carry about 10% of the proton energy. Assuming that the energy of accelerated protons is comparable to the observed photon energy, and given that luminous GRBs account for significant fraction of the total GRB energy production they get an observed neutrino energy flux from all GRBs combined:

$$\epsilon_\nu^2 \Phi_\nu(\epsilon_\nu \approx 10^{15} \text{ eV}) \sim 10^{-9} \text{ GeV/cm}^2\text{/s/sr}. \tag{2}$$

An additional neutrino production site is the reverse shock[39]. The process here is similar to that in the internal shocks, with the main difference being that the reverse shock emits mostly optical-UV photons[40]. As a result, the protons that participate in the photo-meson production have $E_p \sim 10^{20}$ eV and the resulting neutrinos have $\epsilon_\nu \approx 10^{18}$ eV. The efficiency of the process depends linearly on the UV flux. Assuming a very bright reverse shock emission (comparable to the bright optical flash observed in GRB 990123[41]) Waxman & Bahcall[39] estimated an efficiency of $\sim 10\%$ and therefore a comparable flux to that of internal shock neutrinos. However, after a year of Swift[g] observations it seems that GRB 990123 was unique and that early optical-UV emission from the majority of the bursts is fainter by several orders of magnitude. As a result, if neutrinos are produced in reverse shocks, their energy flux is significantly lower than Eq. 2.

The acceleration of a baryonic fireball[42–46] is another phase in which neutrinos can be produced[47]. During the acceleration the protons are coupled to the radiation (through electrons) while the neutrons are coupled to the protons through collisions. If the wind is accelerated to $\Gamma \gtrsim 300$ the protons decouple from the neutrons while still being accelerated and the relative velocity of the two components become relativistic. As a result, inelastic nucleon-nucleon collisions take place leading to pion production[47,48]. Since the nucleons are mildly relativistic in the wind rest frame, the energy of the produced neutrinos is ~ 50 MeV in this frame. In the observer frame the nutrino energy is $\epsilon_\nu \sim 10$ GeV. Since every nucleon in the wind is expected to experience ~ 1 inelastic collision, the efficiency of this process is again $\sim 10\%$. However the low neutrino energy makes them hard to detect.

The progenitors of long bursts are massive stars [49,50]. For a GRB to be observed, the relativistic jetted wind must penetrate the stellar envelope. The propagation of relativistic baryonic winds within the envelopes of massive progenitor stars is yet another phase in which neutrinos may

[g]swift.gsfc.nasa.gov/

be produced[51]. Here, the process is again photo-meson production, where the relativistic protons are accelerated by internal shocks within the wind and the photons are supplied by the wind that is shocked by the interaction with the envelope. Mészáros & Waxman[51] find that the head of the jetted wind is propagating at a mildly relativistic velocity, and the photon temperature in the jet front is ~keV. As a result, pions are produced by protons with $E_p \gtrsim 10^{14}$ eV and the resulting neutrinos have $\epsilon_\nu \gtrsim 5 \cdot 10^{12}$ eV. The efficiency of the process is high (given the high photon density) and thus the energy flux of these neutrinos is comparable to the one from the other processes discussed above. However, this process is unique since here a similar neutrino flux is expected both when the jet succeeds in penetrating the envelope (and a GRB is produced), and when the jet is choked and no γ-ray signal is observed.

To conclude, if the relativistic wind in GRBs is baryonic then various processes can produce neutrinos with energies ranging from 10 GeV to 10^{18} eV. The energy flux in these neutrinos can be as high as $\sim 10\%$ of the energy emitted in soft γ-rays (unless high energy protons carry much more energy than observed in γ-rays[25]). If the wind is Poynting-flux-dominated then none of these processes will take place and no strong neutrino emission is expected. Therefore neutrino detection from GRBs will strongly suggest that the relativistic flow is baryonic.

5. GeV photons

EGRET[h] detected GeV photons from a few GRBs[52] where the most energetic photon (18 GeV) was detected 90min after the burst (during the afterglow phase)[53]. Significant GeV emission during the afterglow is predicted as a result of the synchrotron self Compton (SSC) process[54–57] - the electrons that are accelerated by the external shock Comptonize some of the synchrotron photons emitted by the same electron population. For typical GRB parameters most of the SSC energy is emitted at $\nu_{SSC} \sim \nu_{c,synch}\gamma_c^2 \sim GeV$ where $\nu_{c,synch}$ is the synchrotron cooling frequency (usually between the optical and X-ray bands) and γ_c is the Lorentz factor of external shock electrons that were efficiently cooled (both by synchrotron and SSC) and is typically[i] $\sim 10^3 - 10^4$. The ratio between the energy emitted in the synchrotron and SSC branches depends on the ratio of the magnetic field en-

[h] http://cossc.gsfc.nasa.gov/docs/cgro/egret/

[i] Note that $\nu_{c,synch}$ photons, that are softer by a factor Γ in the shock frame, are typically well below the Klein-Nishina limit of the γ_c electrons

ergy density and the electron energy density in the emitting region. These two quantities are usually measured as fractions from the total internal energy and denoted as ε_B and ε_e [j]. Multi-wavelength modeling of GRB afterglows[58,59] shows that $\varepsilon_e \sim 10\varepsilon_B$ in which case the energy in the SSC branch is larger by a factor of $\sim (\varepsilon_e/\varepsilon_B)^{1/2} \approx 3$. Therefore, a clear prediction of the current modeling of the synchrotron radiation is that during the afterglow phase, GeV emission with energy that is comparable to or larger than the prompt emission is emitted (the observed energy in the synchrotron afterglow emission is comparable to the prompt emission energy). Note that SSC emission that arises from the external shock does not depend on the composition of the relativistic wind.

Many other processes are suggested to produce GeV emission and discussing them all in detail is beyond the scope of this short review. These processes involve comptonization of soft photons or production of hard photons by elements heavier than electrons (pions or nucleons). Among the comptonization models are: prompt emission photons on reverse shock electrons[60] (GeV-TeV); Reverse shock photons on forward shock electrons[61] (GeV-TeV); X-ray flare photons on forward shock electrons[62] (GeV-TeV); Hypothesized UV flare photons on forward shock electrons[63] (sub-GeV). Emission from heavy elements is expected in a baryonic flow due to acceleration of protons and production of pions. Some of the discussed processes are synchrotron emission of accelerated protons and produced pions[64,65] and photons emitted by pion decay as a result of inelastic neutron-proton collisions during the acceleration of the wind[47,48].

Finally, while TeV emission is expected to be produced in some of the processes discussed above, only once every several years a bright long GRB is expected to occur at a distance from earth within which the IR background is optically thin[66] to TeV photons. The local rate of low luminosity long GRBs and short GRBs is higher[7,10,13]. However, both are much less energetic than bright long GRBs and in most of the processes discussed above the high energy emission depends strongly on the total burst energy.

6. Summary

The predictions for high energy emission from GRBs depend strongly on the composition and the conditions of the relativistic wind. If the wind is

[j] ε_B here is measured in the shocked downstream and not to be confused with the magnetic energy density in the unshocked upstream which we considered in §2. The two quantities are not necessarily related as magnetic fields may be generated in the shock.

composed of mildly magnetized baryons then it is very plausible that GRBs are the source of UHECRs. If the wind is composed of weakly magnetized baryons then cosmic rays are expected to be accelerated but not to energies of 10^{20} eV. In any case of baryonic wind GRBs are also expected to produce neutrinos over a wide range of energies which might very well be detectable by next generation neutrino detectors. Under these conditions the energy output rate of GRBs in soft ($\sim$ MeV) γ-rays, hard ($\sim$ GeV) γ-rays and in cosmic-rays is expected to be comparable ($\sim 10^{53}$erg/Gpc^3/yr in each component) while the energy output in neutrinos is $\sim 10\%$ of this rate. If however the relativistic wind is dominated by Poynting flux then no neutrino component is predicted while the cosmic-ray component is currently unexplored. In this case only soft ($\sim$ MeV) and hard ($\sim$ GeV) γ-rays with comparable energies are predicted.

Acknowledgments

I am grateful to A. Gal-Yam and M. Lyutikov for helpful comments.

References

1. T. Piran, *Reviews of Modern Physics* **76**, 1143(January 2005).
2. P. Meszaros, *ArXiv Astrophysics e-prints* (May 2006).
3. C. Kouveliotou, et. al., *ApJ* **413**, L101(August 1993).
4. D. B. Fox, et. al., *Nature* **437**, 845(Octtober 2005).
5. J. Hjorth, et. la., *Nature* **437**, 859(Octtober 2005).
6. E. Berger, et. al., *Nature* **438**, 988(December 2005).
7. D. Guetta, T. Piran and E. Waxman, *ApJ* **619**, 412(January 2005).
8. C. Meegan, et. al., A Review of Gamma Ray Bursts, in *AIP Conf. Proc. 410: Proceedings of the Fourth Compton Symposium*, 1997.
9. M. Schmidt, *ApJ* **559**, L79(Octtober 2001).
10. E. Nakar, A. Gal-Yam and D. Fox, *ArXiv Astrophysics e-prints* (Nov. 2005).
11. D. Guetta and T. Piran, *ArXiv Astrophysics e-prints* (November 2005).
12. B. E. Cobb, et. al., *ArXiv Astrophysics e-prints* (March 2006).
13. A. M. Soderberg, et. al., *ArXiv Astrophysics e-prints* (April 2006).
14. Y. Lithwick and R. Sari, *Apj* **555**, 540(July 2001).
15. M. J. Rees and P. Meszaros, *ApJ* **430**, L93(August 1994).
16. R. Sari and T. Piran, *ApJ* **485**, 270(August 1997).
17. M. Lyutikov and R. Blandford, *ArXiv Astrophysics e-prints* (December 2003).
18. M. Lyutikov, *ArXiv Astrophysics e-prints* (December 2005).
19. A. Dar, *ArXiv Astrophysics e-prints* (January 2006).
20. E. Nakar and T. Piran, *MNRAS* **330**, 920(March 2002).
21. G. Ghirlanda, G. Ghisellini and A. Celotti, *A&A* **422**, L55(July 2004).
22. E. Waxman, *Physical Review Letters* **75**, 386(July 1995).
23. M. Milgrom and V. Usov, *ApJ* **449**, L37+(August 1995).

24. M. Vietri, *ApJ* **453**, 883(November 1995).
25. S. Wick, C.. Dermer and A. Atoyan, *Astroparticle Physics* **21**, 125(May 2004).
26. L. O. Drury, *Reports of Progress in Physics* **46**, 973(August 1983).
27. R. Blandford and D. Eichler, *Physics Reports* **154**, 1(Octtober 1987).
28. A. Achterberg, et. al., *MNRAS* **328**, 393(December 2001).
29. D. Ellison and G. Double, *Astroparticle Physics* **18**, 213(December 2002).
30. U. Keshet and E. Waxman, *Phys. Rev. Letters* **94**, 111102(March 2005).
31. B. Zhang, S. Kobayashi and P. Mészáros, *ApJ* **595**, 950(Octtober 2003).
32. P. Kumar and A. Panaitescu, *MNRAS* **346**, 905(December 2003).
33. A. R. Bell and S. G. Lucek, *MNRAS* **321**, 433(March 2001).
34. M. Milosavljevic and E. Nakar, *ArXiv Astrophysics e-prints* (December 2005).
35. E. Waxman and J. Bahcall, *Physical Review Letters* **78**, 2292(March 1997).
36. E. Waxman and J. Bahcall, *Phys. Rev. D* **59**, 023002(January 1999).
37. J. P. Rachen and P. Mészáros, *Phys. Rev. D* **58**, 123005(December 1998).
38. C. D. Dermer and A. Atoyan, *Phys. Rev. Letters* **91**, 071102(August 2003).
39. E. Waxman and J. N. Bahcall, *ApJ* **541**, 707(Octtober 2000).
40. R. Sari and T. Piran, *ApJ* **520**, 641(August 1999).
41. C. Akerlof, et. al., *Nature* **398**, 400 (1999).
42. B. Paczynski, *ApJ* **308**, L43(September 1986).
43. J. Goodman, *ApJ* **308**, L47(September 1986).
44. T. Piran, A. Shemi and R. Narayan, *MNRAS* **263**, 861(August 1993).
45. P. Meszaros, P. Laguna and M. J. Rees, *ApJ* **415**, 181(September 1993).
46. E. Nakar, T. Piran and R. Sari, *ApJ* **635**, 516(December 2005).
47. J. N. Bahcall and P. Mészáros, *Phys. Rev. Letters* **85**, 1362(August 2000).
48. E. Derishev, V. Kocharovsky and V. Kocharovsky, *ApJ* **521**, 640(Aug. 1999).
49. J. Hjorth, et. al., *Nature* **423**, 847(June 2003).
50. K. Z. Stanek, et. al., *ApJ* **591**, L17(July 2003).
51. P. Mészáros and E. Waxman, *Phys. Rev. Letters* **87**, 171102(Octtober 2001).
52. B. L. Dingus, J. R. Catelli and E. J. Schneid, Bursts Detected and not Detected by EGRET Imaging Spark Chambers, in *American Institute of Physics Conference Series*, eds. C. A. Meegan, R. D. Preece and T. M. Koshut1998.
53. K. Hurley, *Nature* **372**, 652(December 1994).
54. P. Meszaros, M. J. Rees and H. Papathanassiou, *ApJ* **432**, 181(Sept. 1994).
55. C. D. Dermer, J. Chiang and K. E. Mitman, *ApJ* **537**, 785(July 2000).
56. R. Sari and A. A. Esin, *ApJ* **548**, 787(February 2001).
57. B. Zhang and P. Mészáros, *ApJ* **559**, 110(September 2001).
58. A. Panaitescu and P. Kumar, *ApJ* **560**, L49(Octtober 2001).
59. S. A. Yost, et. al., *ApJ* **597**, 459(November 2003).
60. A. M. Beloborodov, *ApJ* **618**, L13(January 2005).
61. A. Pe'er and E. Waxman, *ApJ* **633**, 1018(Novmber 2005).
62. X.-Y. Wang, Z. Li and P. Mészáros, *ApJ* **641**, L89(April 2006).
63. Y. Fan and T. Piran, *ArXiv Astrophysics e-prints* (January 2006).
64. M. Bottcher and C. D. Dermer, *ApJ* **499**, L131+(June 1998).
65. T. Totani, *ApJ* **502**, L13+(July 1998).
66. P. S. Coppi and F. A. Aharonian, *Apj* **487**, L9+(September 1997).

EXPLOSION MECHANISM OF CORE-COLLAPSE SUPERNOVAE AND COLLAPSARS

S. NAGATAKI

Yukawa Institute for Theoretical Physics, Kyoto University, Oiwake-cho Kitashirakawa Sakyo-ku, Kyoto 606-8502, Japan
E-mail: nagataki@yukawa.kyoto-u.ac.jp

We have done 2-dimensional MHD simulation of collapsars with magnetic fields and neutrino cooling/heating processes. It is found that explosion energy of a hypernova is not obtained from the neutrino heating process. However, strong jet is found when magnetic fields are included, and total energy of the jet component can be of the order of 10^{52} erg, which is comparable to the one of a hypernova.

1. INTRODUCTION

There has been growing evidence linking long gamma-ray bursts (GRBs; in this study, we consider only long GRBs, so we call long GRBs as GRBs hereafter for simplicity) to the death of massive stars. In fact, direct evidences of some GRBs accompanied by supernovae have been reported such as the association of GRB 980425 with SN 1998bw and that of GRB 030329 with SN 2003dh. It should be noted that these supernovae are categorized as a new type of supernovae with large kinetic energy ($\sim 10^{52}$ ergs), nickel mass ($\sim 0.5 M_{\odot}$), and luminosity, so these supernovae are sometimes called as hypernovae. Also, since GRBs are considered to be jet-like phenomena, it is natural to consider the accompanying supernova to be jet-induced explosion.

The central engine of GRBs accompanied by hypernovae is not known well. But it is generally considered that normal core-collapse supernovae can not cause an energetic explosion of the order of 10^{52} erg. So another scenario has to be considered to explain the system of GRBs associated with hypernovae. One of the most promising scenario is the collapsar scenario (MacFadyen & Woosley 1999). In the collapsar scenario, a black hole is formed as a result of gravitational collapse. Also, rotation of the progenitor plays an essential role. Due to the rotation, an accretion disk is formed around the equatorial plane. On the other hand, the matter around the rotation axis falls into the black hole. It was pointed out that the jet-induced explosion along to the rotation axis occurs due to the heating through neutrino anti-neutrino pair annihilation that are emitted from the accretion disk. MacFadyen and Woosley (1999) demonstrated the numerical simulations of the collapsar, showing that the jet is launched $\sim$ 7 sec after the gravitational collapse and the duration of the jet is about 10 sec, which is comparable to the typical observed duration of GRBs. However, detailed neutrino heating process is not included inMacFadyen & Woosley 1999. Also, it is pointed out that effects of magnetic fields may be so important (Proga et al. 2003; Fujimoto et al. 2005). So in this study, we solved the dynamics of collapsars with neutrino cooling/heating processes and magnetic fields.

2. MODELS AND NUMERICAL METHODS

Our models and numerical methods of simulations in this study are shown in this section. First we present equations of ideal MHD, then initial and boundary conditions are explained. Micro physics included in this study, equation of state (EOS), nuclear reactions, and neutrino processes are also explained.

2.1. *Magnetohydrodynamics*

We have done two-dimensional MHD simulations taking account of self-gravity and gravitational potential of the central point mass. The calculated region corresponds to a quarter of the meridian plane under the assumption of axisymmetry and equatorial symmetry. The spherical mesh with $150(r)\times$ $30(\theta)$ grid points is used for all the computations. The radial grid is nonuniform, extending from 1.0×10^6 cm to 1.0×10^{10} cm with finer grids near the center, while the polar grid is uniform.

The basic equations in the following form are finite differenced on the spherical coordinates:

$$\frac{D\rho}{Dt} = -\rho\nabla\cdot\mathbf{v} \tag{1}$$

$$\rho\frac{D\mathbf{v}}{Dt} = -\nabla p - \rho\nabla\Phi + \frac{1}{4\pi}(\nabla\times\mathbf{B})\times\mathbf{B} \tag{2}$$

$$\rho\frac{D}{Dt}\left(\frac{e}{\rho}\right) = -p\nabla\cdot\mathbf{v} - \mathbf{L}_\nu^- + \mathbf{L}_\nu^+ + \mathbf{L}_{\rm nucl} \tag{3}$$

$$\frac{\partial\mathbf{B}}{\partial t} = \nabla\times(\mathbf{v}\times\mathbf{B}), \tag{4}$$

where ρ, $\mathbf{v}$, P, Φ, e, $L_\nu^\pm$, and $L_{\rm nucl}$ are density, velocity, pressure, gravitational potential, internal energy density, heating/cooling rates due to neutrino processes, and energy gain (loss) rate due to nuclear reaction. The Lagrangian derivative is denoted as D/Dt. The ZEUS-2D code developed by Stone and Norman (1992) has been used to solve the MHD equations.

2.2. *Initial Conditions*

We adopt the Model E25 in Heger et al. (2000). The star in this model has $25M_\odot$ initially with solar initial metallicity, but lose its mass and becomes to be $5.45M_\odot$ as a Wolf-Rayet star at the final stage. This model seems to be a good candidate as a progenitor of a GRB since losing their envelope will be suitable to make a baryon poor fireball. The mass of iron core is $1.69M_\odot$ that is covered with Si layer whose mass is $0.55M_\odot$. So we assume that the iron core has collapsed and formed a black hole at the center.

Angular momentum was distributed so as to provide a constant ratio of 0.04 of centrifugal force to the component of gravitational force perpendicular to the rotation axis at all angles and radii, except where that prescription resulted in j_{16} greater than a prescribed maximum value,10. This treatment is exactly same with MacFadyen and Woosley (1999). Total initial rotation energy is 5.7×10^{48} erg that corresponds to initial ratio of the rotation energy to the gravitational energy, $T/W = 8.3\times10^{-3}$.

Configuration and amplitude of the magnetic fields in a progenitor prior to collapse are still uncertain. So in this study we choose a simple form of the initial configuration and the amplitude is changed parametrically. Initial configuration of the magnetic fields is chosen as follows:

$$\vec{B}(\vec{r}) = \frac{1}{3}B_0\left(\frac{r_0}{r}\right)^3(2\cos\theta\vec{e_r}+\sin\theta\vec{e_\theta}) \quad \text{for } \mathrm{r}\geq\mathrm{r_0} \tag{5}$$

$$= \frac{2}{3}B_0(\cos\theta\vec{e_r}-\sin\theta\vec{e_\theta}) \quad \text{for } \mathrm{r}<\mathrm{r_0}. \tag{6}$$

This configuration represents that the magnetic fields are uniform in a sphere ($r < r_0$), while dipole at outside of the sphere. We set r_0 to be the boundary between CO core/Si layer. B_0 corresponds to the strength of the magnetic field in the sphere. We have chosen B_0 to be 0, 10^8G, 10^9G, and 10^{10}G.

2.3. *Micro Physics*

2.3.1. *Equation of State*

The equation of state (EOS) used in this study is the one developed by Blinnikov et al. (1996). This EOS contains an electron-positron gas with arbitrary degeneracy, which is in thermal equilibrium with blackbody radiation and ideal gas of nuclei.

2.3.2. *Nuclear Reactions*

Although the contribution of ideal gas of nuclei to the total pressure is negligible, effects of energy gain/loss due to nuclear reactions are important. In this study, nuclear statistical equilibrium (NSE) was assumed for the region where $T \geq 5 \times 10^9$ [K] is satisfied Nagataki et al. 2003, while no nuclear reaction occurs for the region where $T < 5 \times 10^9$ [K].

2.3.3. *Neutrino Processes*

Neutrino cooling processes due to pair capture on free nucleons, pair annihilation, and plasmon decay are included in this study. Since photoneutrino and bremsstrahlung processes are less important ones at $10^9 < T < 10^{11}$ [K] and $\rho < 10^{10}$ [g cm^{-3}] where effects of neutrino cooling are important in our calculations, we do not include these processes.

Neutrino heating process due to ν_e and $\bar{\nu}_e$ captures on free nucleons and neutrino pair annihilation with blocking factors of electrons and positrons are included in this study. The ν_e and $\bar{\nu}_e$ captures on free nucleons are inverse processes of electron/positron captures. As for the neutrino pair annihilation process, the formulation of Goodman et al. (1987) is adopted. We assume that the matter is optically thin against neutrinos to obtain the neutrino heating rate as mentioned above.

3. RESULTS

In Figure 1, density contour with velocity fields at t-2.5sec after the collapse. The case without magnetic fields is shown in the left panel, while the case

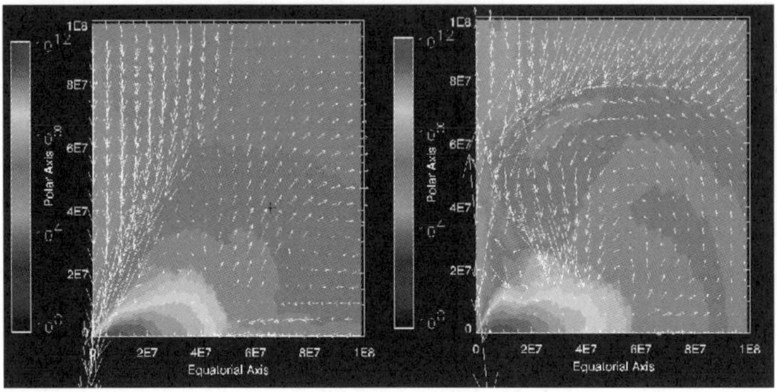

Figure 1. Density contour with velocity fields at t-2.5sec after the collapse. The case without magnetic fields is shown in the left panel, while the case with magnetic fields (10^9G) is shown in the right panel.

with magnetic fields (10^9G) is shown in the right panel. It is clearly shown that a jet propagates along to the rotation axis for the case with magnetic fields. The total energy of the jet component can be of the order of 10^{52} erg, which is comparable to the one of a hypernova.

In Figure 2, total emitted energy by neutrino processes as a function of time (solid line), total absorbed energy by neutrino anti-neutrino pair annihilation (dashed line), and total absorbed energy by neutrino capture on nucleons (dotted line) are shown. It is noted that total absorbed energy is so little that this effect can not explain the explosion energy of a hypernova.

As for the dependence of the strength of the initial magnetic fields, we found that strong jet is launched when strong initial magnetic fields are assumed initially.

4. SUMMARY AND DISCUSSION

We have done 2-dimensional MHD simulation of collapsars with magnetic fields and neutrino cooling/heating processes. It is found that explosion energy of a hypernova is not obtained from the neutrino heating process. However, strong jet is found when magnetic fields are included, and total energy of the jet component can be of the order of 10^{52} erg, which is comparable to the one of a hypernova.

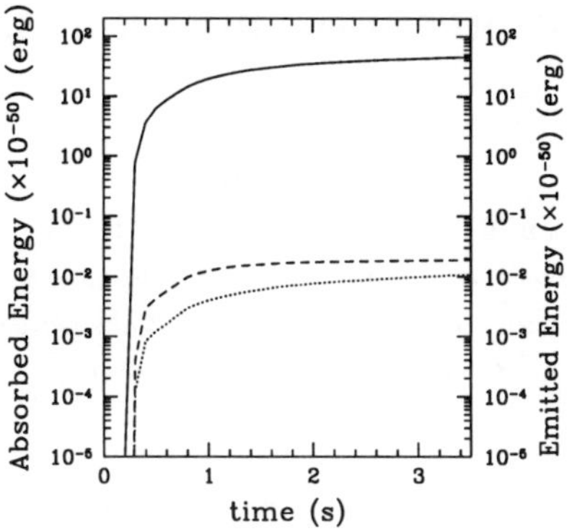

Figure 2. Solid line: total emitted energy by neutrino processes as a function of time. Dashed line: total absorbed energy by neutrino anti-neutrino pair annihilation. Dotted line: total absorbed energy by neutrino capture on nucleons.

5. Acknowledgments

S.N. are also grateful to M. Watanabe and S. Yamada for useful discussion. The computation was partly carried out on NEC SX-5 and SX-8, SGI Altix3700 BX2, and Compaq AlphaServer ES40 at Yukawa Institute for Theoretical Physics, and Fujitsu VPP5000 at National Astronomical Observatory of Japan. This work is partially supported by Grants-in-Aid for Scientific Research from the Ministry of Education, Culture, Sports, Science and Technology of Japan through No. 14102004, 14079202, and 16740134.

References

Blinnikov et al. 1996. Blinnikov, S.I., Dunina-Barkovskaya, N.V., Nadyozhin, D.K. 1996, ApJ, 106, 171

Fujimoto et al. 2005. Fujimoto, S., et al. ApJ, in press (astro-ph/0602457)

Heger et al. 2000. Heger, A., Langer, N., Woosley, S.E. 2000, ApJ, 528, 368

MacFadyen & Woosley 1999. MacFadyen, A.I., Woosley, S.E., ApJ, 1999, 524, 262

Nagataki et al. 2003. Nagataki, S., Mizuta, A., Yamada, S., Takabe, H., Sato, K., ApJ, 2003, 596, 401

Proga et al. 2003. Proga, D., MacFadyen, A.I., Armitage, P.J., Begelman, M.C. 2003, ApJ, 599, L5

RECENT RESULTS FROM CANGAROO

MASAKI MORI
FOR THE CANGAROO TEAM*
Institute for Cosmic Ray Research, University of Tokyo
5-1-5 Kashiwanoha, Kashiwa, 277-8582 Chiba, Japan
E-mail: morim@icrr.u-tokyo.ac.jp

The CANGAROO-III telescope system for very-high-energy gamma-ray astrophysics consists of four 10 m atmospheric Cherenkov telescopes located near Woomera, South Australia. The construction of the fourth telescope was completed in summer 2003, and stereoscopic observations has been in progress since March 2004. Here we report on the status of the system and some recent results from CANGAROO-III observations.

1. Introduction

CANGAROO is an acronym for the Collaboration of Australia and Nippon (Japan) for a GAmma Ray Observatory in the Outback. After successful operation of the 3.8m imaging Cherenkov telescope (CANGAROO-I) for 7 years, which was the first of this kind in the southern hemisphere, we constructed a new telescope of 7m diameter [1] (CANGAROO-II) in 1999 next to the 3.8m telescope near Woomera, South Australia (136°47′E, 31°06′E, 160m a.s.l.). Then the construction of an array of four 10m telescopes (CANGAROO-III) was approved and as the first step the 7m telescope was upgraded to 10m diameter in 2000, with this becoming the first telescope of the CANGAROO-III array [2,3,4,5,6]. Results from observations with this first 10m telescope have been reported in publications (see, e.g. [7]).

In the following years, we have constructed an additional three 10m telescopes located at the corners of a diamond of 100m sides with improved mirrors, cameras and electronics. After tuning, we have started observation with the full system in stereo mode in March 2004 (Fig.1) [8].

The major parameters of the CANGAROO-III telescopes are summarized in Table 1. The detailed design of reflectors [9,10], cameras [11,12,13],

*see http://icrhp9.icrr.u-tokyo.ac.jp for the full collaboration list.

Figure 1. The CANGAROO-III telescopes in Woomera, South Australia, as of March 2004. From the left to right, they are called T2, T3, T4 and T1 in the order of construction. T1 was called CANGAROO-II before.

electronics [14,15], and telescope control system [16] are described elsewhere.

Table 1. Parameters of the CANGAROO-III telescopes.

	T1	T2, T3, T4
Mount	Alt-azimuth	
Focal length	8m	
Number of mirrors (area)	114 (57m^2 in total)	
Reflector type	Parabola	
Number of PMTs	552 (1/2")	427 (3/4")
Camera pixel size	0.115°	0.168 °
Readout	TDC(CAMAC) & ADC	TDC(VME) & ADC
Point image size (FWHM)	0.20°	$0.14 \sim 0.21°$
Completion	2000.3	2002.3 (T2), 2002.11 (T3), 2003.7 (T4)

2. Stereo analysis: the case of the Crab nebula

The Crab nebula is an established TeV gamma-ray source and is used as a calibration source to check performance of a Cherenkov telescope. However, from Woomera, it can be observed only at large zenith angles ($> 53°$). For stereo observations, the threshold energy of T1 is higher than other telescopes and thus we used the newer three telescopes for analysis. Because of the geometrical arrangement of the array, the effective baseline for large zenith angle observations becomes short which makes stereo reconstruction of images difficult.

To overcome the unfortunate situation described above, we developed new analysis methods [17]. To avoid the increased uncertainty of the intersection points, we introduced a new parameter, "IP distance" (D_{IP}), which is defined as the distance between the intersection point and centroid of images. Then we searched best intersection points which minimizes the image widths and the difference between distance and D_{IP}. This results in better angular resolution as seen in the θ^2 distribution in Monte Carlo simulations, where θ is the space angle between the source direction and the reconstructed arrival direction: gamma-ray signals should be seen as a peak toward $\theta^2 = 0$, whose sharpness depends on the angular resolution and the angular extent of a gamma-ray source.

We observed the Crab nebula in December 2003 in so-called wobble mode, changing the pointing directions $\pm 0.5°$ in declination from the target every 20 minutes. After basic data quality check, such as rejecting runs affected by clouds, a total of 890 minutes data were used for further analysis.

In addition to the conventional square cuts method using image parameters to enhance gamma-ray fractions, we applied two different analyses: the Likelihood method [18,19] and the Fisher Discriminant method [17,20]. In the latter method, effectiveness of the parameters for the gamma-ray-like event selection is evaluated using the simulation, and we can optimize the weights of the parameters in estimating the probability of gamma-ray-like events. Finally we obtained the spectrum of the Crab nebula in the energy range from 2 to 20 TeV [20], which is consistent within the statistical and systematic errors with other measurements [21,22] (Fig.2).

3. Recent results

3.1. *Pulsar PSR 1706-44*

A detection of a gamma-ray signal from PSR 1706-44, which was one of the EGRET-detected pulsars, was reported using the data acquired by CANGAROO-I 3.8m telescope [23]. The Durham group also reported a detection with their Mark 6 telescope [24]. H.E.S.S., however, claimed no detection from that direction [25]. We observed this source for 27 hours (ON) and 29 hours (OFF) with CANGAROO-III in May 2004. Preliminary analyses using T2 and T3 telescope pair did not show a peak in the θ^2 distribution [26]. The upper limit from this result is shown in Fig. 3, which is lower than the flux reported by CANGAROO-I. Further analysis is underway and the details will be reported elsewhere.

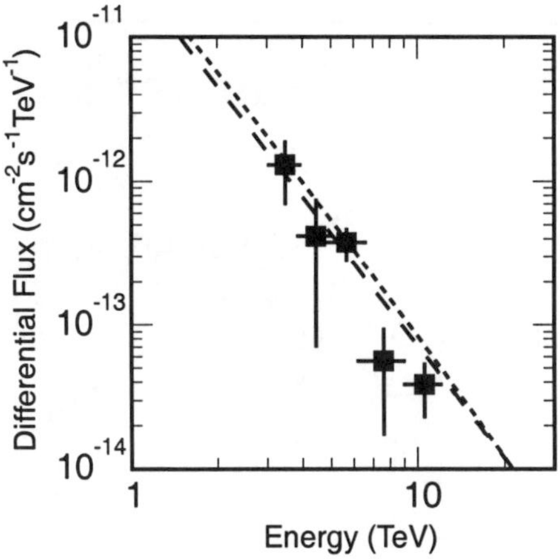

Figure 2. Differential gamma-ray flux from the Crab Nebula as a function of energy. The dashed line is the HEGRA result [21] and the dotted one is the Whipple result [22].

3.2. *Supernova remnant SN1006*

A detection of a gamma-ray signal from SN1006, which was shown to be a source of high-energy electrons through observation of non-thermal X-rays with ASCA [27], was reported using the data acquired by CANGAROO-I [28]. H.E.S.S., however, claimed no detection from that direction [29]. We observed this source for 27 hours (ON) and 29 hours (OFF) with CANGAROO-III in May 2004. Preliminary analyses using T2 and T3 telescope pair did not show any peaks in the θ^2 distribution for the NE-rim point which was the maximum point of the gamma-ray emission in the CANGAROO-I data [28]. The upper limit from this result is lower than the flux reported by CANGAROO-I. Further analysis is underway to check the possible extended emission, and the details will be reported elsewhere.

3.3. *Vela pulsar and nebula*

The Vela pulsar was observed in January/February 2004. After basic data quality check, a total of 1311 minutes data were used for further analysis [20], where the minimum elevation angle was set at 60°. The mean elevation angle was 70.9°, corresponding to an energy threshold of 600 GeV. The

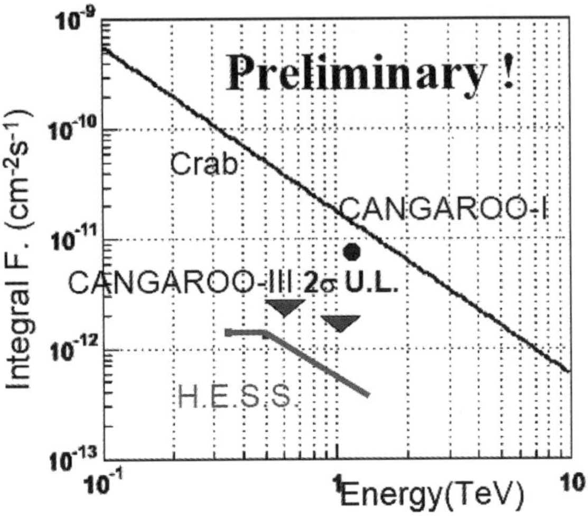

Figure 3. Upper limits on gamma-ray flux from PSR 1706-44 from CANGAROO-III observations (triangle) [26]. The CANGAROO-I result is shown by a filled circle [23] and the H.E.S.S. limits [25] are also shown.

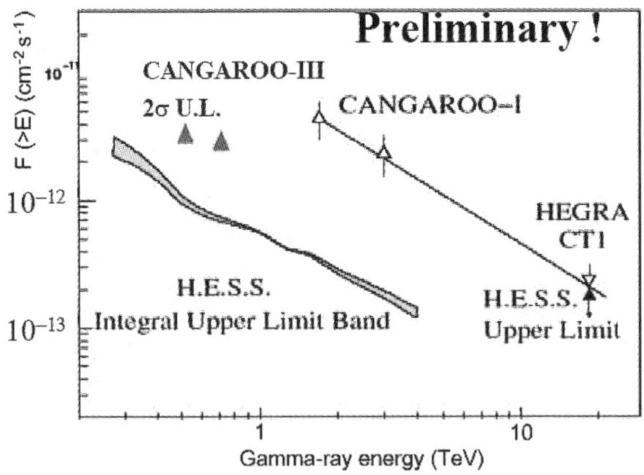

Figure 4. Upper limits on gamma-ray flux from NE-rim of SN1006. The CANGAROO-I results are shown by open triangles [28] and the HEGRA CT1 result by inverted open triangle [30]. The CANGAROO-III upper limits [26] are shown by filled triangles with H.E.S.S. limits [34].

observations were carried out using the same wobble mode as for the Crab nebula observations. In this period, T2 and T3 were in operation, and we analyzed the stereo data from these two telescopes. For Vela, at a declination of $-45°$, the relative orientation of the two telescopes does not present any problems.

We used the optimized analysis procedure used for the Crab nebula analysis described above. The resulting θ^2 distribution for the Vela pulsar position showed no significant gamma-ray signal, giving upper limits as shown in Fig.5, which are consistent with H.E.S.S. results [34]. Also we did not see excess from the point offset by 0.13° from the pulsar, which was the maximum of the excess detected with the CANGAROO-I telescope [31].

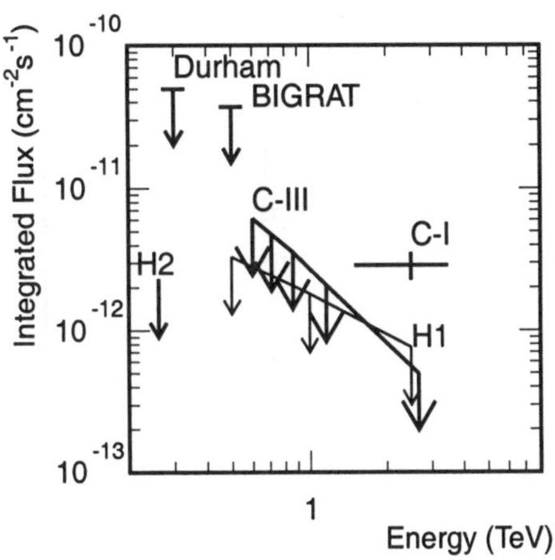

Figure 5. The 2σ upper limits for the gamma-ray flux from the Vela pulsar by CANGAROO-III (C-III) [20]. C-I represents the CANGAROO-I excess from the point offset by 0.13° from the pulsar [31]. Also shown are upper limits reported by the Durham group [32], BIGRAT [33] and H.E.S.S. [34].

The H.E.S.S. group detected a gamma-ray excess from the Vela X nebula, extended over a 0.6° radius from the center of the emission [(R.A., decl.) = ($8^h35^m, -45°36'$)] [34]. In order to analyze extended emission, we applied the following method. Gamma-ray-like events can be extracted by fitting position-by position F (Fischer discriminant) distributions under the assumption that gamma rays obey the Monte Carlo predictions, the proton background follows the average F distribution of all directions, and the total distribution is a linear combination of those two. We chose the background region to be more than 0.8° from the center, since we do not have sufficient statistics for off-source regions for these observations. The result of fitting is shown in Fig. 6. An excess was observed at $\theta^2 < 0.6$ deg^2 around the center of the Vela X region. The excess radius is marginally consistent with H.E.S.S. considering our angular resolution. The total number of gamma-ray-like events is 561 ± 114. Though the statistical significance is below the 5σ level, this could be a supporting evidence of the H.E.S.S. detection.

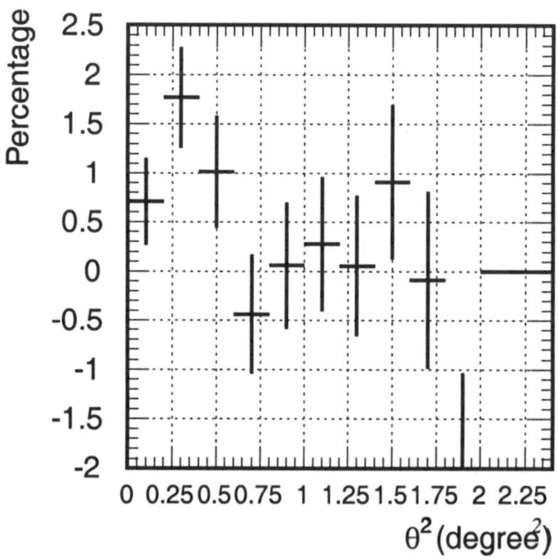

Figure 6. Wide-range θ^2 plot for the Vela X region [20], where θ is a space angle of an event direction from (R.A., decl.) $= (8^{\rm h}35^{\rm m}, -45°36')$, i.e., the peak of the emission detected by H.E.S.S. [34].

The differential fluxes were obtained and compared with H.E.S.S. result in Fig. 7, which are in general agreement considering our poor statistics.

3.4. *SNR RX J0852.0-4622*

We reported a gamma-ray signal from this SNR using observations by CANGAROO-II [35]. We applied the Fisher discriminant method to the stereo data for RX J0852.0-4622 observed in January and February 2004 using T2 and T3 taken in the wobble mode for 2,197 minutes in total. We used the northwest rim as a target point in the wobble mode. After

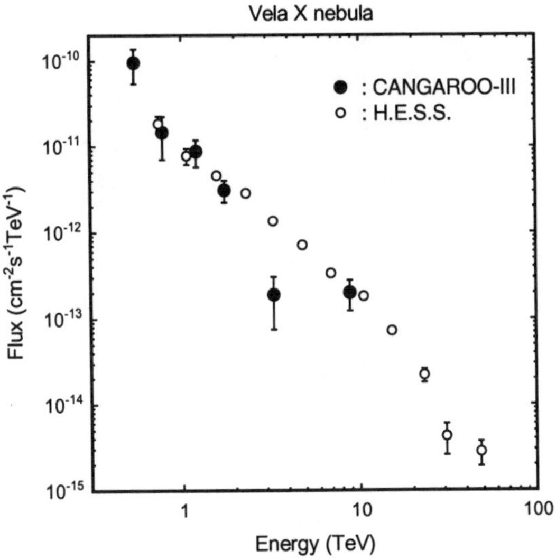

Figure 7. Gamma-ray spectra in the Vela X region observed by CANGAROO-III [20] compared with those reported by H.E.S.S. [34].

the coarse selections, 1,204 minutes data were available. For the Fisher discriminant, we used four image parameters, lengths and widths, determined with each telescope independently. Finally the gamma-ray events were extracted by comparing the Fisher discriminant values between the SNR region and the background region. The excess count map is shown in Figure 8. Te region inside the solid arcs shows the maximum acceptance region, which is an overlap of the two field-of-views in the wobble mode. Also the one-degree arc from the SNR center is indicated by the dotted line. The strong gamma-ray emission from the NW rim is obviously seen, which was first reported by CANGAROO-II. This maximum acceptance region covers about a half of the whole SNR, and the integrated flux above 0.81 TeV is about 60% of the H.E.S.S. result [36], which value is reasonable considering our coverage of the SNR [26,37].

4. Summary

We have been carrying out stereo observations of sub-TeV gamma-rays with CANGAROO-III since March 2004. Results from stereo observations were presented: PSR 1706-44 and SN1006, from which gamma-ray signals were reported by CANGAROO-I, were not confirmed by CANGAROO-III

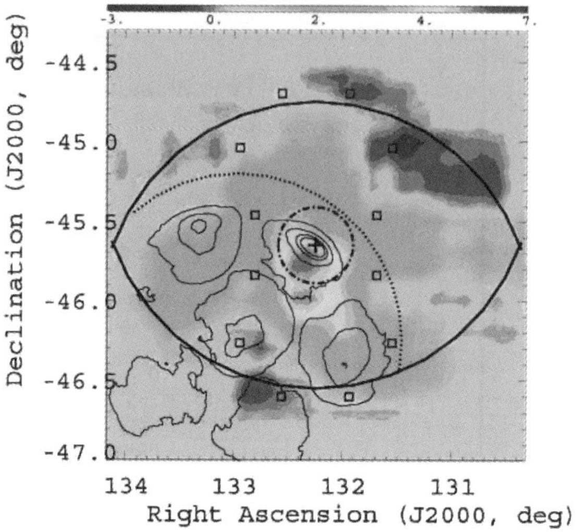

Figure 8. Excess event map around the SNR RX J0852.0-4622 obtained from the CANGAROO-III stereo observations in 2004 (preliminary) [37].

observations. For two supernova remnants, the Vela SNR and RX J0852.0-4622, our results are consistent with the recent H.E.S.S. results.

The distribution of 'gamma-ray SNRs' is important in the quest for the origin of cosmic-rays and the high-energy content of the Universe: we will continue systematic study of SNRs in high-energy gamma-rays in the Galaxy.

References

1. T. Tanimori et al., in *Proc. 26th ICRC (Salt Lake City)* (University of Utah, Utah), **5**, 203–206 (1999).
2. M. Mori et al., in "GeV-TeV Gamma Ray Astrophysics Workshop: Towards a Major Atmospheric Cherenkov Detector V", (eds. B.L.Dingus, M.H.Salamon and D.B.Kieda, AIP Proceedings, New York, 2000), pp. 485–491.
3. T. Tanimori et al., in "Very High Energy Phenomea in the Universe", eds. M.Boer and J. Tran Than Van, GIOI Publishers, France, 2001), pp.105–108.
4. M. Mori et al., in *Proc. 27th ICRC (Hamburg)* (Copernicus Gesellshaft, Berlin, 2001), pp. 2831–2834.
5. R. Enomoto et al., in *Proc. 28th ICRC (Tsukuba)* (Universal Academy Press, Tokyo, 2003), pp. 2807–2810.
6. H. Kubo et al., *New Astronomy Reviews* **48**, 323–329 (2004).
7. M. Mori et al., in *AIP Conference Proceedings* 745 (eds. F.A. Aharonian, H.J.Völk and D. Horns, AIP, New York, 2005), pp. 639-644.

8. K. Nishijima et al., in *Proc. 29th ICRC (Pune)* (Tata Inst. Fundamental Research, India, 2005), Vol. 5, pp. 327–330.
9. A. Kawachi et al., *Astropart. Phys.* **14**, 261–269 (2001).
10. M. Ohishi et al., in *Proc. 28th ICRC (Tsukuba)* (Universal Academy Press, Tokyo, 2003), pp. 2855–2858.
11. F. Kajino et al., in *Proc. 27th ICRC (Hamburg)* (Copernicus Gesellshaft, Berlin, 2001), pp. 2909.
12. S. Kabuki et al., in *Proc. 28th ICRC (Tsukuba)* (Universal Academy Press, Tokyo, 2003), pp. 2859–2862.
13. S. Kabuki et al., *Nucl. Instr. Meth.*, **A500**, 318–336, 2003.
14. H. Kubo et al., in *Proc. 27th ICRC (Hamburg)* (Copernicus Gesellshaft, Berlin, 2001), pp. 2900–2903.
15. H. Kubo et al., in *Proc. 28th ICRC (Tsukuba)* (Universal Academy Press, Tokyo, 2003), pp. 2863–2866.
16. S. Hayashi et al., in *Proc. 28th ICRC (Tsukuba)* (Universal Academy Press, Tokyo, 2003), pp. 2867–2870.
17. T. Nakamori et al., in *Proc. 29th ICRC (Pune)* (Tata Inst. Fundamental Research, India, 2005), Vol. 4, pp. 203–206.
18. R. Enomoto et al., in *Proc. 27th ICRC (Hamburg)* (Copernicus Gesellshaft, Germany), Vol.5, pp. 2477–2480 (2001).
19. R. Enomoto et al., *Nature* **416**, 823–826 (2002).
20. R. Enomoto et al., *Astrophys. J.* **638**, 397–408 (2006).
21. F.A. Aharonian et al., *Astrophys. J.* **539**, 317–324 (2000).
22. A.M. Hillas et al., *Astrophys. J.* **503**, 744–759 (1998).
23. T. Kifune et al., *Astrophys. J.* **438**, L91–94 (1995).
24. P.M. Chadwick et al., *Astropart. Phys.* **9**, 131–136 (1998).
25. F.A. Aharonian et al., *Astron. Astrophys.* **432**, L9–L12 (2005).
26. T. Tanimori et al., in *Proc. 29th ICRC (Pune)* (Tata Inst. Fundamental Research, India, 2005), Vol. 4, pp. 215–218.
27. K. Koyama et al., *Nature*, **278**, 255–228 (1995).
28. T. Tanimori et al., *Astrophys. J.* **497**, L25–28 (1998).
29. F.A. Aharonian et al., *Astron. Astrophys.* **437**, 135–139 (2005).
30. V. Vitale et al., in *Proc. 28th ICRC (Tsukuba)* (Universal Academy Press, Tokyo, 2003), pp. 2889–2892.
31. T. Yoshikoshi et al., *Astrophys. J.* **487**, L65–68 (1997).
32. P.M. Chadwich et al., *Astrophys. J.* **537**, 414–421 (2000).
33. S.A. Dazeley, Ph.D. thesis, University of Adelaide (1999).
34. F.A. Aharonian et al., *Astron. Astrophys.* **448**, L43–47 (2006).
35. H. Katagiri et al., *Astrophys. J.* **619**, L163–L165 (2005).
36. F. Aharonian et al., *Astron. Astrophys.* **437**, L7–10 (2005).
37. R. Enomoto et al., to be submitted for publication (2006).

OBSERVATIONS OF GALACTIC GAMMA-RAY SOURCES WITH H.E.S.S.

D. BERGE FOR THE H.E.S.S. COLLABORATION
Max-Planck-Institut für Kernphysik
P.O. Box 103980
D-69029 Heidelberg
Germany
berge@mpi-hd.mpg.de

H.E.S.S. results from the first three years of nominal operation are presented. Among the many exciting measurements that have been made, most gamma-ray sources are of Galactic origin. I will concentrate here on an overview of Galactic observations and summarise and discuss observations of selected objects of the different source types.

1. Introduction

The High Energy Stereoscopic System (H.E.S.S.) is a system of four imaging atmospheric Cherenkov telescopes which commenced full operation in the Khomas Highland of Namibia in December 2003 [1], at an altitude of 1800 m. The experiment is run by an international collaboration of mostly European institutes. It is built for very-high-energy (VHE) gamma-ray astronomy, exploiting the energy range above 100 GeV up to several tens of TeV. Gamma rays are measured by means of Cherenkov light emitted in air showers of secondary particles that form whenever a primary gamma ray hits the earth's atmosphere and is being absorbed. Using Cherenkov images of air showers one can deduce the energy and direction of the primary particle.

The H.E.S.S. Cherenkov telescopes are operated in moonless nights yielding a total observation time of roughly 1000 h per year. In normal data taking mode, five to ten objects are tracked per night with a typical cosmic-ray event rate of 300 Hz. The observations proceed in stereoscopic mode: events are recorded if at least two out of the four telescopes have triggered on the same air shower [3]. The telescopes itself have a 60-t steel structure with *altitude-azimuth* mount. Each has a tessellated mirror surface

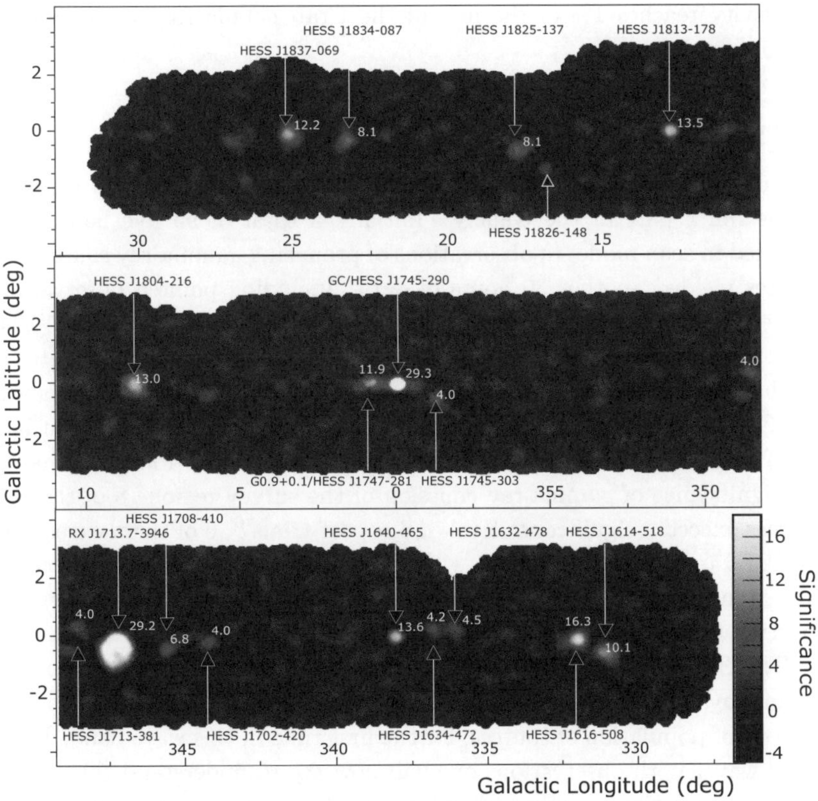

Figure 1. Significance map of the H.E.S.S. Galactic plane survey in 2004 [2]. The data include re-observations of gamma-ray candidates as well as pointed observations of known gamma-ray sources. The gamma-ray sources of the survey region are labelled and the significance of the signal is given for all of them. Note that the colour scale is truncated at 18 σ.

consisting of 380 single round facets, comprising a total area of 107 m^2 [4]. With a focal length of 15 m, the Cherenkov light is imaged onto 960-photomultiplier cameras with integrated fast readout electronics [5]. Each camera covers a large field of view of 5°. The resulting $FWHM \approx 4°$ of the system field-of-view response makes H.E.S.S. the currently best suited experiment in the field for the study of extended VHE gamma-ray sources and the search for unknown sources in surveys.

At zenith, the energy threshold of the system is about 100 GeV and for point sources an energy resolution of 15% is achieved. The angular resolution for individual gamma rays is better than 0.1° and the point source sensitivity reaches 1% of the flux of the Crab nebula for long exposures ($\approx$ 25 hours).

2. The H.E.S.S. Survey of the Inner Galaxy

One of the first observation campaigns of H.E.S.S. in 2004 was a survey of the inner part of the Galaxy. Initially a total of 95 live hours were recorded in scan mode, re-observations of promising gamma-ray source candidates yielded another 30 hours of data. Including pointed observations of the Galactic-centre region and the supernova remnant RX J1713.7–3946 (which will both be discussed below), the H.E.S.S. data set accumulates to 230 hours and reaches an average sensitivity of 2% of the Crab flux above 200 GeV. In the region covered ($\pm 30°$ in Galactic longitude, $\pm 3°$ in latitude) 14 previously unknown sources were detected. Fig. 1 shows a map of the significance of gamma-ray emission of the survey region. 8 of the new sources exceed a significance level of 6 σ post-trials [6], 6 of them exceed the level of 4 σ [2]. They all line up with the Galactic plane, except for one all are extended at the 2 to $3'$ level and reveal hard power-law type energy spectra with a mean photon index of 2.3.

The H.E.S.S. survey is a major breakthrough for the field of gamma-ray astronomy. The increased number of sources allows to consider the behaviour of population of sources, for the first time in this wave band. Using multi-wavelength observations one will now try to understand the physics of the acceleration processes that lead eventually to the emission of VHE gamma radiation. The sources in the survey region might be associated with four source classes:

- **Pulsar Wind Nebulae (PWNe):** HESS J1825–137, HESS J1747–281 (G0.9+0.1), HESS J1702–420, and HESS J1616–508.
- **X-ray binaries:** HESS J1826–148 (LS 5039).
- **Supernova remnants (SNRs):** HESS J1834–087, HESS J1813–178, HESS J1804–216, RX J1713.7–3946, HESS J1713–381, and HESS J1640–465.
- **Unknown nature:** HESS J1837–069, HESS J1745-290 (Galactic centre), HESS J1745-303, HESS J1708–410, HESS J1634–472, HESS J1632–478, HESS J1614–518.

I will step now sequentially through the source classes and discuss examples of H.E.S.S. measurements.

3. Pulsar Wind Nebulae

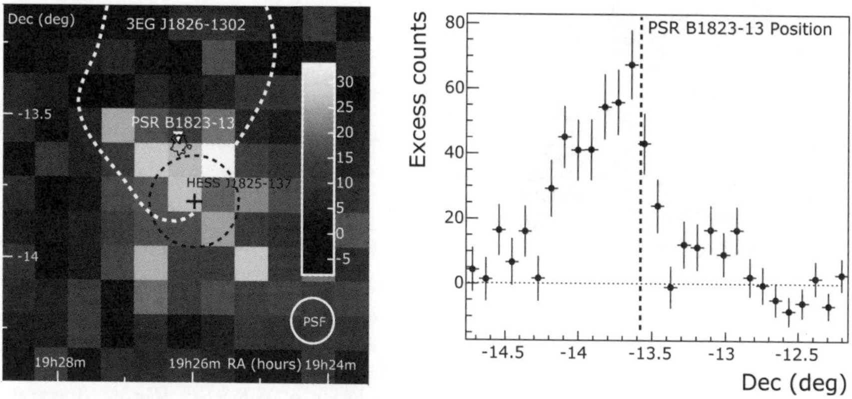

Figure 2. **Left:** Gamma-ray excess image of the region surrounding PSR B1823-13 (marked with triangle) in uncorrelated bins [7]. The H.E.S.S. best-fit position is shown with error bars together with the emission-region size. The black contours denote the XMM measurement, the dotted white line the unidentified EGRET source. **Right:** Excess slice (0.4° wide) through the H.E.S.S. data taken along the north-south direction. The one-sided nature of the emission with respect to the pulsar is clearly seen.

Energetic pulsars dissipate rotational energy in form of relativistic outflows. Confinement of these winds by the ambient medium leads to the formation of PWNe which can emit X-rays via Synchrotron radiation and gamma rays via the Inverse Compton mechanism. One of the four PWN candidates in the H.E.S.S. survey region is HESS J1825–137, shown in Fig. 2 [7]. The source is probably associated with PSR J1826–334, a 2.1×10^4 years old pulsar. As can be seen from the figure, the emission region is offset from the pulsar and extends asymmetrically to the south. The reason for this asymmetric PWN, which is also seen in the X-ray measurement, is the reverse shock from the northern side, where an increased density of the interstellar medium is encountered. The shock presumably crashed into the PWN and pushed it to the south. Note that follow-up observations of this object have been performed with H.E.S.S. and more detailed analyses, including spatially resolved energy spectra, will be published very soon.

Another example of a PWN measured in VHE gamma rays with H.E.S.S. (which is *not* in the survey region) is Vela X, the nebula associated with PSR B0833–45. Fig. 3 shows the combined image from the 2004 and 2005 H.E.S.S. data [8]. The gamma-ray emission region is extended, roughly ellipsoidal in shape, and coincides well with ROSAT and Chandra X-ray

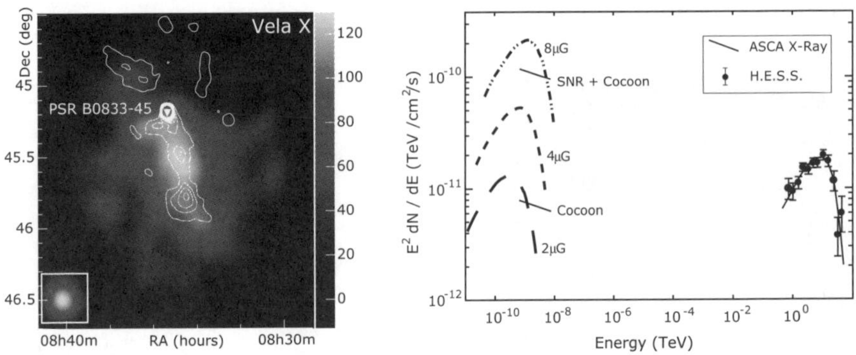

Figure 3. **Left:** Gaussian smoothed gamma-ray image of the region surrounding the Vela pulsar [8] (the pulsar position is marked with a triangle). The white contours are the ROSAT X-ray measurement of this region. In the bottom left-hand corner, a simulated point source is shown and demonstrates the resolution of H.E.S.S.. **Right:** Spectral energy distribution using H.E.S.S. and ASCA data. The black lines show one-zone model fits with different synchrotron flux predictions for different magnetic fields (see publication [8] for details).

measurements. Also here, the emission is offset from the pulsar, again due to an asymmetric reverse shock from the northern side. The energy spectrum measured with H.E.S.S. is well explained by a one-zone Inverse Compton model, as is shown in Fig. 3 (right). The measured photon index is very hard, 1.45 ± 0.09, with an exponential cutoff of 13.8 ± 2.3. This is actually the first measurement of a complete VHE gamma-ray peak in a spectral energy distribution.

4. X-ray Binaries

A point-like gamma-ray source was found close to HESS J1825–137 in the H.E.S.S. survey, HESS J1826–148, likely to be associated with a system called LS 5039. This system is an X-ray binary, a companion star orbiting around a compact object. Radio and X-ray observations of relativistic outflows of some X-ray binaries have led to the term *Microquasar*, suggesting that they behave as scaled-down active galactic nuclei. The H.E.S.S. measurement is shown in Fig. 4 [9]. It is noteworthy that this is the only point-like source in the whole survey region. The positional coincidence with LS 5039 led to the identification of the gamma-ray source with the microquasar, and it is the first detection of such an object in VHE gamma rays. The spectrum of HESS J1826–148 is shown in Fig. 4 (right), it follows a power law and suggests an association with the EGRET source

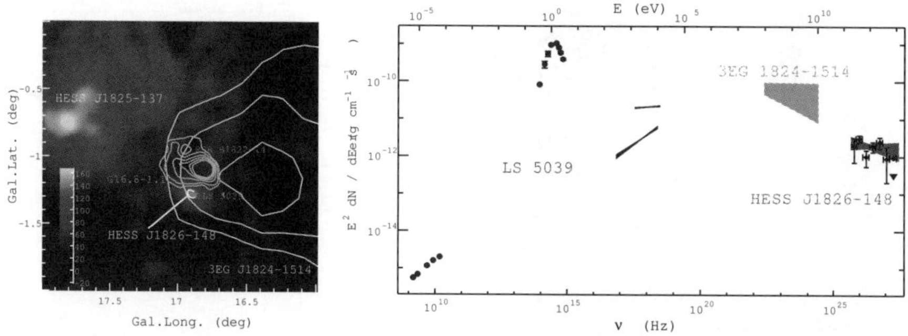

Figure 4. **Left:** Smoothed excess image of the region around LS 5039 [9]. The H.E.S.S. position is indicated, overlaid are gray contours of radio emission and yellow contours of EGRET data. **Right:** Spectral energy distribution of LS 5039. H.E.S.S. data (black points) are compared to optical and X-ray data. Shown in gray is the EGRET measurement suggesting an association of the H.E.S.S. and the EGRET source.

3EG 1824–1514, despite a spatial separation of $\approx 0.5°$. More H.E.S.S. data from follow-up observations in 2005 exist and allow to search for orbital modulations. Detailed results will be published soon.

5. Supernova Remnants

SNRs are the best source candidates for cosmic rays in our Galaxy. The standard notion of particle acceleration is the diffusive shock acceleration of charged particles in the shells of SNRs. The source with the largest extension in the survey region is such a shell-type SNR, RX J1713.7–3946. It has an apparent diameter of $\approx 1°$, twice the size of the full moon. The remnant was discovered with ROSAT in X-rays [13], follow-up observations with ASCA revealed a dominantly non-thermal X-ray continuum without line emission [14,15], most plausibly explained by Synchrotron emission of multi-TeV electrons. The presumed acceleration of electrons to TeV energies in the expanding shell of RX J1713.7–3946, together with indications of interactions of the shock with molecular clouds [16], made this SNR a prime target for H.E.S.S. to look for gamma rays from interactions of accelerated cosmic rays with ambient matter.

After the first detection of VHE gamma rays from this object with CANGAROO [17,18], H.E.S.S. has indeed confirmed gamma-ray emission with its 2003 observation campaign. It revealed the first ever resolved image of an astronomical source in VHE gamma rays [12]. Follow-up observations allowed for detailed analyses with unprecedented precision [11], the resulting gamma-

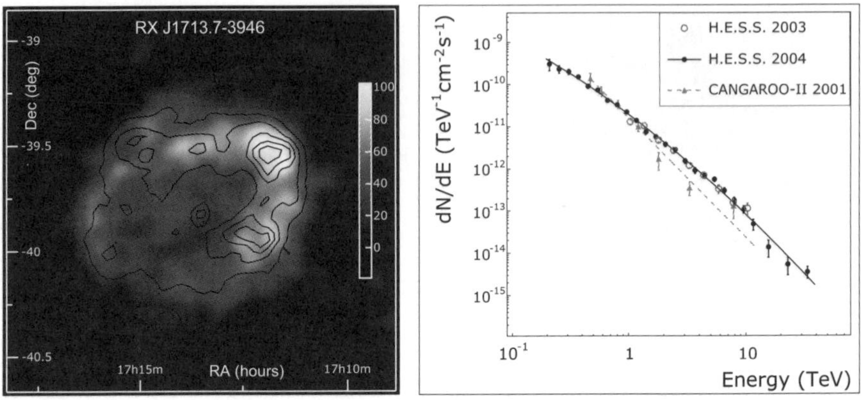

Figure 5. **Left:** Smoothed gamma-ray excess image of RX J1713.7–3946, produced from H.E.S.S. data of 2004 and 2005 [10]. Note the angular resolution of 3.6′ achieved here. Overlaid as black contours is the ASCA 1-3 keV X-ray measurement. **Right:** H.E.S.S. gamma-ray spectrum of the whole SNR [11]. The black line is the best fit of a power law with photon index that depends logarithmically on energy, determined from the 2004 data set. The 2003 H.E.S.S. data [12] shown as blue points are in good agreement. The CANGAROO-II data are also drawn.

ray image is shown in Fig. 5 [10]. It shows a clear shell structure, brighter in the northwest, resembling very much the picture seen in X-rays. In fact a detailed correlation study revealed a striking correspondence between keV

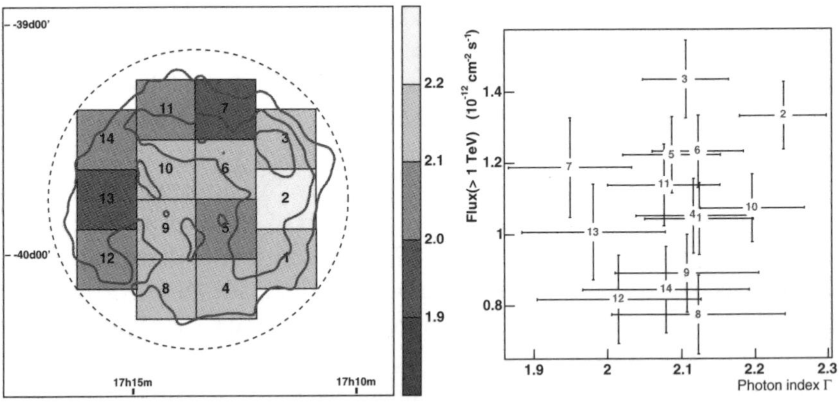

Figure 6. **Left:** Gamma-ray excess contours are shown in red, superimposed are 14 boxes (each $0.26° \times 0.26°$ in dimension) for which spectra were obtained independently [11]. The photon index obtained from a power-law fit in each region is colour coded in bins of 0.1. **Right:** Integral flux above 1 TeV versus the photon index, for the 14 regions shown left. The error bars are $\pm 1\sigma$ statistical errors.

and TeV energies. The differential energy spectrum of the whole remnant is shown in Fig. 5 (right). It extends over more than two decades well beyond 10 TeV and is well described by power-law type spectral shapes, albeit with deviations from a pure power law at large energies. The spectrum reported by the CANGAROO-II collaboration, also shown in the figure, is in marginal agreement with the H.E.S.S. measurement.

The H.E.S.S. 2004 data of RX J1713.7–3946 enabled us to perform a spatially resolved spectral analysis, to look for spectral variation on scales down to $\approx 0.3°$. The result is shown in Fig. 6. When determining spectra in 14 boxes arranged to cover the whole SNR, no significant index variation is found, the spectral shape is the same everywhere, only the flux varies by more than a factor of two.

The key issue from the interpretation side for the RX J1713.7–3946 data is the identification of the particle population responsible for gamma-ray emission. While with the H.E.S.S. measurement it is clear that primary particles are accelerated in the shock wave to energies beyond 100 TeV, it remains difficult to say whether these particles are electrons or protons, in other words, if we really have the proof at hands that this SNR is a source of nucleonic cosmic rays. A broadband approach to answer this question is shown in Fig.7. A one-zone electron model fails to reproduce the spectral shape measured with H.E.S.S., in a hadronic scenario on the other hand the spectral shape seen in gamma rays is qualitatively as expected from

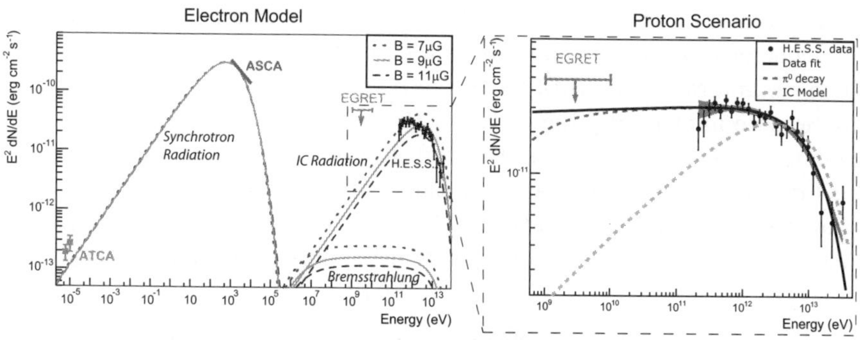

Figure 7. **Left:** Spectral energy distribution of RX J1713.7–3946. Shown are broadband data together with model curves obtained from a one-zone electron model [11]. Curves are plotted for three assumed magnetic field values. **Right:** Blow-up view of the high-energy part showing H.E.S.S. data together with the fit of a power law with exponential cutoff, extrapolated to small energies. Moreover, a curve taking the gamma-ray suppression due to the π^0-decay kinematics into account is indicated and one of the Inverse Compton model curves from the left-hand side.

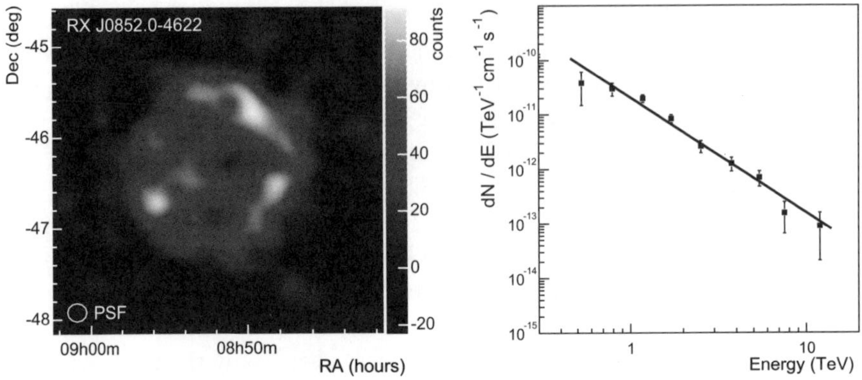

Figure 8. **Left:** H.E.S.S. gamma-ray excess image of RX J0852.0–4622 from 2004 and 2005 data [19]. The image is smoothed with a Gaussian of $\sigma = 0.1°$. The point-spread function (PSF) of this data set is shown in the bottom left corner. **Right:** H.E.S.S. spectrum of the whole SNR from 2004 data, determined from only 3.2 h live time (corresponding to 700 ± 60 excess events) [20]. The best-fit of a power law is shown as black line.

theory. In that sense the hadronic scenario is favoured by the H.E.S.S. data, although the correlation between X-rays and gamma rays is then challenging and so far not well understood.

Another prominent SNR that was detected with H.E.S.S. in 2004 is RX J0852.0–4622 [20], sometimes called *Vela Junior* (it is close to the PWN Vela X, discussed above). Also first discovered with ROSAT [21], this object is in many regards similar to RX J1713.7–3946. It is largely extended with a diameter of almost 2° and reveals a shell structure, correlated in X-rays and gamma rays. The H.E.S.S. image is shown in Fig. 8. It demonstrates once more impressively the ability of H.E.S.S. to map extended objects in gamma rays. The spectrum of the whole SNR is shown in Fig. 8 (right). It extends beyond 10 TeV and is within statistics well described by a pure power law with a photon index of 2.1 ± 0.1. Note that detailed analysis of more data from 2005 is underway and in the pipeline for publication.

6. Sources of Unknown Nature – The H.E.S.S. Galactic Centre Signal

Among all the H.E.S.S. sources in the survey region that so far could not be unequivocally identified the Galactic centre is probably the most exciting one. The point-like VHE gamma-ray emission is coincident with the supermassive black hole Sgr A* and the SNR Sgr A East [23]. The spec-

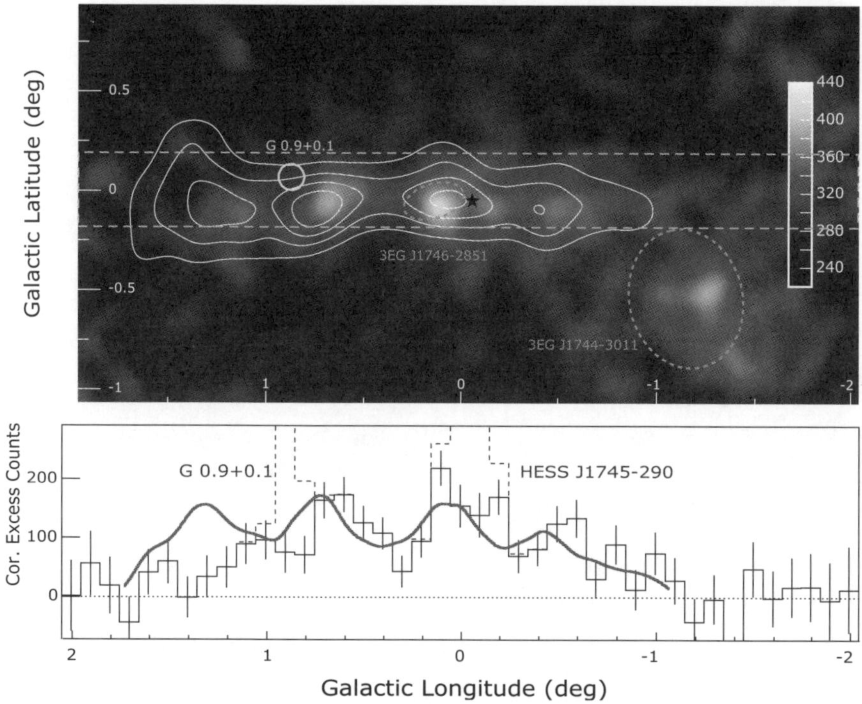

Figure 9. **Upper panel:** Acceptance-corrected smoothed gamma-ray image of the Galactic centre region after subtracting the two dominant point sources in the field of view [22]. White contour lines indicate the density of molecular gas, traced by its CS emission. The dashed gray rectangle shows the 0.4° wide slice region that was used to produce the profile shown in the **lower panel:** Here we show the distribution of gamma-ray counts versus Galactic longitude and compare it to the CS line emission (red line). The signal of the two subtracted point sources is shown as dashed blue lines.

trum is well described by a pure power law with photon index 2.21 ± 0.09. No sign for any time variability of the signal is found. Possible emission processes that have been discussed include electron and proton origin of gamma rays, produced in the vicinity of the black hole or the shocks of the SNR. Moreover, the H.E.S.S. signal has been discussed in the framework of dark matter annihilations [24].

The deep exposure of 2004 revealed not only a second source of VHE gamma rays, G0.9+0.1, but also enabled us to subtract these two strong point sources and search for remaining diffuse emission. The result is shown in the upper panel of Fig. 9 which shows the residual gamma-ray excess after subtraction. Two significant features appear: a region of extended emission

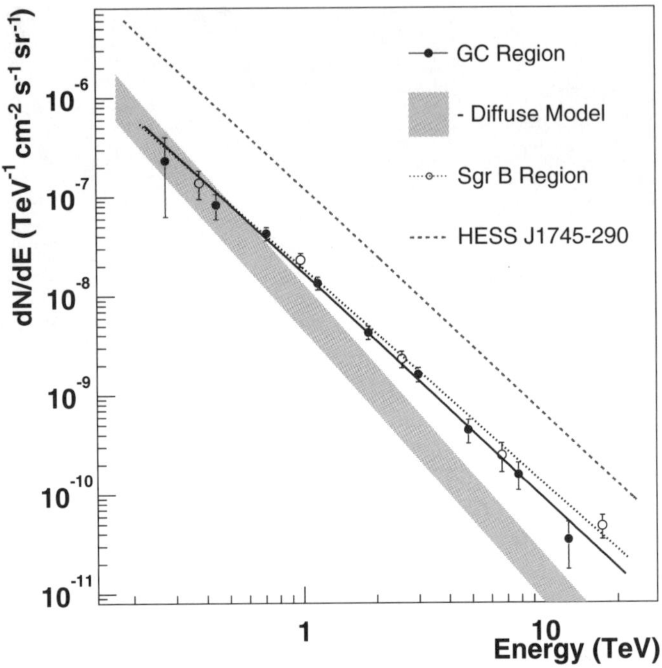

Figure 10. Energy distribution of Galactic cosmic rays per unit angle in the Galactic centre region [22]. The spectrum is well described by a power-law fit (solid line). Data points are compared with the expected flux from π^0-decay assuming the local (solar) cosmic-ray spectrum and a target mass as measured with the CS emission. The open points correspond to the Sgr B complex, the dotted red line gives the spectrum of the bright central source HESS J1745-290.

spatially coincident with the unidentified EGRET source 3EG J1744–3011, and emission extending along the Galactic plane for roughly 2° [22]. Overlaid in the figure are velocity-integrated CS data from the Galactic centre direction which trace molecular gas. There is a close correlation visible between the gamma-ray signal and the molecular gas density. In the lower panel of Fig. 9 the gamma-ray count rate is shown as a profile, plotted versus Galactic longitude, integrated in a 0.4° thick slice. The good match between gamma-ray and CS data suggests a cosmic-ray origin of gamma rays, produced in interactions of cosmic rays with molecular clouds. The similarity in the distributions of CS-line and gamma-ray emission implies a rather uniform cosmic-ray density in this region. Fig. 10 shows the gamma-ray flux measured in this region. The data is well described by a power law with photon index 2.29 ± 0.07. Keeping in mind that in case of a power-law energy dependence, the gamma-ray spectral index closely traces the cosmic-ray index itself, it follows that the measured spectrum is significantly harder than in the solar neighbourhood. If we estimate the gamma-ray flux assuming a target mass as determined from the CS measurement, and the local cosmic-ray flux and spectrum, we obtain the shaded grey band shown in Fig. 10. There is a clear excess measured beyond 500 GeV. This could simply be due to the proximity to the accelerator, meaning that propagation effects, which lead to a steepening of the spectrum, are less pronounced.

7. Summary and Conclusions

During its first three years of operation H.E.S.S. has had a number of significant achievements in the field of VHE gamma-ray astronomy. Only with the sensitivity, the good angular and energy resolution and the large field of view of experiments like H.E.S.S. is it now possible to measure the morphology and spectra of extended gamma-ray sources with great precision. Moreover, the good off-axis sensitivity make H.E.S.S. ideally suited for sky surveys. This was demonstrated here by means of the Galactic plane survey data, which revealed 14 previously unknown VHE gamma-ray sources. The detection of extended emission from SNRs such as RX J1713.7–3946, which resembles indeed as expected a shell structure, proves the existence of highest energy particles in the shocks of SNRs and presents a major step forward towards solving the puzzle of the origin of Galactic cosmic rays. Finally, the detection of a diffuse VHE gamma-ray component from the direction to the Galactic centre provides new vistas of the centre of our Galaxy delivering exciting insights into acceleration and diffusion processes of cosmic rays.

References

1. Hinton, J. A., 2004, New Astronomy Review **48**, 331.
2. Aharonian et al. (H.E.S.S. Collaboration), 2006, The Astrophysical Journal **636**, 777.
3. Funk et al., 2004, Astroparticle Physics **22**, 285.
4. Bernlohr et al., 2003, Astroparticle Physics **20**, 111.
5. Vincent et al. (H.E.S.S. Collaboration), 2003, Proc. 28th ICRC (Tsukuba), 2887.
6. Aharonian et al. (H.E.S.S. Collaboration), 2005, Science **307**, 1938.
7. Aharonian et al. (H.E.S.S. Collaboration), 2005, Astronomy and Astrophysics **442**, L25

8. Aharonian et al. (H.E.S.S. Collaboration), 2006, Astronomy and Astrophysics **448**, L43
9. Aharonian et al. (H.E.S.S. Collaboration), 2005, Science **309**, 746.
10. Aharonian et al. (H.E.S.S. Collaboration), 2006, Astronomy and Astrophysics **in preparation**.
11. Aharonian et al. (H.E.S.S. Collaboration), 2005, Astronomy and Astrophysics **449**, 223.
12. Aharonian et al. (H.E.S.S. Collaboration), 2004, Nature **432**, 75.
13. Pfeffermann, E. & Aschenbach, B., 1996, in Roentgenstrahlung from the Universe, 267.
14. Koyama et al., 1997, Publications of the Astronomical Society of Japan **49**, L7.
15. Slane et al., 1997, The Astrophysical Journal **525**, 357.
16. Fukui et al., 2003, Publications of the Astronomical Society of Japan **55**, L61.
17. Muraishi et al., 2000, Astronomy and Astrophysics **354**, L57.
18. Enomoto et al., 2002, Nature **416**, 823.
19. Komin et al. (H.E.S.S. Collaboration), 2005, Proc. 29th ICRC (Pune).
20. Aharonian et al. (H.E.S.S. Collaboration), 2005, Astronomy and Astrophysics **437**, L7.
21. Aschenbach, B., 1998, Nature **396**, 141.
22. Aharonian et al. (H.E.S.S. Collaboration), 2006, Nature **439**, 695.
23. Aharonian et al. (H.E.S.S. Collaboration), 2004, Astronomy and Astrophysics **425**, L13.
24. Aharonian et al. (H.E.S.S. Collaboration), 2006, Physical Review Letters **in preparation**.

PARTICLE ACCELERATION IN SUPERNOVA REMNANTS AND THE RESULTING NONTHERMAL EMISSION

H.J. VÖLK

Max-Planck-Institut für Kernphysik,
Postfach 103980,
D-69029 Heidelberg, Germany
E-mail: Heinrich.Voelk@mpi-hd.mpg.de

We give an overview on the theoretical description of individual supernova remnants (SNRs) as particle accelerators and report about the latest developments regarding Tycho's SN, SN 1006, Cassiopeia A, and especially SNR RXJ1713.7-3946. In all these objects the nuclear relativistic component dominates over the relativistic electron component. Then we discuss the global nonthermal effects of SNRs. The entire population of SNRs in our Galaxy collectively drives a Galactic Wind that extends halfway to the neighboring galaxies, the CRs being dynamically important. The same is probably true for the diffuse intergalactic medium around field galaxies, especially starburst galaxies. The intracluster medium in galaxy clusters contains a significant nonthermal energy fraction, more or less in equal share the result of star formation, AGN activity and cluster accretion. It should be observable in high energy gamma rays. To wherever the nonthermal component extends, it is expected to be in rough equipartition with the thermal gas. This "Nonthermal Universe" is the result of cosmic structure formation.

1. SNRs: Observations and acceleration theory

Empirically the nonthermal characteristics of individual SNRs are characterized through three different types of observational information (i) the gamma-ray emission at energies comparable to the expected particle energies (ii) the broadband synchrotron emission from radio to hard X-ray energies, and (iii) additional multi-wavelength information on the angular size/morphology and the expansion rate. In the future we should be able to add also the high-energy neutrino flux.

If we want to understand the meaning of such measurements, we need to compare them with theory. It describes the system by kinetic equations for the particle distribution functions $f_{\mathrm{protons}}(p, r, t)$ and $f_{\mathrm{electrons}}(p, r, t)$, nonlinearly coupled with the gas dynamics of the thermal plasma through the mass velocity, the gradient of the CR pressure, and the dissipation of wave energy. In spherical symmetry these transport equations have been solved in their full time dependence (1) and coupled with the gamma-ray emission (2), approximating the diffusion mean free path by the particle gyro radius. However, this theory contains several dynamical variables which can at present not be quantitatively determined very well at a deeper level: the effective (amplified) magnetic field strength B_{eff} (3), the injection rate of suprathermal ions into the acceleration process, and the amplitude of the electron distribution $f_{\text{electrons}}$. We therefore need observational input to determine them as "parameters of the theory". This input is the measured synchrotron spectrum in form and amplitude (for reviews, see (4) (5)), and it is important that the parameters can be derived in a form that is consistent with the solution of the kinetic equation for the nonthermal electron component that produces the synchrotron emission. In addition, we can for almost all sources determine B_{eff} from X-ray filaments which are assumed to delineate the outer shock and whose thickness is interpreted as the synchrotron cooling length of the radiating electrons (Fig.1) (6).

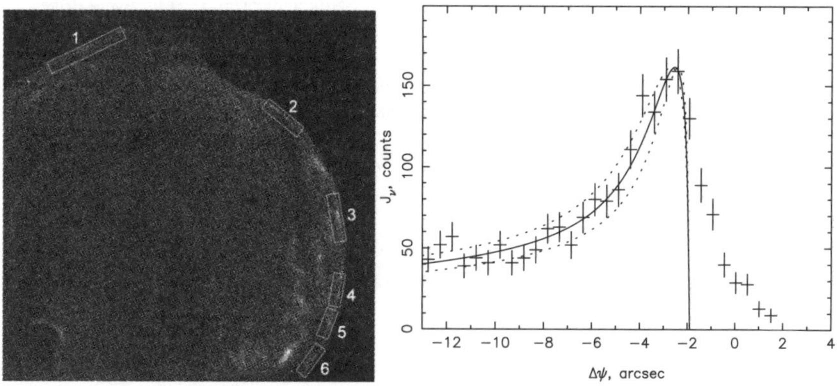

Figure 1. Tycho's SNR in the 4-6 keV band, as imaged by *Chandra*. The angular profiles of the filaments are determined for six *rectangular boundary boxes* (left figure). The average profile is shown in the right figure. (From (6)).

There are also "physical parameters" which can in principle be furnished by models of the SN explosion process itself. These are: the SN type (nuclear deflagration/explosion or core collapse), the amount of mechanical energy E_{sn} released, and the ejected mass M_{ej}. In practice, for core collapse events, the explosion parameters have to be plausibly assumed from circumstantial information.

Such a theoretical model has still not a unique solution for a given SNR unless we know what are best called the "environmental parameters": the source distance and age, the topology of the magnetic field in the external circumstellar medium and its gas density, and the mass of the progenitor star. These environmental parameters are part of the "astronomical landscape"; in other words, they cannot be deduced from theory and must be determined independently from observations.

Also deviations from spherical symmetry necessarily occur because the magnetic field is divergence-free. Globally, this effect can be taken into account for the hadronic gamma-ray emission by reducing the spherically symmetric flux by a renormalization factor $f_{re} \approx 0.2$ (7).

After this construction of a model, the solutions of the dynamical equations are first of all the time dependencies of outer shock radius and expansion velocity, as well as the overall gas dynamic morphology, and the distribution functions of the accelerated charged particles. From these we then obtain the nonthermal X-ray and gamma-ray morphologies, and the local energy spectra of the various forms of gamma-ray emission, i.e. nonthermal Bremsstrahlung (NB), inverse Compton (IC) collisions, and π^0-decay from hadronic collisions. This can be compared with the observations and with the general expectations we have on the CR sources in the Galaxy and beyond.

Let us conclude that SNRs are at the same time thermal and nonthermal astrophysical objects whose detailed theoretical understanding is well possible, even though a SNR is something much more complex than a star which is always thermally dominated. Prerequisite is a detailed observational knowledge of the individual source and its environment, and that knowledge is clearly in most cases not available in complete form. Nevertheless we have a number of objects which are described quite satisfactorily. We will discuss several of them in the next sections.

2. Individual SNRs

2.1. *Tycho's SNR*

Being a historical event, observed by Tycho Brahe in 1572 AD, its age is known as 433 yr. From X-ray spectroscopy Tycho's SN is now generally believed to be a SN type Ia (e.g. (8) (9)). Therefore M_{ej} should equal the Chandrasekhar mass. Interpreting the dynamics of the gas dynamic discontinuities (10) we obtain $f_{re} \approx 0.2$ (11); from SN explosion modeling the mechanical energy release has recently been estimated to be $E_{sn} = 1.2 \times 10^{51}$ erg (12). We can also use radio measurements of the angular size and expansion velocity (13) to put constraints on the distance and the ambient gas density, if the hadronic gamma-ray flux is measured or has an upper limit.

Anticipating a strong amplification of the magnetic field by the accelerating CRs to a downstream value of 300 ± 60 μG, and using initially a source distance $d = 2.3$ kpc and ISM density $N_H = 0.5$ H-atoms cm^{-3}, one can fit the observations of the gas dynamics with a total hydrodynamic explosion energy $E_{sn} = 0.27 \times 10^{51}$ erg (14) (see also Fig.2).

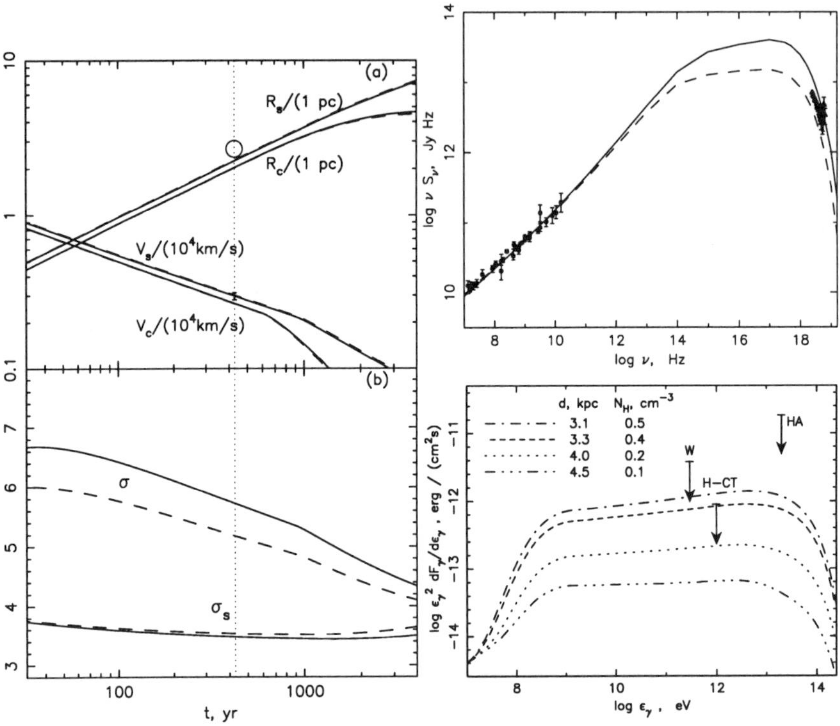

Figure 2. Time evolution of the outer blast wave position R_s and velocity V_s, as well as contact discontinuity radius R_c and velocity V_c in Tycho's SNR (*upper left panel*), together with the overall shock (σ) and subshock (σ_s) compression ratios (*lower left panel*). The *solid* and *dashed* lines correspond to the internal magnetic field strengths $B_d = 240\mu$G and $B_d = 360\mu$G, respectively. The *dotted vertical line* marks the current epoch. The observed mean size and speed of the shock are given from radio measurements. The *upper right panel* shows the theoretical overall synchrotron spectrum for the two field strengths, in comparison to the radio and X-ray measurements. For $E_{sn} = 1.2 \times 10^{51}$ erg the *lower right panel* gives the gamma-ray energy flux predictions for a source distance $3.1 < d < 4.5$ kpc. Observational upper limits are from the *Whipple* (W (15)) and *HEGRA* (H-CT (16), HA (17)) experiments, respectively (11).

The SNR is not very far anymore from entering a quasi-selfsimilar Sedov phase. Yet it is interesting that the relative thickness $(R_s - R_c)/R_s \leq 0.1$ of the shocked layer of circumstellar gas is very small. This implies that the downstream medium is unusually compressible. This is only possible if a substantial part of the shock energy goes into relativistic particles with their small adiabatic index of $4/3$. The fact that the overall compression ratio σ is significantly larger than 4 (Fig.2) is consistent with this conclusion.

The solution for the accelerated electrons has to be compared with the synchrotron data. The difference in slope of the relatively flat radio spectral energy density to that of a test particle spectrum up to about 10 GHz is a consequence of the nonlinear modification of the shock by the dominantly accelerating nuclear CRs and determines the injection rate of these nuclear particles. In addition, the radiating electrons at ≈ 10 GHz must have energies $\leq m_{\rm p}c^2$ which determines the effective downstream magnetic field strength; it has the value $B_{\rm d} = 360 \pm 60 \mu$G. A field strength that is consistent with this magnitude, up to a distance $d < 4$ kpc, is obtained if the thickness of the filamentary structure of the outer shock in hard X-rays is interpreted as the synchrotron loss length length of the radiating particles (Fig.1).

Finally, for given values of $E_{\rm sn}$ and distance d, we can predict the gamma-ray energy flux from the acceleration of nuclear particles. Using $E_{\rm sn} = 1.2 \times 10^{51}$ erg, this flux is shown in Fig.2 in comparison with several upper limits from observations. The hadronic gamma rays turn out to dominate by one order of magnitude. The prediction is then – selfconsistent with gas dynamics and sychrotron spectrum – that a very hard gamma-ray spectral energy distribution of strength $(2 \text{ to } 5) \times 10^{-13} \mathrm{erg}/(\mathrm{cm}^2\mathrm{s})$ should extend up to almost 100 TeV if the source lies in a distance range between 3.3 and 4 kpc, and if $E_{\rm sn}$ has indeed the above value. The flux should be detectable by the new Northern Hemisphere TeV detectors *VERITAS* and *MAGIC* in their full stereo complements, and then also give a consistent value for the ambient density $0.2 < N_{\rm H} < 0.5 \ \mathrm{cm}^{-3}$.

2.2. *SN 1006 and Cas A*

These two objects have been much discussed in the past years. Cas A is a weak gamma-ray source detected with the *HEGRA* stereoscopic array in a very deep observation of 232 hr (18). SN 1006 had originally been claimed in TeV gamma rays by the *CANGAROO* collaboration (e.g. (19)) and was for a long time considered by many the prototype VHe gamma-ray-SNR.

Despite its clear detection as a synchrotron source in hard X-rays (20), SN 1006 could not be detected in TeV gamma rays with the *H.E.S.S.* telescope system (21). Subsequent stereo observations by the *CANGAROO III* telescope system (22) confirmed this result. In all probability the very low gamma-ray flux is the result of a very low gas density of this type Ia supernova, cf. Fig.3 (23). Note that this is also consistent with $E_{\rm sn} < 1.9 \times 10^{51}$ erg from explosion theory.

It seems that in our cosmic neighborhood SN 1006 is the simplest and best-documented, almost prototypical explosion into a uniform interstellar medium with a uniform B-field. This is also indicated by the clear dipolar structure of its

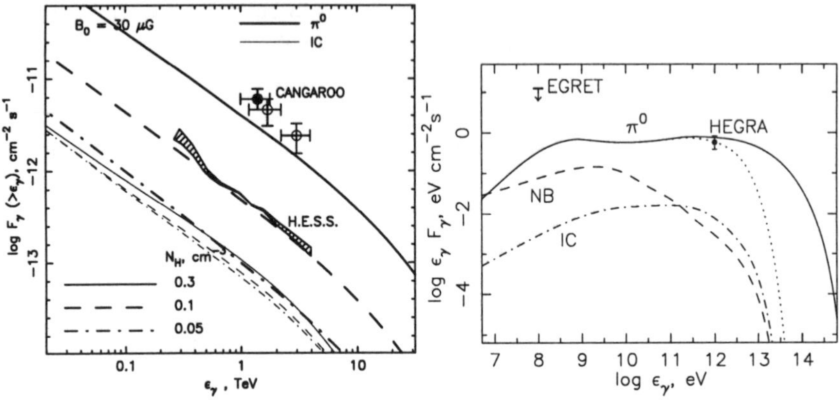

Figure 3. (*Left:*) Integral TeV gamma-ray flux measurements and theoretical estimates for the NW rim of SN 1006. The old *CANGAROO* data from 2001 lie about one order of magnitude above the *H.E.S.S.* upper limit band. The theoretical curves show the strong dependence of the hadronic flux (*thick lines*) on the gas density $N_{\rm H}$. For densities decreasing below 0.05cm^{-3} the hadronic flux tends to reach the IC flux in the upstream field of 30μG. (*Right:*) Predicted spectral energy distributions of gamma rays from Cas A for the π^0-decay hadronic, NB and IC channels. The *HEGRA* point may already indicate the escape of particles at the highest energies (*dotted line*).

X-ray synchrotron emission (24). It would therefore be of basic interest to finally detect this source in a deep gamma-ray observation.

In contrast to SN 1006, Cas A is the result of the core collapse of a massive star, probably a Wolf-Rayet star that had shaped its environment by several successive phases of mass loss. And therefore the explosion occurred into the wind bubble of the progenitor. Cas A is the strongest radio source in the sky.

The spatially integrated synchrotron emission has a spectrum that is – like in the cases of Tycho's SNR (Fig.2) and SN 1006 – nonlinearly modified at low frequencies and is characterized by synchrotron losses in the amplified field at optical frequencies. In the case of Cas A the magnetic field strength reaches the enormous value of 500 μG, consistent with the value derived from filamentary structures in hard X-rays. Therefore the basic features of the acceleration process are analogous to the ones described before when considering Tycho's SNR. And they exist in similar form in SN 1006.

If we use the specific model parameters of Borkowski et al. (25) who describe the thermal X-ray emission of Cas A in terms of a Wolf-Rayet progenitor, it is possible to predict the gamma-ray emission spectrum (Fig.3 (26)). It is characterized by a preponderance of the hadronic π^0-decay emission by two orders of magnitude at 1 TeV; up to $\sim$ 10 GeV the Nonthermal Bremsstrahlung is still comparably strong within a factor of a few, albeit with a somewhat different energy spectrum. A clear distinction between hadronic and leptonic emission is therefore only possible above $\sim$ 100 GeV, that means primarily with ground-based instruments. (This argument favoring detections at very high energies is different from that of Drury et al. (27) who discussed the *diffuse* Galactic gamma-ray background as a mounting obstacle for the detection of extended sources, when going to GeV energies and below). It is also interesting that this youngest Galactic SNR may already show signs of escape of the highest energy particles from the accelerator. In other words, Cas A may be "old" in an evolutionary sense, because the outer shock is presently already moving through the radially decreasing magnetic fi eld drawn out by the progenitor's Red Super Giant wind. The *HEGRA* detection signifi cance is slightly below 5σ. An eventual confi rmation by the Northern arrays would be an important measurement.

2.3. *SNR RX J1713.7-3946*

Like Cas A this source is most probably due to the core collapse of a massive progenitor in a dense interstellar cloud. The existence of a X-ray bright central point source as the probable compact relic also supports this view (e.g. (28)). The diffuse X-ray emission seems to be entirely nonthermal (29) (30) (31) (28) (32). From their TeV detection of part of the source and their phenomenological fi t to the spectrum Enomoto et al. (33) concluded that the gamma-ray emission was dominantly hadronic. Reimer and Pohl (34) and others argued that the upper limits from *EGRET*, supposedly from the same source, violated this interpretation. In fact, the data available at the time, both in TeV gamma rays and in other wavelength ranges, were scarce and the source distance ambiguous. As a consequence the alternative of hadronic vs. IC emission remained undecided.

Subsequently the H.E.S.S. experiment obtained a resolved gamma-ray image of the source (35). It showed clearly the morphology expected for a shell-type SNR (Fig. 4 (left)). The latest gamma-ray spectrum (36) can be phenomenologically fi tted as $dN_\gamma/dE = I_0 E^{-\Gamma} \exp(E/E_c)$ cm^{-3} s^{-1} erg^{-1}, with a photon index $\Gamma = 1.96 \pm 0.04$ and a cutoff energy $E_c = 14.7 \pm 2.7$ TeV. This is unambiguous proof for the acceleration of charged particles to energies beyond 100 TeV. (For further details see the article by D. Berge in these Proceedings.)

The radio faintness of this 1°–diameter source and the correspondingly not well defi ned slope of the observed radio synchrotron spectrum unfortunately does not permit a determination of the proton injection rate. Also the effective magnetic fi eld can only be approximately obtained from an overall comparison of the accelerated electron distribution (Fig. 4 (right)) to the observed synchrotron spectrum (Fig. 5). The upstream fi eld strength obtained is 20 μG. With a total shock

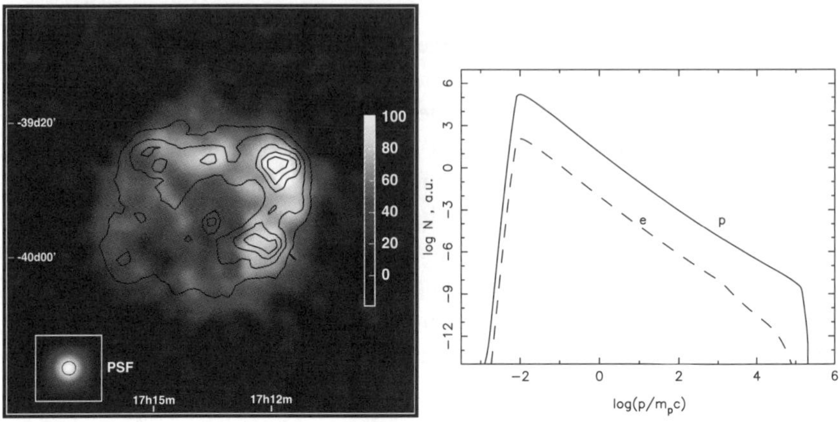

Figure 4. (*Left:*) *H.E.S.S.* image of SNR RX J1713.7-3946 at energies > 800 GeV with *ASCA* 1-3 keV X-ray intensity contours superposed. (*Right:*) Theoretical, spatially integrated CR spectra as function of particle momentum (38). *Solid* and *dashed* lines correspond to protons and electrons, respectively. The proton number exceeds the electron number by a factor $\sim 10^3$).

compression ratio of 6.3 this implies a downstream field strength of 126 μG. Such an amplified field is nevertheless consistent with the lower limit $B_d \approx 65\ \mu$G that one can derive from a radial X-ray profile recently published (32).

On the other hand, the original argument (30) for a distance $d \approx 1$ kpc, is now largely agreed upon. It is quite consistent with an age of 1612 yr, according to historical Chinese records (39). The faintness in thermal X-rays suggests that the object had a fairly massive progenitor ($15 < M_*/M_\odot < 20$) into whose rarefied wind bubble the star exploded (38). An acceleration model with these characteristics is able to reproduce the synchrotron and gamma-ray observations remarkably well (Fig. 5). It assumes a proton injection rate $\eta = 3 \times 10^{-4}$. The resulting gamma-ray spectral energy density is dominated by π^0-decay and is nevertheless very hard at gamma energies ≤ 1 TeV on account of the nonlinear shock modification, to be expected for a source of the Galactic CRs. The theoretical spectrum agrees with the *H.E.S.S.* spectrum and stays below the *EGRET* upper limit, estimated for the region of the source. A *GLAST* measurement would be very useful to confirm the flux level for energies $> 10^9$ eV.

The fact that we have here the first case of a TeV-extended shell-type SNR which could be detected with high significance, makes the question of its theoretical understanding in terms of a hadronic gamma-ray source particularly critical. The need to assume a hadronic injection rate – as a result of the inadequate observational situation at radio wavelengths – spoils the predictive power of the theory in its standard from. However, this is removed if we invoke in addition a

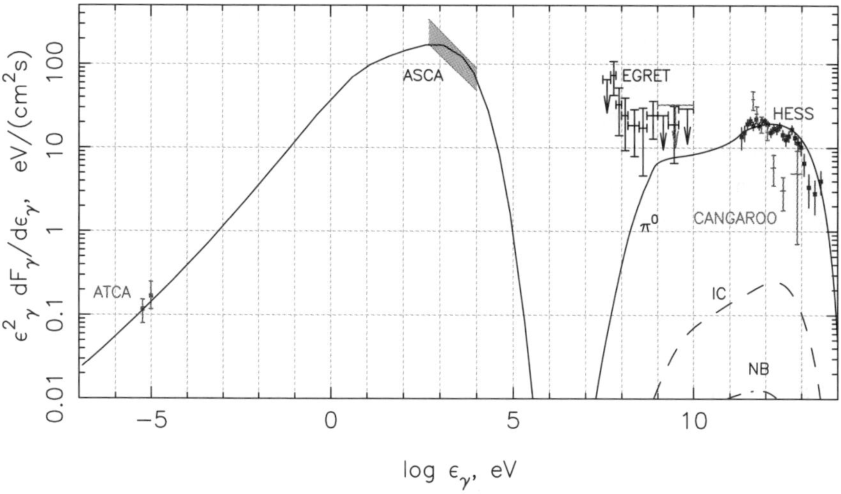

Figure 5. Spatially integrated spectral energy distribution of SNR RX J1713.7-3946(38). Among other data also the *EGRET* upper limit for the RX J1713.7-3946 position is shown (*in red color*). (From (38)).

semi-empirical relation $B_d^2/(8\pi P_c) \approx 5 \times 10^{-3}$ between the amplified magnetic field energy and the energetic particle pressure P_c whose gradient drives the field amplification; P_c is clearly dominated by the proton component. Such a plasma theoretical relation can be traced back to Bell & Lucek (2001) (3). For a given field strength B_d it requires the proton injection rate to be consistent with the value of P_c. With this extension the injection parameter is determined. The result is still basically consistent with the observations. For details, see the recent paper by Berezhko and Völk (2006) (38).

An object similar to RX J1713.7-3946 seems to be the even more extended SNR RX J0852.0-4622, also called "Vela Jr.". It was recently detected in TeV gamma rays by both *CANGAROO III* (40) and *H.E.S.S.* (41). Again the synchrotron emission has a rather uncertain spectral form if compared to, say, Cas A or Tycho's SNR. However, the TeV emission is incontrovertible, and thus Vela Jr. constitutes the second confirmed TeV detection of a shell-type SNR. It will be of basic importance to see, whether and to which extent also this source can be shown to be a hadronically dominated Galactic CR source.

3. Nonthermal energetics due to SNRs in the Universe

The escaping energetic particles from the collection of SNRs in normal star forming galaxies like the Milky Way establish a large-scale pressure gradient above/below the gas disk which can carry thermal gas along with the CRs as a result of pitch angle scattering (42). In concrete terms this flow reaches supersonic velocities at distances $\sim$ 20 kpc (43) (44).

The asymptotic wind speed amounts to about 300 km/s and is only reduced to subsonic values in a termination shock at distances of some 100 kpc (see Fig. 6 (left)). The CR transport properties in this diffusive/convective wind halo agree with the basic constraint of the rigidity-dependent "grammage" encountered by the particles that are observed in the disk (45). It is therefore quite plausible that the internal energy densities of the CRs, the thermal gas, and magnetic fi eld remain in rough equipartition ($U_{\rm CR} \sim U_{\rm gas} \sim U_{\rm B}$) also far from the thin gas disk, where this equipartition is observed (Fig. 6). At least in groups or clusters these "galactic wind bubbles" should essentially touch each other in pressure equilibrium.

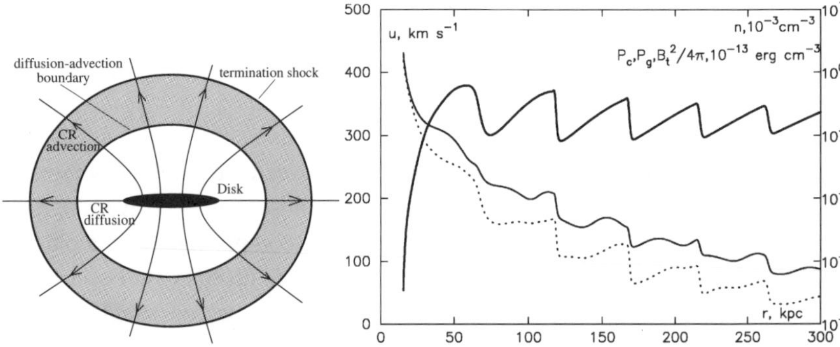

Figure 6. (*Left:*) Cartoon of the Galactic Wind from a disk galaxy like the Milky Way. Particles accelerated at the outer termination shock (46) can practically not return to the disk due to the modulating effect of the wind fbw. (*Right:*) Spiral density wave modification of the Galactic wind velocity u as a function of distance from the disk (*upper solid curve*). The gas pressure (*dotted curve*) is increasingly dominated by the CR pressure (*lower solid curve*) that drives the wind. This results from the softer equation of state of the CRs. Particles can be re-accelerated in this large-scale sawtooth wave and can return to the disk at energies $\geq$ knee-energies (47).

Apart from our Galaxy only crude estimates of CR energetics are possible. I will give some in the following.

As far as the local intergalactic medium (IGM) in the Milky Way's Local Group is concerned, its dynamics may be comparable to that in the Leo group, where the internal gas pressure of an HI cloud of $p_{\rm cloud}/{\rm k} \sim 5$ K cm^{-3} has been argued to be the result of a recent galaxy-galaxy interaction and therefore the cloud may be in expansion against the total inter-group pressure (48). We can then ask ourselves to which extent this inter-group pressure is produced by CRs from the ensemble of group galaxies. Putting the total Galactic CR energy production rate $\sim 2 \times 10^{41}$ erg/s into kinetic energy of the CR-driven wind and dissipating again 10 percent of this kinetic energy into CRs at the termination shock, to be released into the local IGM, the CR energy accumulates over 10^{10} yr in a volume of size l^3, where l is the present mean distance between galaxies. The resulting CR pressure is $p_{\rm CR}^{\rm IG}/{\rm k} \approx 0.6/(l/{\rm Mpc})^3$ Kcm^{-3}, where k denotes Boltzmann's constant. In the Local Group $l \approx 0.5$ Mpc, and thus $p_{\rm CR}^{\rm IG}/{\rm k} \approx 4.8$ Kcm^{-3}. This rough calculation ignores all evolutionary effects which might in fact lead to a higher CR pressure as the result of a higher star formation rate in the past.

For the environment of field galaxies, with $l \approx 1$ Mpc, $p_{\rm CR}^{\rm IG}/{\rm k} \approx 6 \times 10^{-1}$ Kcm^{-3}. A lower limit to the ambient gas density there is the present mean cosmological baryon density $\rho_b \approx 2 \times 10^{-7}$cm^{-3}, but the cosmic web of walls and filaments containing the field galaxies should have substantially higher gas densities ρ. With a gas temperature of $\sim 5 \times 10^4$ K, roughly characteristic for a Lyman-α forest gas, one obtains $p_{\rm gas}^{\rm IG}/{\rm k} \approx (\rho/\rho_b) \times 10^{-2}$ Kcm^{-3}. With $(\rho/\rho_b) \sim 100$ (e.g. (49)), it is again likely that the CRs from galaxies are a significant dynamical element in the evolution of the IG gas.

In starburst galaxies the pressure effect from nonthermal particles is enhanced and in fact superpowered by the expansion of the hot thermal gas that cannot cool radiatively fast enough any more before being set in motion, producing a "superwind". The main difference to normal galaxies is the fact that the wind energy is now supplied by the total energy released by the SN population – a factor ~ 10 increase in efficiency – and that these objects have a ~ 10 times higher SN rate to begin with. Their CR output at the wind termination shock can therefore be up to a factor ~ 100 higher than normal galaxies, with corresponding consequences on the environment.

Finally, we may consider the nonthermal evolution of rich clusters of galaxies. The main sources of CRs - and probably comparable in strength - are the accretion shocks from the infall of cold IG gas, the relativistic particle production in AGNs, and the strong early starbursts and galaxy mergers that led to the present-day predominance of gas-poor elliptical and lenticular galaxies. The large geometrical size of clusters ensures the confinement of all but the most energetic nonthermal particles over the age of the Universe (50) (51). Clusters are therefore reservoirs of "cosmological CRs", accumulating them since the beginning of cosmic structure formation. For large clusters the total CR energy content in the intracluster medium from SNRs and wind termination shocks alone is roughly estimated as $E_{\rm CR} \sim 10^{62}$ erg (52), which is higher than the thermal energy content of all stars from all the cluster galaxies together.

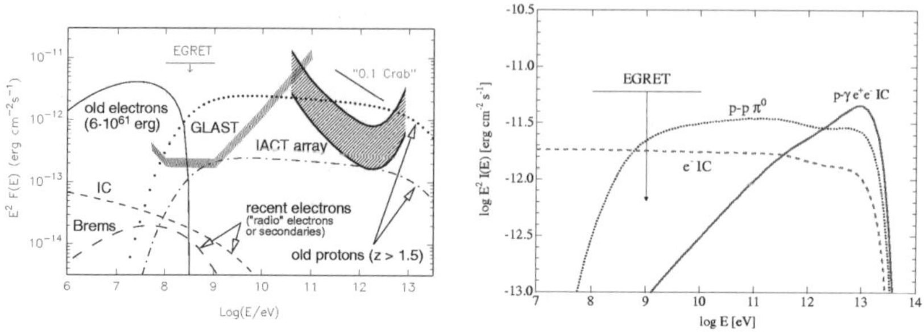

Figure 7. The COMA cluster in gamma rays (*Left:*) Gamma-ray spectral energy densities from protons (*dotted and dash-dotted curves*) and electrons (*solid curve*) over the internal merger history, and from recent electrons, plus instrumental sensitivities (adapted from (52)). (*Right:*) The same (53) from p-p collisions and IC collisions of primary (e^-) and secondary electrons from $p-\gamma$ interactions (54), due to particles accelerated in accretion shocks.

It would be extremely interesting to detect high energy gamma rays from clusters. Fig. 7 shows gamma-ray flux estimates for internally produced and cluster accretion shock accelerated nonthermal particles. The fluxes are comparable. Deep observations with instruments like H.E.S.S. or GLAST might lead to detections which could unravel the nonthermal history of these objects.

In conclusion, it seems appropriate to summarize the sources of these processes in terms of a *Nonthermal Universe* whose components are interacting with the thermalized structures we are used to think of. The origin of the nonthermal particles is intimately connected with cosmic structure formation in its most energetic events. These tend to result in rough energy equipartition between the thermal and the nonthermal particle fraction.

References

1. E.G. Berezhko,, V.K. Elshin and L.T. Ksenofontov, *JETPh* **82**, 1 (1996).
2. E.G. Berezhko and H.J. Völk, *Astropart. Phys.* **7**, 183 (1997).
3. A.R. Bell and S.G. Lucek, *MNRAS* **321**, 433 (2001).
4. H.J. Völk, in it Frontiers of Cosmic Ray Science, Proc. 28th ICRC (Tsukuba), ed. Kajita et al. (Tokyo, Japan: Universal Press, Inc.) (Invited papers) **8**, 29 (2004); arXiv:astro-ph/0312585.
5. E.G. Berezhko, *Adv. Space Res.* **35**, 1031 (2005).
6. H.J. Völk, E.G. Berezhko and L.T. Ksenofontov, *A&A* **433**, 229 (2005).
7. H.J. Völk, E.G. Berezhko and L.T. Ksenofontov, *A&A* **409**, 563 (2003).
8. A. Decourchelle, J. Sauvageot, M. Audard et al., *A&A* **365**, L218 (2001).
9. U. Hwang, A.Decourchelle, S. Holt et al., *ApJ* **581**, 1101 (2002).

10. J.S. Warren, J.P. Hughes, C. Badenes et al., *ApJ* **634**, 376 (2005).
11. H.J. Völk, E.G. Berezhko and L.T. Ksenofontov, (2006); arXiv:astro-ph/05/12086 v2.
12. C. Badenes, K.J. Borkowski, J.P. Hughes et al., *ApJ* submitted (2006); arXiv:astro-ph/0511140.
13. S.M. Tan and S.F. Gull, *MNRAS* **216**, 949 (1985).
14. H.J. Völk, E.G. Berezhko, L.T. Ksenofontov et al., *A&A* **396**, 649 (2002).
15. J.H. Buckley, C.W. Akerlof, D.A, Carter-Lewis, et al., *A&A* **329**, 639 (1998).
16. F.A. Aharonian et al. (HEGRA Collaboration), *A&A* **373**, 292 (2001).
17. J. Prahl and C. Prosch, in *Proc 25th ICRC (Durban)* **3**, 217 (1997).
18. F.A. Aharonian et al. (HEGRA Collaboration), *A&A* **370**, 112 (2001).
19. T. Tanimori, T. Naito, T. Yoshida et al., *Proc. 27th ICRC (Hamburg)* **6**, 2465 (2001).
20. K. Koyama, R. Petre, E.V. Gotthelf, et al. *Nature* **378**, 255 (1995).
21. F.A. Aharonian et al. (H.E.S.S. Collaboration), *A&A* **437**, 135 (2005).
22. M. Mori, in B. Degrange, G. Fontaine (eds.), *Cherenkov 2005*, Proc. Workshop: "Towards a Network of Atmospheric Cherenkov Detectors VII", Palaiseau, April 2005, p.19 ff. (2005).
23. L.T. Ksenofontov, E.G. Berezhko and H.J. Völk, *A&A* **443**,973 (2005).
24. R. Rothenflug, J. Ballet, G.Dubner, et al., *A&A* **425**, 121 (2004).
25. K.J. Borkowski, A.E. Szymkowiak, J.M. Blondin et al., *ApJ* **466**, 866 (1996).
26. E.G. Berezhko, G. Pühlhofer and H.J. Völk, *A&A* **400**, 971 (2003).
27. L. O'C. Drury, F.A. Aharonian and H.J. Völk, *A&A* **287**, 959 (1994).
28. G. Cassam-Chenaï, A. Decourchelle, J. Ballet, et al., *A&A* **427**, 199 (2004).
29. E. Pfeffermann and B. Aschenbach, in *Roentgenstrahlung from the Universe*, ed. H.H. Zimmermann, J. Trümper and H. Yorke, *MPE Rep.***263**; Garching: MPE, 267 (1996).
30. K. Koyama, K. Kinugasa, K. Matsuzaki et al., *PASJ* **49**, L7 (1997).
31. P. Slane, B.M. Gaensler, T.M. Dame, et al.*ApJ* **527**, 357 (1999).
32. J.S. Hiraga, Y. Uchiyama, T. Takahashi, et al., *A&A* **431**, 953 (2005).
33. R. Enomoto, T. Tanimori, T. Naito et al., *Nature* **416**, 823 (2002).
34. O. Reimer, and M. Pohl, *A&A* **390**, L43 (2002).
35. F.A. Aharonian et al. (H.E.S.S. Collaboration), *Nature* **432**, 75 (2004).
36. F.A. Aharonian et al. (H.E.S.S. Collaboration), *A&A* **449**, 223 (2006).
37. D. Berge, *these Proceedings*.
38. E.G. Berezhko, H.J. Völk, *A&A* **451**, 981 (2006).
39. Z.R. Wang, Q.-Y. Qu, Y. Chen, *A&A* **318**, L59 (1997).
40. H. Katagiri, R. Enomoto, L.T. Ksenofontov, et al., *ApJ* **619**, L163 (2005).
41. F.A. Aharonian et al. (H.E.S.S. Collaboration), *A&A* **437**, L7 (2005).

42. F. Ipavich, *ApJ* **196**, 107 (1975).
43. D. Breitschwerdt, J.F. McKenzie, H.J. Völk, *A&A* **245**, 79 (1991).
44. V.N. Zirakashvili, D. Breitschwerdt, V.S. Ptuskin, et al., *A&A* **311**, 113 (1996).
45. V.S. Ptuskin, H.J. Völk, V.N. Zirakashvili, et al., *A&A* **321**, 434 (1997).
46. J.R. Jokipii and G.E. Morfi ll, *ApJ* **290**, L1 (1985).
47. H.J. Völk and V.N. Zirakashvili, *A&A* **417**, 801 (2004).
48. H.J. Rood and B.A. Williams, *A&A* **288**, 535 (1985).
49. R. Cen, J. Miralda-Escudé, J.P. Ostriker, et al., *ApJ* **437**, L9 (1994).
50. H.J. Völk, F.A. Aharonian, D. Breitschwerdt, *Space Sci. Rev.* **75**, 279 (1996).
51. V.S. Berezinsky, P. Blasi and V.S. Ptuskin, *ApJ* **487**, 529 (1997).
52. A.M. Atoyan and H.J. Völk, *ApJ* **535**, 45 (2000).
53. F.A. Aharonian, S. Gabici, S. Inoue, *Private communication* (2005).
54. S. Inoue, F.A. Aharonian, N. Sugiyama, *ApJ* **628**, L9 (2005).

RECENT RESULTS FROM THE MAGIC PROJECT AND OUTLOOK

J.A. COARASA
MAGIC Group, Max-Planck-Institut für Physik, Foehringer Ring 6, D-80805 Munich, Germany

FOR THE MAGIC COLLABORATION

The 17 m diameter Major Atmospheric Gamma Imaging Telescope (MAGIC) for gamma ray astronomy from 50 GeV, the lowest threshold achievable by such kind of telescopes, has been taking data regularly with a very high duty cicle since its commissioning phase. The status of the telescope, the results for the 1st year of regular data taking and the outlook of the project will be reviewed.

1. Introduction

The MAGIC telescope [1] is the largest γ-ray imaging atmospheric Cherenkov telescope (IACT) in the world. It is operating at the *Roque de Los Muchachos observatory* (28.75° N, 17.9° W, 2200 m above see level), on the Canary island of La Palma. After the commissioning phase, it started regular observations in October 2004. It yields the lowest threshold (~50 GeV) of the current existing IACTs giving reach to an explored energy range and sensitivity without precedent for short-term variation signals.

MAGIC has a 17 m diameter, f /D = 1, parabolic reflector covering a total surface of 234 m^2. The reflector dish is composed of 956 (0.495×0.495 m^2) diamond milled aluminum mirrors. The reflector shape is parabolic to minimize the time spread of Cherenkov light flashes on the focal plane, thus preserving the time structure of the Cherenkov flashes and allowing for a better signal-to-noise ratio with respect to the Night Sky Background light due to a reduced integration window. Aluminum mirrors (reflectivity of about 80%) were chosen to reduce the weight of the reflecting surface and allow a fast slewing of the telescope. For the same reason the telescope frame is made of carbon fiber tubes and has a weight of only about 8 tons.

A very short slewing time is needed to catch the γ-ray prompt emission of Gamma Ray Bursts (GRBs). MAGIC can fast slew to any part of the sky in less than 60 s. The focusing of the mirrors is corrected for mechanical deformations of the frame at different positions using an "Active Mirror Control System". Thanks to this system, the RMS of the Point Spread Function is less than 0.1°.

The reflected Cherenkov photons are recorded by a 3.5°–3.8° field of view (FOV) hexagonal camera, composed by 397 0.1° FOV photomultiplier tubes (PMTs), surrounded by 180 0.2° FOV PMTs. The PMTs have hemispherical windows and only 6 dynodes to minimize the time response width. The PMT photo conversion efficiency has been enhanced up to 30% and extended to the UV by coating the window with wavelength shifter [2].

The PMT signals are amplified at the camera, converted into optical signals by Vertical Cavity Emitting Lasers and transmitted over 162 m long optical fibers to the counting house [3]. This reduces drastically the weight and size of the cables and protects the signals from electromagnetic pick-up.

The signals in the electronic room are split and sent to trigger and digitizing systems. The trigger decision is generated by a 2-level system using the signals of the 325 innermost PMTs. Only signals above an adjustable threshold are considered. The first level trigger requires a four-fold next neighbour coincidence within a 5 ns window while the second level imposes topological constraints on the event images [4]. The analog signals are continuously digitized by 8 bit 300 MHz Flash ADCs. If the trigger condition is fulfilled the signals stored in FADC ring buffers are written to a FIFO buffer and saved by the Data Acquisition System (DAQ) [5].

The low energy threshold together with the fast repositioning of the telescope allows closing two observational gaps, giving us access to physics that was unexplored so far: the energy gap between (10-300 GeV) the satellite-borne detectors and ground-based telescopes, and the time gap between the actual observation and the time of the prompt emission detection of the GRBs.

2. Summary of observations by the MAGIC Telescope during the first year of regular operations

The performance of the telescope has been experimentally evaluated and found in good agreement with the expectations and Monte Carlo simulations. For the time being we are routinely performing analyses above 100 GeV, where the performance of our instrument is fully understood.

Since fall 2004 the first MAGIC telescope is regularly collecting data from a long list of astrophysical objects. Four galactic and four extra-galactic objects have been seen so far by MAGIC, one of which is a new discovery. We review shortly in the following sections part of this results, starting with the galactic ones, followed by the extragalactic ones and by the observation of the prompt emission of the Gamma Ray Burst GRB050713A.

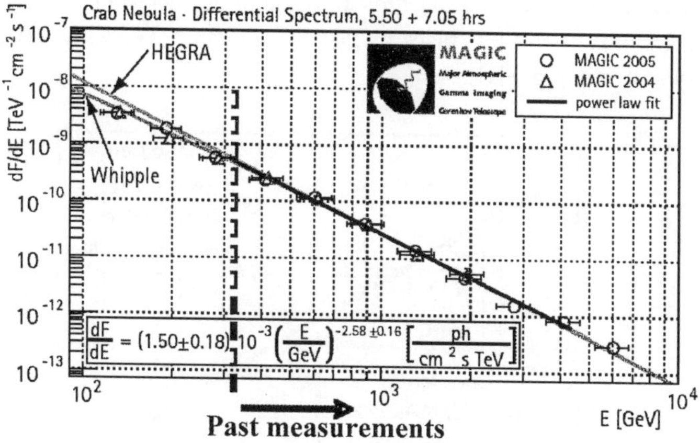

Figure 1. Crab Nebula differential energy spectrum down to 100 GeV as measured by MAGIC in different hardware situations (years 2004 and 2005 respectively) compared to the one measured previously by Whipple and HEGRA (which have been extrapolated down to the same energies). The plot shows good agreement among all of them.

2.1. *Crab Nebula*

The Crab Nebula is a steady emitter at GeV and TeV energies, what makes it into an excellent calibration candle. It was detected by MAGIC soon during the commissioning phase. MAGIC measured with high precision and for the first time, the spectrum down to 100 GeV, as shown in Fig. 1 with a reduced data set. The analysis of the whole combined data set will give more information on the existence of the inverse Compton peak, which is expected to be close to 100 GeV. It serves to evaluate the performance of the telescope [6, 7] as well.

We have also carried out a search for pulsed γ-ray emission from Crab and two millisecond pulsars [8], albeit without positive result.

2.2. *Galactic Center*

MAGIC measured also, albeit a difficult observation due to the high zenith angle, the VHE γ-ray flux from the Galactic Center (GC) [9], whose high-energy emission has been of very much interest during the last years. The energy spectrum of these three sources measured by MAGIC is shown in Fig. 2. All spectra are well described by unbroken power laws. In the case of the GC, the result disfavours dark matter annihilation as the main origin of the detected flux.

Furthermore, there is neither evidence for variability of the flux on hour/day time scales nor on a year scale (by comparison with HESS results obtained one year before). This disfavours the typical variable acceleration observed so far from VHE γ-ray emitting black holes in favour of steady acceleration mechanisms seen in e.g. SNRs.

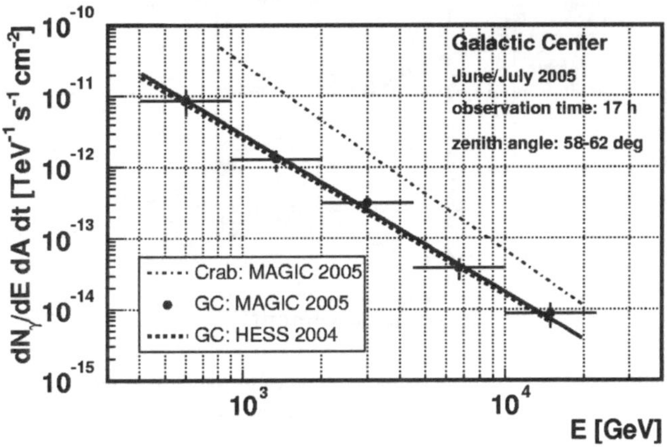

Figure 2. Differential energy spectrum for the region of the Galactic Center as measured by MAGIC compared to the one measured by HESS in 2004. The plot shows good agreement between them. The Crab Nebula one is also shown for comparison.

2.3. *HESS J1813-178 and J1834-087*

Within its program of observation of galactic sources, MAGIC has confirmed the VHE γ-ray emission from the supernova remnants (SNRs) HESS J1813-178 [10] and HESS J1834-087 [11] shortly after their discovery by HESS [12]. Our observations have confirmed SNRs as a well established population of VHE γ-ray emitters. In Fig. 3 the differential energy spectrum for HESS J1813-178 is shown. The measurement was also made under high zenith angle around 40°-50° which implies a higher energy threshold E_{Th} = 300 GeV but also a larger effective area.

2.4. *1ES1959+650*

MAGIC has measured for the first time the spectrum of the Blazar 1ES1959+650 (z=0.047) in a low energy range (from ~200 GeV to ~2 TeV). This source is especially interesting because it has shown in the past flares without any counterpart on other wavelenghts (orphan flares) and there are hints for detection of neutrinos and UHECR from its direction, and therefore might provide a connection between these three kinds of astronomies. The low flux measured by MAGIC points to detection in the low state (Fig. 4) [13] and shows the difficulty of previous telescopes in selecting a true "low state" due to the long window integration time.

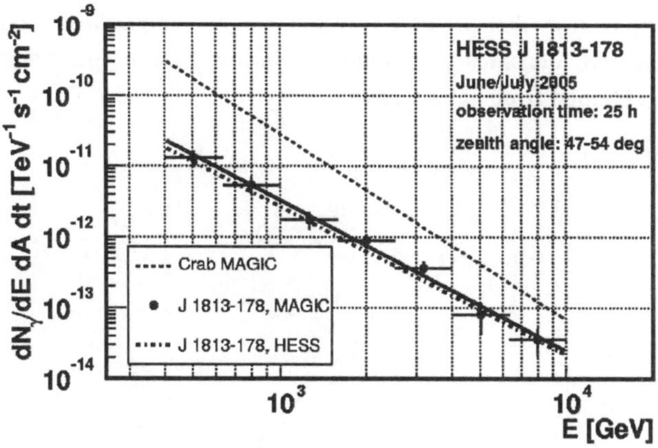

Figure 3. Differential energy spectrum for HESS J1813-178 as measured by MAGIC compared to the one measured by HESS. The plot shows good agreement between them. The Crab Nebula one is also shown for comparison.

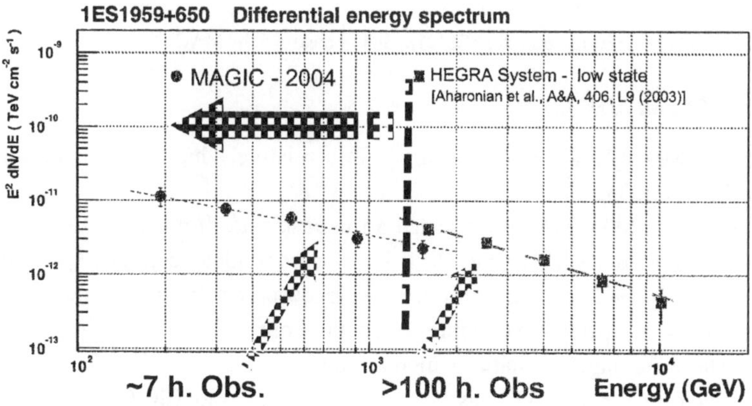

Figure 4. Flux energy spectrum for 1ES1959+650 as measured by MAGIC compared to the "low state" one measured by HEGRA.

2.5. *1ES1218+304*

MAGIC has added a new member to the list of VHE γ-ray emitters: the distant AGN 1ES1218+304 (z = 0.182), the second most distant gamma-ray source. This discovery enlarges the so-called γ-horizon at these energies and has allowed constraining models of the Extragalactic Background light (EBL). This can be done by unfolding the effect of γ-ray absorption by pair production in the EBL from the measured spectrum. The conclusions point to a more transparent universe to gamma rays that predicted by most of the theoretical models and with very small EBL densities with saturate the lower limits provided by the determinations obtained from galaxy counts.

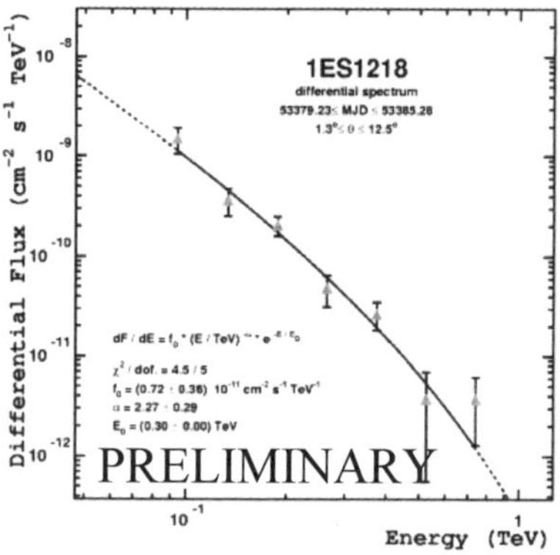

Figure 5. Differential energy spectrum for 1ES1218+304 as measured by MAGIC.

2.6. *Mrk 501 and Mrk 421*

Mrk 501 and Mrk 421 were both observed in a relatively high state of flaring during 2004 and 2005.

On June 30th 2005, MAGIC detected an intense flare from Mrk 501 2005, while monitoring this source and alerted the astronomical community (IAU circular 8562). Very large flux variations were observed on short time scale. There are hints of spectral hardening when the source is in high flux state. Mrk 501. There are ongoing studies which seem to point to an orphan flare.

The spectrum for Mrk 421 has been measured down to 100 GeV [14] for the first time with an IACT. flux variations of a factor of 2 between successive days are observed while the intra-night light curve does not show significant variations below 1 hour (we note that MAGIC is sensitive to detect variability on the order of 10 minutes time scale for the moderate flux level involved).

There is no evidence of spectral index variations associated to the flux variations. A clear correlation between X-rays and γ-rays was found (Fig. 6).

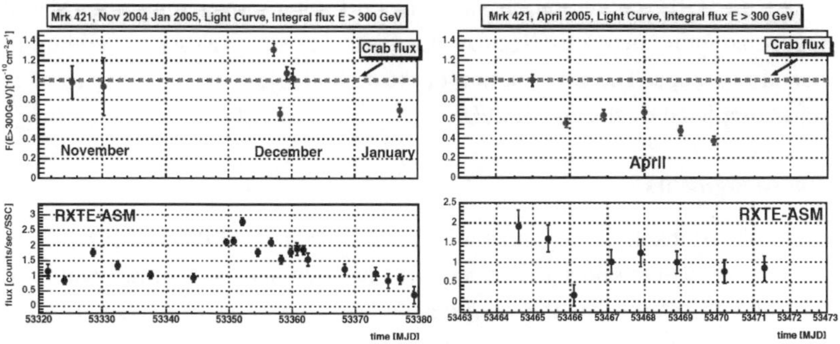

Figure 6. Light curve for the emission of Mrk 421 as measured by MAGIC and RXTE-ASM.

2.7. *Prompt GRB follow-up by MAGIC for GRB050713A*

The extremely fast repositioning of the MAGIC telescope allowed the follow-up of the Gamma Ray Burst GRB050713A only 13 seconds after the reception of the alert provided by the SWIFT satellite and only ~40 seconds after the actual gamma ray burst happened [15]. MAGIC observed the GRB in coincidence with the BAT and XRT instruments and while the x-ray activity was still high. So far no significant gamma ray emission was detected in the recorded MAGIC data and this will provide already strong constraints in many GRB models. MAGIC has observed more than 10 GRB since April 2004.

3. Conclusions and Outlook

MAGIC has been taking data regularly since Fall 2004. Known sources are confirmed in a few hours of observation time, and new results already emerge from a broad observation program (more than 50 possible γ-ray sources so far). Many galactic and extragalactic sources are under study, out of which nine have been detected: MAGIC detected γ-rays from 1ES 1218+304 ($z = 0.182$) the most distant γ-ray source (for a few hours only) detected so far; MAGIC confirmed emission from two sources recently discovered by HESS. MAGIC is currently producing high quality spectra above 100 GeV. This limit will be lowered further by improving the data analysis technique, but has produced physics output for known sources such as Mrk 421, Mrk 501 and 1ES 1959+650 that allows physics studies without precedent such as testing the existence of the Inverse Compton peak, time variability studies...

The MAGIC collaboration is actively further developing the MAGIC telescope adopting new 2.5 GHz FADCs for the existing telescope and building a second telescope. The construction of the second telescope already started and will be completed by the end of 2007.

Acknowledgments

I would like to thank the organization of the *International Workshop on Energy Budget in the High Energy Universe* for the nice stay and organization. We would like to thank also the *Instituto de Astrofísica de Canarias* (IAC) for excellent working conditions. The support of the German *Bundesministerium für Bildung und Forschung* (BMBF) and *Max-Planck-Gesellschaft* (MPG), the Italian *Istituto Nazionale di Fisica Nucleare* (INFN) and the Spanish *Comisión Interministerial de Ciencia y Tecnología* (CICYT) is gratefully acknowledged.

References

1. E. Lorenz, Procs. of Workshop: Towards a Major Atmospheric Cherenkov Detector IV. Ed. Cresti M. (1995) 277; E. Lorenz, *Nucl. Phys.* **B Proc. Suppl. 48**, 494 (1996); J. A. Barrio *et al.*, 1998, Max-Planck-Institut Report MPI-PHE/98-5. March 1998.
2. D. Paneque *et al.*, *Nucl. Instrum. Meth.* **A518,** 619 (2004).
3. D. Paneque *et al.*, Procs. of 28th Int. Cosmic Ray Conference (2003) 2927.
4. M. Meucci *et al.*, *Nucl. Instrum. Meth.* **A518,** 554 (2004).
5. J. Cortina *et al.*, Procs. of the 2002 Int. Science Symposium on The Universe Viewed in Gamma-Rays (2001).
6. R. Wagner *et al.*, Procs. of the 29th Int. Cosmic Ray Conference (2005).
7. J. Cortina *et al.*, Procs. of the 29th Int. Cosmic Ray Conference (2005); M. Gaug *et al. Ibid.*; E. Domingo *et al. Ibid.;* J. Rico *et al. Ibid.;* F. Goebel *et al. Ibid.;* P. Majumdar *et al. Ibid*
8. M. López *et al.*, Procs. TAUP (2005); E. de Oña-Wilhelmi *et al.,* Procs. of the 29th Int. Cosmic Ray Conference (2005).
9. J. Albert *et al.*, Astrophys. Journal, 638, L101-L104 (2006).
10. J. Albert *et al.*, Astrophys. Journal, 637, L41-L44 (2006).
11. J. Albert *et al.*, accepted in Astrophys. Journal. astro-ph/0604197
12. F. Aharonian *et al.*, Science 307, 1938 (2005).
13. J. Albert *et al.*, Astrophys. Journal, 639, 761-765 (2006).
14. D. Mazin *et al.*, Procs. of the 29th Int. Cosmic Ray Conference (2005).
15. J. Albert *et al.*, submitted for publication. astro-ph/0602231.

All-SKY SURVEY HIGH RESOLUTION AIR-SHOWER DETECTOR (ASHRA)

MAKOTO SASAKI*
on behalf of the Ashra collaboration
ICRR, Univ. Tokyo, Kashiwa, 277-8582, Japan
**E-mail: sasakim@icrr.u-tokyo.ac.jp*

Ashra (*All-sky Survey High Resolution Air-shower detector*) [1–3] is a project to build an unconventional optical telescope complex that images very wide field of view, covering 80% of the sky, yet with the angle resolution of 1.2 arcmin, sensitive to the blue to UV light with the use of image intensifier and CMOS technology. The project primarily aims to observe Cherenkov and fluorescence lights from the lateral and longitudinal developments of very-high energy cosmic rays in the atmosphere. It can also be used to monitor optical transients in the wide field of sky. In 2004 we built prototype telescopes to verify and develop techniques at Haleakala in Hawaii, needed for the development of the full-scale telescopes. Construction of the main detector station has begun at Mauna Loa on the Hawaii Island in the summer of 2005.

Keywords: Ashra; all-sky; air-shower; very-high-energy gamma; very-high-energy neutrino; optical transient

1. Project

Main Technical Features

The observatory will fi rstly consist of one main station having 12 detecor units and two sub-stations having 8 and 4 detector units. One detector unit has a few light collecting systems with segmented mirrors. The features of the system were studied with a prototype detector unit located at Haleakala. The main station is being constructed at Mauna Loa (3,300 m).

The key technical feature of the Ashra detector rests on the use of electrostatic lenses to generate convergent beams rather than optical lens systems. This enables us to realise a high resolution over a wide fi eld of view. This electron optics requires:

- *image pipeline;* the image transportation from imaging tube (image intensifi er) to a trigger device and image sensors of fi ne pixels (CCD+CMOS), with high gain and resolution, and

- *parallel self-trigger:* the systems that trigger separately for for atmospheric Cherenkov and fluorescence lights.

Observational Objectives

optical transients; Ashra will acquire optical image every 5 s after 4-s exposure. This enables us to explore optical transients, possibly associated with gamma ray bursts (GRBs), flares of soft gamma-ray repeaters (SGRs), supernovae explosion, and so on, in so far as they are brighter than $B \simeq 15$ mag, for which we expect 3-σ signals (Fig. 1). The unique advantage is the on-time detection of the events without resorting to usual satellite alerts. 10$\sim$20 events per year are expected in coincodence with the Swift gamma-ray events. The fi eld of view that is wider than satellite instruments allows to detect more optical transients, including an interesting possibility for an optical flash, not visble with gamma-rays.

TeV gamma rays; Atmospheric Cherenkov radiation will be imged by Ashra. Requiring the signal-to-noise ratio (SNR) $>$5, the system will allow to explore VHE gamma-ray sources with the energy threshold of 2 TeV at the limiting flux sensitivity of 5% Crab for 1-year observation.

EeV cosmic rays; For fluorescence lights from VHE cosmic rays the effective light gathering effi ciency is comparable with that of the High Resolution Fly's Eye detector (HiRes). The arcmin pixel resolution of Ashra provides fi ner images of longitudinal development profi les of EeV cosmic ray (EeV-CR) air-showers. The resolution of arrival direction with the stereo reconstruction is thus signifi cantly improved and it is better than one arcmin for the primary energy of EeV and higher [7]. This is useful to investigate events clustered around the galactic and/or extragalactic sources. This in turn would give us information as to the strength and coherence properties of the magnetic fi eld[4].

PeV-EeV neutrinos; Ashra may detect Cherenkov and/or fluorescence signals generated from tau-particle induced air-showers that is generated from interactions of tau neutrinos with the mountain and/or the earth. This is identifi ed by peculiar geometry of the air-shower axis. The 1-year detection sensitivity with the full confi guration of Ashra is 5 and 2 times larger than the Waxman-Bahcall limit for mountain-produced event (Cherenkov) and earth-skimming event (fluorescence),

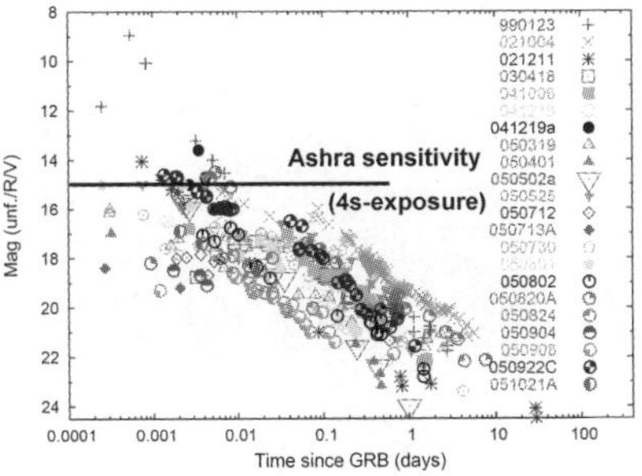

Fig. 1. Early light curves (unfiltered, R and V) for a set of GRBs with detections within minutes of the gamma-ray event. All the measurements shown have been taken from GCN circulars (Guidorzi et al. astro-ph/0511032). The limiting magnitude expected with Ashra (4s-exposure) is shown as a horizontal line.

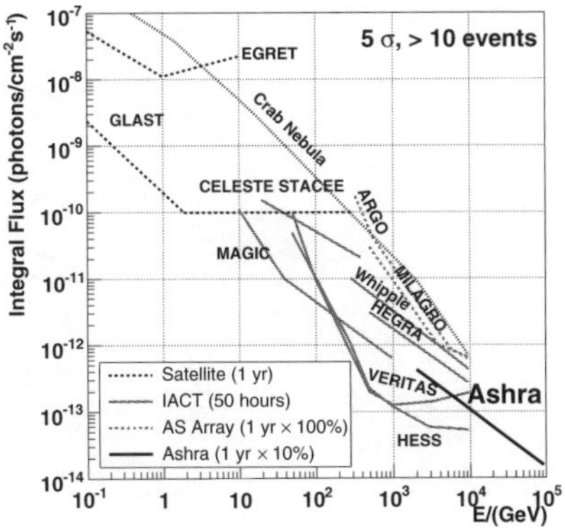

Fig. 2. Sensitivity of gamma ray telescopes (Surveys in High Energy Physics, vol.16, 255-244, 2002). All sensitivities are at 5σ. The sensitivity of Imaging Atmospheric Cherenkov Telescopes (IACTs) other than Ashra is for a 50 hour exposure on a single source. Satellites, AS arrays, and Ashra sensitivity is shown for one year of all sky survey. The Ashra duty of 10% is assumed.

respectively [5,6]. The most sensitive energy of around 100 PeV is suitable for the GZK neutrino detection.

Station Layout

From the viewpoint of funding strategy, the Ashra project is subdivided into two project phases. The first phase (Ashra-1) will include a main station and a substation for higher field of view (subH) at Mauna Loa (3,300m) and another substation for lower field of view (subL) at Camp Kilohana to demonstrate the all-sky monitor of optical transients and the stereoscopic Cherenkov and fluorescence survey observation in a cost effective manner (Fig. 3). The centers of the main and subH stations are separated by 80m to maximize the stereoscopic observation efficiency for TeV-gamma rays (Fig. 4). The main, subH, and subL stations consist of 12, 4, 8 detector units, respectively.

2. Detector

light collector optics; The Ashra light collector, as shown in Fig. 5, uses a modified Baker-Nunn optics[7,8]. It utilizes 2.3-m projected diameter spherical reflector for viewing a 0.42 sr (full FOV 42°) region of the sky. The ray tracing study shows the RMS point spread of 0.8 arc-minutes for incident parallel lights stably within

Fig. 3. Ashra-1 station layout on the Hawaii Island.

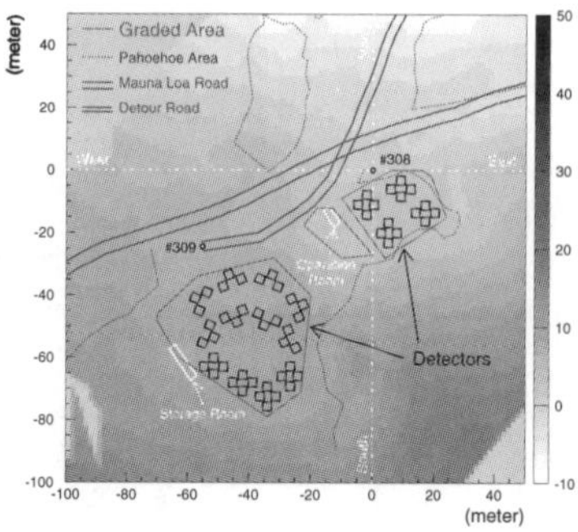

Fig. 4. Detector layout plan at the Mauna Loa site.

the full FOV of 42°. It does not strongly depend on the spectrum shapes of the Cherenkov, fluorescence, and star lights. The 1m-diameter correcting aspherical lens system composed of three acrylic direct cut plates can signifi cantly reduce spherical and coma aberrations. The reflector has seven segmented glass-mirrors coated with aluminum and aulminum oxide. The RMS point spread and reflectivity have been evaluated to be better than 0.4 arc-minutes and 90% respectively.

photoelectric lens image-intensifier tube; At the end of the light collector, a 20-inch diameter photoelectric lens image-intensifi er tube (PLIT) [13,14] is equipped. The cathode sphere is aligned to be combined with the surface of the sharp focus in the layout of the optical system. The PLIT makes input images scaled to the 35 mm diameter of the output fi ber optic (FOP) window with the electrostatic lens effect.

photoelectric image pipeline and readout; The photoelectric image pipeline (PIP) consists of relay lens systems, an amp II, a delay II, and half-mirror light splitters. It transports images to trigger processing, recording on 4-mega pixel CMOS sensors with trigger controls, and recording on 4-mega pixel CCD sensors for images of B-UV star lights periodically (Fig. 6). The delay II just before the CMOS sensor makes time delay for trigger controls utilizing sustained afterglow in the output phosphor screen.

trigger; The Ashra trigger system [15] is composed by two separated hybrid photosensors with 64×64 silicon pixel array (SiPix) and processors for Cherenkov and fluorescence signals transmitted through 64×64 FOP bundle. Each SiPix pixel is

Fig. 5. Ashra light collector and photoelectric lens image-intensifier tube mounted on the light collector.

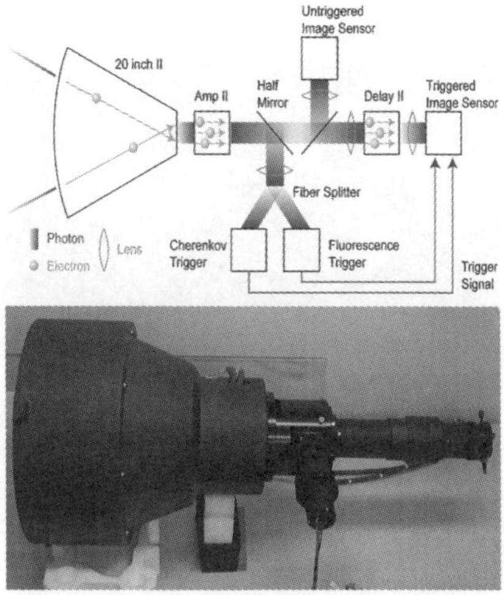

Fig. 6. Block diagram (top) and prototype (bottom) of the Ashra photoelectric image pipeline (PIP) system.

responsible for $0.67° \times 0.67°$ FOV, in which the night sky background flux corresponds to ~ 0.3 photoelectrons / 100 ns in the first cathode on the input window of PLIT. Two types of trigger (L1 and L2) control exposure and readout timings of local regions on the CMOS sensors, respectively. For long-duration fluorescence tracks, the L1 trigger starts the exposure to local regions one by one along the development of the track image on the CMOS sensor.

3. Test Observation

We have constructed a 2/3-scale prototype Ashra detector and a 3m-diameter altazimuth Cherenkov telescope at Haleakala to verify the optical and trigger performances. From October 2004 to August 2005 at the observatory, We made good observations for 844 hours out of 1,526 hours of the moonless night time. The efficiency is 55% of the moonless night time and 11% of entire time. This inefficiency is due primarily to bad weather.

The fine resolution (arc-minutes) in the ultra wide field of view (0.5 sr) has already been demonstrated using a 2/3-scale model. Fig. 7 shows an example of a 50-degree FOV image in which the constellations Taurus and Orion can be clearly identified with the 2/3-scale prototype. The inset, a two-degree square window, shows a close-up view of the Pleiades.

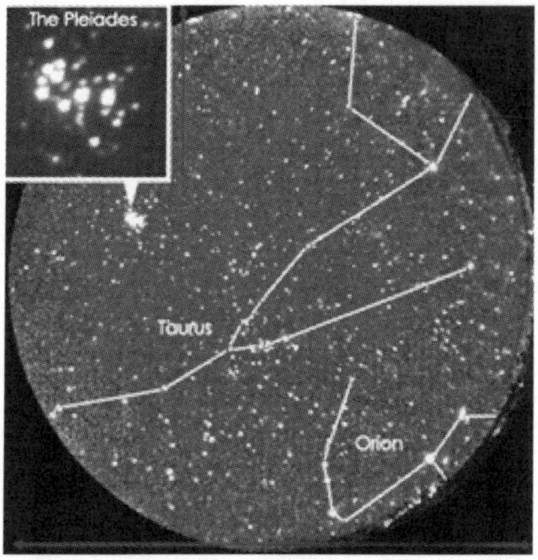

Fig. 7. Example of a 50-degree FOV image taken by the Ashra prototype. The solid lines are drawn to indicate the constellations Taurus and Orion. The inset, a two-degree square window, shows a close-up view of the Pleiades.

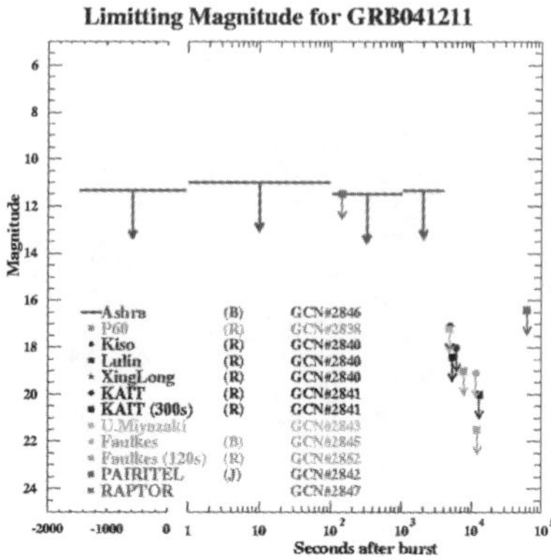

Fig. 8. 3-sigma limiting magnitudes of the test observation with the Ashra prototype and comparison with other observations for GRB041211 as a function of time after GRB. Note that the horizontal axis unavoidably stands for time (s) in logarithmic scale after the burst (positive) and in linear scale before the burst (negative).

Our wide fi eld ovservation covered the HETE-2 WXM error box at the time of GRB041211. 2,000 images were taken every 5 s with 4-s exposure from the time 1h7m before GRB041211 to 1h41m after GRB041211. We detected no objects showing time variation in the WXM error box. It indicates the 3-sigma limiting magnitudes of B$\sim$11.5 magnitude [9–11]. This is compared with other observations in Fig. 8 [16,17]. We also successfully performed two more observations coincident with Swift: GRB050502b [18] and GRB050504 [19].

A demonstration of air Cherenkov imaging of high-energy gamma/cosmic ray is shown in Fig. 9 which was taken during observing Mkr501. Separately, we have confi rmed the alpha-parameter[12] peak of TeV γ-rays from the Crab nebula to be greater than 5 σ.

4. Current Status and Plan

After fi nishing the grading work for the area of 2,419 m^2 at the Mauna Loa site at the end of July 2005, installation of electrical power lines and transformers was performed until the beginning of September. We started the construction of the detector in October 2005 after receiving materials from Japan. Currently, (mid December 2005) a few shelters having motorized rolling doors, acrylic plate windows to maintain air-tightness, and heat-insulating walls and floors have been con-

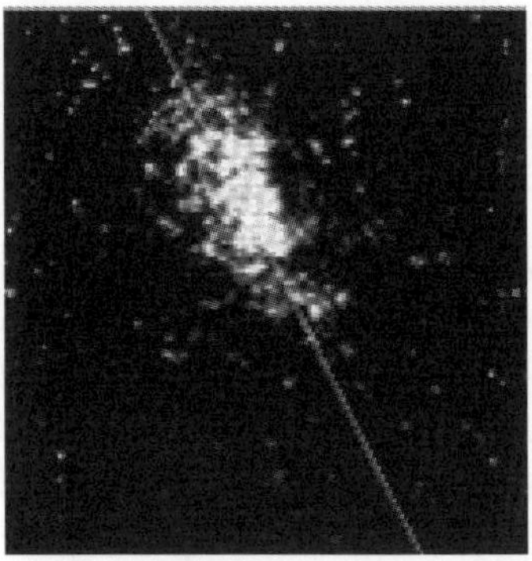

Fig. 9. Self-triggered Cherenkov image of air-shower detected by using the Ashra photoelectric image pipeline and a prototype trigger sensor system. This image was taken during tracking Mkr501.

Fig. 10. Photograph of an Ashra light collector installed in a shelter. A motorized aluminum rolling door is equipped on the roof. Under the rolling door, an acrylic plate is placed over a 1,800-mm-square window.

structed and positioned on eight construction piers of concrete blocks at the Mauna Loa site as shown in Fig. 10. In the shelters, the optical elements of the light collectors have been already installed. The optical performance were checked and adjusted to be optimum with star light images from the pilot observation.

In December 2005, we evaluated the night sky background flux at Mauna Loa using the Ashra light collector installed and aligned in a shelter. The result is fairly consistent with the background in La Palma and Namibia by the HESS group [20]. From the star light observations, our understanding of the light correction effi - ciency to be accurate within 5% level.

In this Ashra-1 experiment, we are performing device installation and specifi c observation in a step-by-step way to enhance the scientifi c impacts as follows.

- All-sky monitor of optical transients will start in FY2006 just after installation of shelters and optical parts of light collectors,
- Installation of trigger and test observation for Cherenkov lights will be repeated in FY2006. The all-sky imaging air Cherenkov monitor of TeV-gamma rays and Mountain-neutrino search will be started in FY2007.
- Installation of trigger and observation for fluorescence lights will be alternatively repeated on each light collector in FY2007. The all-sky fluorescence monitor and Earth-skimming-neutrino search will be started in FY2008.
- After acquiring the additional budget, we will improve the Cherenkov and fluorescence ovservations into stereoscopic ones. Installation of

SubH and SubL stations at Mauna Loa and the second Ashra site Camp Kilohana respectively will be completed in FY2008. The wide angle stereoscopic monitor of TeV-gamma, EeV-cosmic rays, and Earth-skimming-neutrinos will be started in FY2008.

The full Ashra observatory (Ashra-2) will consist of three experimental sites separated by about 30 km at Mauna Loa (3,300 m), Camp Kilohana (2,014 m) on the side of Mauna Kea, and Hualalai (2,320 m) on the island of Hawaii. The full confi guration emphasizes the stereoscopic observation Cherenkov and fluorescence lights from air-showers with two or three stations at separated sites as well as the effective detection area for air-showers. The parallax observation for optical transients with two or more stations is also useful for rejecting local background events.

Acknowledgment

This study was supported mainly by the Coordination Fund for Promoting Science and Technology (157-20004100) and by a Grant-in-Aid for Scientifi c Research (16740130) from the Ministry of Education, Science, Sports and Culture of Japan.

References

1. http://www.icrr.u-tokyo.ac.jp/∼ashra
2. Sasaki, M., *Progress of Theoretical Physics Supplement,* vol. 151, pp. 192–200, 2003.
3. Sasaki, M., et al., 2005. Proc. 29th Int. Cosmic Ray Conf. Pune 8, 17-200.
4. D. Harari, S. Mollerach and E. Roulet, astro-ph/0205484.
5. G.W.S. Hou and M.A. Huang, astro-ph/0204145.
6. E. Waxman, J. N. Bahcall, Phys. Rev. Lett., **78** (1997) 2292.
7. Sasaki,M., *et al., Nucl. Instrum. Methods,* vol. A492, pp. 49–56, Oct. 2002.
8. J. Baker, U.S. Patent 3022708, Feb. 27, 1962.
9. Dezalay, J-P, et al., 2004. GCN Circ. 2839.
10. Wozniak, P., et al., 2004. GCN Circ. 2847.
11. Sasaki, M., et al., 2004. GCN Circ. 2846.
12. Hillas, A. M., Proc. 19th Int.Cosmic Ray Conf. (La Jolla), 3, (1985) 445.
13. Sasaki,M., *et al., Nucl. Instrum. Methods,* vol. A501, pp. 359–366, Apr. 2003.
14. Asaoka,Y., Aita,Y., Aoki,T., Sasaki,M., IEEE Trans. Nucl. Sci., IEEE, 52, 1773-1778, 2005.
15. Arai,Y., *et al., Proc. 28th Intl. Cosmic Ray Conf. (Tsukuba),* pp. 961–964, 2003.
16. Sasaki,M., *et al.*., 2005. Proc. 29th Int. Cosmic Ray Conf. Pune 8, 17-200.
17. Sasaki, M., et al., 2005. Proc. 29th Int. Cosmic Ray Conf. Pune 5, 319-322.
18. Sasaki, M., et al., 2005. GCN Circ. 3499.
19. Sasaki, M., Manago, N., Noda, K., Asaoka, Y., 2005. GCN Circ. 3421.
20. Preuß, et al., *Nucl. Instrum. Methods,* **A481** (2002) 229.

TEV GAMMA-RAYS FROM OLD SUPERNOVA REMNANTS

RYO YAMAZAKI

Department of Physics, Hiroshima University,
Hiroshima 739-8526, Japan
E-mail: ryo@theo.phys.sci.hiroshima-u.ac.jp

KAZUNORI KOHRI

Institute for Theory and Computation,
Harvard-Smithsonian Center for Astrophysics,
MS-51, 60 Garden Street, Cambridge, MA 02138, USA

AYA BAMBA

RIKEN (The Institute of Physical and Chemical Research) 2-1,
Hirosawa, Wako, Saitama 351-0198, Japan

TATSUO YOSHIDA

Faculty of Science, Ibaraki University,
Mito 310-8512, Japan

TORU TSURIBE AND FUMIO TAKAHARA

Department of Earth and Space Science, Osaka University,
Toyonaka 560-0043, Japan

We study the emission from an old supernova remnant (SNR) with an age of around 10^5 yrs. When the SNR age is around 10^5 yrs, hadron acceleration is efficient enough to emit TeV γ-rays at the shock of the SNR. The maximum energy of primarily accelerated electrons is so small that TeV γ-rays and X-rays are dominated by hadronic processes, π^0-decay and synchrotron radiation from secondary electrons, respectively. However, if the SNR is older than several 10^5 yrs, there are few high-energy particles emitting TeV γ-rays because of the energy loss effect and/or the wave damping effect occurring at low-velocity isothermal shocks. It is found that the ratio of TeV γ-ray (1–10 TeV) to X-ray (2–10 keV) energy flux can be more than $\sim 10^2$. Such a source showing large flux ratio may be a possible origin of recently discovered unidentified TeV sources.

1. Introduction

The most probable cosmic-ray accelerator in our Galaxy is the young supernova remnant (SNR). The detection of synchrotron X-rays from shells of young SNRs provides us the strong evidence for electron acceleration up to more than ~ 10 TeV[7,8,9]. So far, the evidence for hadron acceleration, however, has not yet obtained. High energy γ-ray observations may give us important information on the accelerated protons[17,12]. For example, TeV γ-rays are detected from the SNRs RX J1713.7−3946[14,2] and RX J0852.0−4622[15,5], which can be originated in either the decay of neutral pion, arising from the collision of high energy protons and interstellar matter, or CMB photons up-scattered by accelerated electrons. At present, the leptonic process is not yet ruled out.

Recently, a survey of the inner part of our Galaxy has revealed several new TeV γ-ray sources[1,3,4,6]. For some of them, no counterpart has been found in any other wave lengths yet. They should be galactic origin because all are located along the galactic plane. One of them, HESS J1303−631[4], was observed by *Chandra*, and no obvious counterpart is revealed[16]. Then, the flux ratio, $R_{\rm TeV/X} = F_\gamma(1-10~{\rm TeV})/F_X(2-10~{\rm keV})$, is more than ~ 2. On the other hand, young SNRs RX J1713.7−3946 and RX J0852.0−4622 have $R_{\rm TeV/X} = 0.51$ and 2.1, respectively. Although at present the lower limit of $R_{\rm TeV/X}$ for HESS J1303−631 is comparable to that of young SNRs, it may become much larger with forthcoming deeper X-ray observations. If so, unlike young SNRs, the unidentified TeV sources may show the evidence for hadron acceleration because inverse-Compton scenario requires unusually small magnetic field strength ($\ll 1~\mu$G).

The large value of $R_{\rm TeV/X}$ is expected for old SNRs. As the SNR ages, the shock velocity decreases. In general, primary electron acceleration is limited by synchrotron cooling. Then, the roll-off energy of electron synchrotron radiation is much smaller than that of typical young SNRs, so that the small synchrotron X-ray flux is expected[19]. It is also important to consider the association with a giant molecular cloud (GMC). Because of large volume, old SNRs may encounter the GMC. Here we study how large $R_{\rm TeV/X}$ can become for the single SNR. More detailed discussions are seen elsewhere[22], where the SNR-GMC interacting system is also investigated.

2. Evolution of SNR

We consider a simple analytical model of the shock dynamics of SNRs expanding into the uniform ambient medium with the density n_0. We assume the shock velocity v_s as a function of the SNR age $t_{\rm age}$ as $v_s(t_{\rm age}) = v_i$ for the free expansion phase ($0 < t_{\rm age} < t_1$), $v_s(t_{\rm age}) = v_i(t_{\rm age}/t_1)^{-3/5}$ for the Sedov-Taylor phase ($t_1 < t_{\rm age} < t_2$), and $v_s(t_{\rm age}) = v_i(t_2/t_1)^{-3/5}(t_{\rm age}/t_2)^{-2/3}$ for the radiative phase ($t_2 < t_{\rm age}$), where $t_1 = (3E/2\pi m_H n_0 v_i^5)^{1/3} = 2.1 \times 10^2 (E_{51}/n_0)^{1/3} v_{i,9}^{-5/3}$ yrs and $t_2 = 4 \times 10^4 E_{51}^{4/17} n_0^{-9/17}$ yrs, and $v_i = v_{i,9} 10^9$ cm s^{-1} and $E = E_{51} 10^{51}$ ergs are the initial velocity, and the initial energy of the ejecta, respectively. Note that for $t_{\rm age} > t_2$, $v_s(t_{\rm age})$ is rewritten as $v_s = 2.3 \times 10^7 E_{51}^{11/51} n_0^{-4/17} (t_{\rm age}/10^5 \text{ yrs})^{-2/3}$ cm s^{-1} and is independent of v_i. We adopt $E_{51} = v_{i,9} = n_0 = 1$ as a fiducial case. Then, we find $v_s = 1.1 \times 10^7$ cm s^{-1} at $t_{\rm age} = 3 \times 10^5$ yrs.

Before entering the radiative phase, the shock of an SNR is strong and adiabatic, so that the density compression ratio, r, is taken as 4. In the radiative phase, the cooling effect is important, and r is given as $r = M^2$, where $M \sim 20(t_{\rm age}/t_2)^{-2/3}$ is the Mach number[10].

3. Emission from an SNR

For the given SNR dynamics, the maximum energy of accelerated particles is calculated. The maximum energy of accelerated protons, $E_{\rm max,p}$, is determined by $t_{\rm acc} = \min\{t_{\rm age}, t_{\rm pp}\}$, while that of accelerated electrons, $E_{\rm max,e}$, is determined by $t_{\rm acc} = \min\{t_{\rm age}, t_{\rm synch}\}$, where $t_{\rm age}$, $t_{\rm acc}$, $t_{\rm synch}$, and $t_{\rm pp}$ are the age of the SNR, the acceleration time scale, the synchrotron loss time scale, and the pion-production loss time scale, respectively. Particles get their energy via diffusive shock acceleration. The acceleration time is given as $t_{\rm acc} = 20hcE_{\rm max}/eB_{\rm d}v_s^2$, where $h = 0.05r(f + rg)/(r - 1)$ and r is the compression ratio, and f and g are functions of the shock angle θ and gyro-factors $\eta_{\rm u}$ and $\eta_{\rm d}$[21]. The downstream magnetic field is given by $B_{\rm d} = rB_{\rm ISM}$, where $B_{\rm ISM} = 10B_{{\rm ISM},-5}$ μG is the ISM magnetic field and we adopt $B_{{\rm ISM},-5} = 1$ as a typical value. As long as $t_{\rm age} < t_{\rm pp}$, the proton acceleration is age-limited, while the electron acceleration is loss-limited when $t_{\rm age} > 10^3$ yrs.

The wide-band radiation spectrum is simulated for given $E_{\rm max,p}$ and $E_{\rm max,e}$ at a certain $t_{\rm age}$. The power-law index of accelerated particles is fixed as $p = 2.2$. We assume the electron-proton ratio, that is the ratio of the number density of electrons to that of protons at relativistic regimes, of 1×10^{-3}. We consider radiation processes from primary electrons: the synchrotron, inverse-Compton (IC), and bremsstrahlung emissions, and from primary protons: π^0 decay γ-rays and synchrotron radiation from secondary

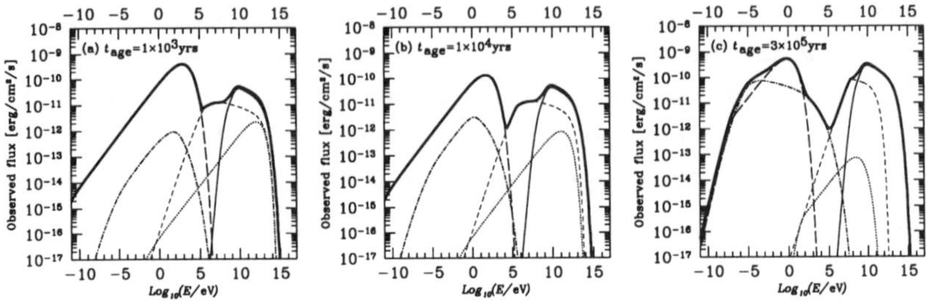

Figure 1. Emission spectra of a single SNR. The thick-solid line shows the total non-thermal flux. The thin-solid, dot-dashed, long-dashed, dotted, and short-dashed lines represent the fluxes of π^0-decay γ-rays, synchrotron radiation from secondary electrons produced by charged pion, synchrotron radiation from primary electrons, inverse-Compton radiation, and bremsstrahlung emission, respectively. The SNR age is (a) 1×10^3 yrs, (b) 1×10^4 yrs, and (c) 3×10^5 yrs, respectively.

electrons arising from the decay of charged pions. Primarily accelerated electrons produce synchrotron emission with the roll-off frequency $\nu_{\rm roll}$[18]. In the loss-limited case, $\nu_{\rm roll}$ can be rewritten as

$$\nu_{\rm roll} \sim 3.2 \times 10^{14} h^{-1} v_{s,7}^2 \ {\rm Hz} \quad , \tag{1}$$

where $v_{s,7} = v_s/10^7 {\rm cm\ s^{-1}}$. When h is around unity, we can reproduce the observed value of $\nu_{\rm roll}$ for young SNRs[7,8,9]. Hence we adopt $h = 1$ as a fiducial value.

When $t_{\rm age} = 1 \times 10^3$ yrs (Fig. 1a), we find $E_{\rm max,p} = 96$ TeV, $E_{\rm max,e} = 27$ TeV and $\nu_{\rm roll} = 4.8 \times 10^{17}$ Hz for the fiducial parameters, and nonthermal X-rays are dominated by the synchrotron emission from primary electrons. The flux ratio is $R_{\rm TeV/X} = 7.6 \times 10^{-2}$, which is consistent with observations for young SNRs. When $t_{\rm age} = 1 \times 10^4$ yrs (Fig. 1b), we find $E_{\rm max,p} = 61$ TeV and $E_{\rm max,e} = 6.9$ TeV for the fiducial parameters. The flux ratio becomes large, $R_{\rm TeV/X} = 1.6$.

When the SNR is in the radiative phase, the compression ratio is large, and strong downstream magnetic field is obtained. The maximum energies are calculated as

$$\begin{aligned} E_{\rm max,p} &= 1.2 \times 10^2 \ h^{-1} E_{51}^{9/34} n_0^{-8/17} B_{\rm ISM,-5} v_{s,7}^{3/2} \ {\rm TeV} \\ E_{\rm max,e} &= 0.30 \ h^{-1/2} E_{51}^{1/34} n_0^{1/17} B_{\rm ISM,-5}^{-1/2} v_{s,7}^{-1/2} \ {\rm TeV} \quad . \end{aligned} \tag{2}$$

Figure 1c shows the spectrum for $t_{\rm age} = 3 \times 10^5$ yrs with fiducial parameters. Then the flux ratio is $R_{\rm TeV/X} = 23$. Since $E_{\rm max,e}$ is small,

TeV γ-rays via IC and bremsstrahlung are suppressed, and therefore, the TeV γ-rays come from the π^0-decay. The roll-off frequency of synchrotron radiation from primary electrons is so small, $\nu_{\rm roll} \sim 4.0 \times 10^{14}$ Hz, that the secondary synchrotron radiation dominates the X-ray band. If $t_{\rm age} > 3 \times 10^5$ yrs, i.e., $v_s < 1.1 \times 10^7$ cm s^{-1}, then, the ionization around the shock front is incomplete[20] and upstream ion-neutral Alfvén wave damping places significant restriction on shock acceleration[13,11]. Once the shock acceleration becomes inefficient, there are few high-energy protons emitting TeV γ-rays around the SNR shell, because they escape the SNR shell due to the diffusion; assuming the Bohm diffusion, the escape time for a particle with an energy $E_{\rm cr} = 10E_{\rm cr,10TeV}$ TeV is estimated as $t_{\rm esc} \sim 4 \times 10^5 \eta^{-1}(B_{\rm d}/2 \times 10^2 \mu{\rm G})E_{\rm cr,10TeV}^{-1}\Delta_{\rm 3pc}^2$ yrs, where η and $\Delta = 3\Delta_{\rm 3pc}$ pc are the gyro factor and the thickness of the shell, respectively. Therefore, when $t_{\rm age} > 3 \times 10^5$yrs, TeV γ-rays are significantly suppressed. Note that for different parameter sets[22], $R_{\rm TeV/X}$ becomes larger than $\sim 10^3$.

4. Discussions

For an old SNR in the radiative phase, the maximum energy of primary electrons is so small that emissions via leptonic processes are vanishingly small both in the X-ray and TeV γ-ray bands. On the other hand, because the strong magnetic field arises downstream of the radiative shock with large compression ratio, there are accelerated protons with energies of $\sim 10^2$ TeV, which is as large as the maximum proton energy for young SNRs. These particles emit the TeV γ-rays via π^0-decay and synchrotron X-rays from secondary electrons generated by charged pions. Such sources may be an origin of recently discovered unidentified TeV sources, and give us the evidence for hadron acceleration. Actually detected TeV sources have the energy flux $F_\gamma(1 - 10 \text{ TeV}) \sim 10^{-12}$–$10^{-11}$erg cm^{-2}s^{-1}. Hence if $R_{\rm TeV/X} < 10^{2-3}$, then $F_X(2 - 10 \text{ keV}) > 10^{-14}$ erg cm^{-2}s^{-1}. Such diffuse, extended source can be detected with current X-ray telescopes.

We can roughly estimate the expected number of observed TeV sources, $N_{\rm TeV}$, which have observed flux larger than 3% of the Crab flux (6×10^{-13} cm^{-2}s^{-1} above 1 TeV)[3,6]. The result is $N_{\rm TeV} \sim 5 \times 10^2 r_{-2} n_{1.4} E_{p,50}$, where r_{-2}, $n_{1.4}$, and $E_{p,50}$ are the SN explosion rate in units of 10^{-2} yr^{-1}, the density of the shell in units of $10^{1.4}$cm^{-3}, and the energy of accelerated protons stored in the SNR in units of 10^{50}ergs, respectively[22]. At present, the number of unidentified TeV sources is a few. If the number of actually detected TeV sources does not increase, then $E_{p,50} \ll 10^{-2}$ is implied, which might suggest the escape of high-energy particles from the SNR begins until $t_{\rm age} \sim 10^5$ yrs.

Acknowledgments

K.K. was supported by NSF grant AST 0307433. A.B. is supported in part by the Grant-in-Aid for young Scientists (B) of the Ministry of Education, Culture, Sports, Science and Technology (No. 17740183).

References

1. F. Aharonian et al., *Astro. Astrophys.* **393**, L37 (2002)
2. F. Aharonian et al., *Nature* **432**, 75 (2004)
3. F. Aharonian et al., *Science* **307**, 1938 (2005)
4. F. Aharonian et al., *Astro. Astrophys.* **439**, 1013 (2005)
5. F. Aharonian et al., *Astro. Astrophys.* **437**, L7 (2005)
6. F. Aharonian et al., *Astrophys. J.* **636**, 777 (2005)
7. A. Bamba, R. Yamazaki, M. Ueno and K. Koyama, *Astrophys. J.* **589**, 827 (2003)
8. A. Bamba, R. Yamazaki, T. Yoshida, T. Terasawa and K. Koyama, *Astrophys. J.* **621**, 793 (2005)
9. A. Bamba, R. Yamazaki and J. S. Hiraga, *Astrophys. J.* **632**, 294 (2005)
10. J. M. Blondin, E. B. Wright, K. J. Borkowski and S. P. Reynolds, *Astrophys. J.* **500**, 342 (1998)
11. A. M. Bykov et al., *Astrophys. J.* **538**, 203 (2000)
12. L. O' C. Drury, F. Aharonian and H. J. Völk, *Astro. Astrophys.* **287**, 959 (1994)
13. L. O' C. Drury, P. Duffy and J. G. Kirk, *Astro. Astrophys.* **309**, 1002 (1996)
14. R. Enomoto et al., *Nature* **416**, 823 (2002)
15. H. Katagiri et al., *Astrophys. J.* **619**, L163 (2005)
16. R. Mukherjee and J. P. Halpern, *Astrophys. J.* **629**, 1017 (2005)
17. T. Naito and F. Takahara, *J. Phys. G: Part. Phys.* **477**, 486 (1994)
18. S. P. Reynolds and J. W. Keohane, *Astrophys. J.* **525**, 368 (1999)
19. S. J. Sturner et al., *Astrophys. J.* **490**, 619 (1997)
20. J. M. Shull and C. F. McKee, *Astrophys. J.* **227**, 131 (1979)
21. R. Yamazaki et al., *Astro. Astrophys.* **416**, 595 (2004)
22. R. Yamazaki et al., *preprint* astro-ph/0601704

THE CALET PROJECT FOR INVESTIGATING HIGH ENERGY UNIVERSE

SHOJI TORII
for the CALET Collboration

Advanced Research Institute for Science and Engineering, Waseda University, 3-4-1,Okubo, Shinjuku-ku, Tokyo, 169-8555, Japan

The CALorimetric Electron Telescope, CALET, mission is proposed for the Japanese Experiment Module Exposed Facility, JEM-EF, of the International Space Station. The mission goal is to reveal high-energy phenomena in the universe by carrying out a precise measurement of the electrons in 1 GeV - 10 TeV and the γ--rays in 20 MeV - several TeV. The CALET has a unique capability to measure the electrons and gamma-rays over 1 TeV since the hadron rejection power might be as much as $\sim 10^6$ and the energy resolution of electromagnetic particles better than a few % over 100 GeV. Therefore, it is promising to detect the change of energy spectra and the γ-ray line expected from candidates of the dark matter. We are expecting to launch the CALET around 2012 by the Japanese H-II Transfer Vehicle, HTV, and to observe for three years.

1. Introduction

We propose CALET as an instrument to observe very high energy electrons and γ-rays on the Japanese Experiment Module Exposure Facility, JEM-EF, on ISS. The objective of the CALET mission is to explore a new frontier at higher energies for the origin of cosmic-rays (CR), the propagation of CR and to search for dark matters. We will measure electrons from 1 GeV to ~10 TeV and γ-rays from 20 MeV to several TeV, free from the hadron backgrounds, with an excellent energy resolution beyond 100 GeV. We are considering using CALET to measure protons and heavy nuclei from several 10 GeV to ~1000 TeV range.

CALET is designed on the basis of experience in balloon observations [1,2,3]. It is a calorimeter, combining an imaging part and a total absorption part, and it will have an excellent capability for proton rejection, $\sim 10^6$, which is necessary to select electrons and γ -rays in the TeV region. It is also suitable for a precise measurement of the energy spectrum, since the energy resolution is better than a few % for energies greater than 100 GeV.

2. Observation and Scientific Objective

2.1. *ELectrons*

As is well known, high-energy electrons lose their energy (per unit time) in proportion to square of the energy, by synchrotron radiation and inverse-Compton scattering. Therefore, in the TeV region, only the electrons at a distance within 1 kpc from the sources and with an age less than $\sim 10^5$ years, can reach the Earth. Since the number of such possible sources are very limited, the energy spectrum observed might have a characteristic structure [4], and the arrival directions are expected to show a detectable anisotropy [5]. The diffusion process in the Galaxy also strongly affects the electron flux. The energy spectrum could, therefore, give a direct evidence of nearby cosmic ray sources and a knowledge of particle diffusion characteristics in interstellar space.

Among the candidates, Vela is the most promising as an observable nearby source since both the distance, ~0.25 kpc, and the age, $\sim 10^4$ years, satisfy the constraints listed above. Figure 1 shows the expected energy spectra calculated by a diffusion model under different assumptions. Several parameters are chosen to reproduce a spectrum consistent with the present data below 100 GeV. These include the injection spectrum of $E^{-2.4}$, the total energy of 10^{48} erg per SN, the size of the Galactic disc, the diffusion coefficient and the energy loss rate [6].

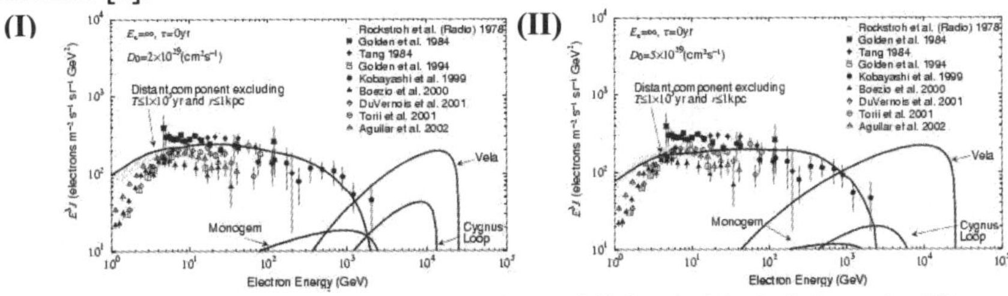

Figure 1. Expected energy spectrum of electrons from a diffusion model calculation under different assumptions, comparing with the present data. The assumptions are following: (I) no cut-off energy ($E_c=\infty$), instant acceleration time (ΔT=0 yr) and the diffusion coefficient at 1 TeV of $D_0= 2 \times 10^{29}$ cm^2/s. (II) $D_0 = 5 \times 10^{29}$ cm^2/s.

There are some expectations that positrons have a line signature around several 100 GeV from the dark matter candidates; Neutralinos in the SUSY theory [7] and Kaluza-Klein (extra dimensional) particles [8]. Although CALET has no capability to distinguish positive and negative charges, an excess of positron and electrons might be detected due to the excellent energy resolution of the instrument and the very high statistics as shown in Figure 2.

2.2. *Gamma-rays*

Since CALET has a large field of view (~2 sr) and a wide effective area (~ 1m^2 @ 10 GeV) for γ -rays , it can observe the whole sky without any attitude control. The coverage per one day is ~70 % and the entire sky can be observed in 20 days. The observation period for point sources is 48 days on average per one year. Most of the GeV sources detected by EGRET have not been observed in the TeV region by Air Cherenkov observations although the detection efficiency should be enough for the case that there is not a break in the spectrum.

CALET will have the ability to detect γ-rays from point sources to fill the energy gap between the EGRET and Air Cherenkov observations. In Table 1, the performance is compared with EGRET and GLAST. The most important targets of observation include: Galactic andextra-Galactic diffuse components, supernova remnants, pulsars, AGNs, and γ-ray bursts. In particular, the diffuse Galactic component of γ-rays above 10 GeV is strongly related to the electron energy spectrum since the γ-rays are mainly produced by inverse-Compton scattering with electrons near the source region [9].

Table 1. Performance for gamma rays compared with other instruments.

	EGRET	GLAST (SRD)	CALET
Energy Range (GeV)	0.02–30	0.02–300	$0.02 - 1\times10^4$
Effective Area (cm^2)	1500	>8000	7.9×10^3 (@10GeV) 4.6×10^3 (>100GeV)
F.O.V. (sr)	0.5	>2	0.5–1.8
Angular Res. (deg)	5.8 (@100MeV)	< 3.5 (@100MeV) < 0.15 (>10GeV)	5.0 (@100MeV) < 0.24 (> 10GeV)
Energy Res. (%)	10	< 10	$7/\sqrt{E/10\text{GeV}}$
Point Source Sensitivity ($cm^{-2}s^{-1}$) (> 100MeV)	5×10^{-8}	6×10^{-9}	1×10^{-8}

Because the energy resolution improves at higher energies, CALET can precisely measure the change in the γ-ray energy spectrum index around from 10 to ~100 GeV. Such changes might be brought by the decrease of acceleration power and/or the absorption by starlight photons in the extra-Galactic space. Finally, observations of gamma-ray lines from the annihilation of SUSY particles [10] are also feasible if such particles exist in sufficient numbers as shown in Figure 3.

2.3. *Protons and Nuclei*

Although CALET is an electromagnetic calorimeter, it can detect protons up to 1000 TeV as the absorber thickness corresponds to 1.8 mean free path for protons as described in next section. Determining the energy spectrum of protons in proximity of the Knee region is very important for resolving the acceleration limits of protons and heavy nuclei. Further, validity of the leaky box model will be tested up to the energy region of 10 TeV by measuring the cosmic ray secondary to primary ratio energy dependence.

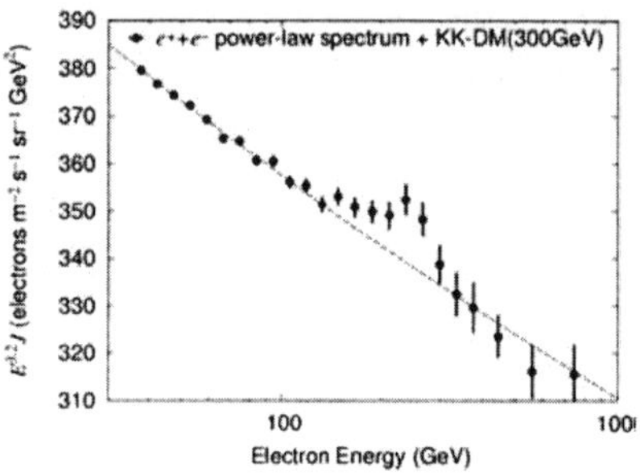

Figure 2. Simulated energy spectrum of e+ + e− power law spectrum with Kaluza-Klein dark matter annihilation for 300 GeV mass.

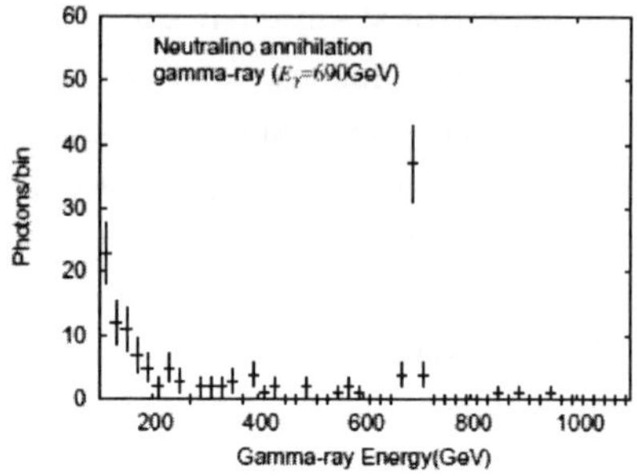

Figure 3. Simulated gamma-ray line at 690 GeV from neutralino annihilation, including the background of the Galactic diffuse emission.

3. Detector

3.1. *Detector Concept*

CALET is a combination of an imaging calorimeter, IMC, with a total absorption calorimeter, TASC. The IMC is used for identification of the incident particle and energy measurement below 10 GeV, while the TASC is for proton rejection in the TeV region and for energy measurements above 10 GeV. The detector weight is nearly 1,760 kg and the effective geometrical factor (for electrons) is ~1 m^2 sr. A schematic side view of the CALET detector is shown in Figure 4.

The IMC consists of 17 layers of lead plates each separated by 2 layers of 1mm square cross section scintillating fiber (SciFi) belts arranged in the x and y direction and is capped by an additional x,y SciFi layer pair. The dimension of the IMC is about 100 cm by 100 cm. While the total thickness of the IMC is 4 radiation length (X_0) , and about 0.13 proton interaction lengths (λ), the first 10 lead – SciFi layers sample the particle at 0.1 X_0, followed by 5 layers that are 0.2 X_0 thick and finally 2 layers that are 1 X_0 deep. This provides the precision necessary to 1) separate the incident particle from backscattered particles, 2) precisely determine the starting point for the electromagnetic shower, and 3) identify the incident particle.

The TASC measures the development of the electromagnetic shower to 1) determine the total energy of the incident particle and 2) separate electrons and gamma-rays from hadrons. The TASC is composed of 14 layers of Bismuth

Germanate (BGO) "logs" where each log has dimensions of 2.5cm×2.5cm× 35cm. There are 56 such logs in each layer. Alternate layers are orientated 90^0 to each other to provide an x,y coordinate for tracking the shower core. The total area of TASC is about 0.5 m^2 and the vertical thickness is 32 X_0 and 1.6 λ .

For low energy gamma-rays (<10 GeV), the IMC is covered by plastic scintillators for anti-coincidence with hadrons. A pixelated silicon detector module will be placed at the top of the IMC for having sufficient charge separation capability and dynamic range to identify relativistic nuclei in the range from proton to Iron and above. It can also afford to identify precise position of an incident particle among the copious backscattered particles at higher energies. Figure 5 presents the conceptual structure of CALET.

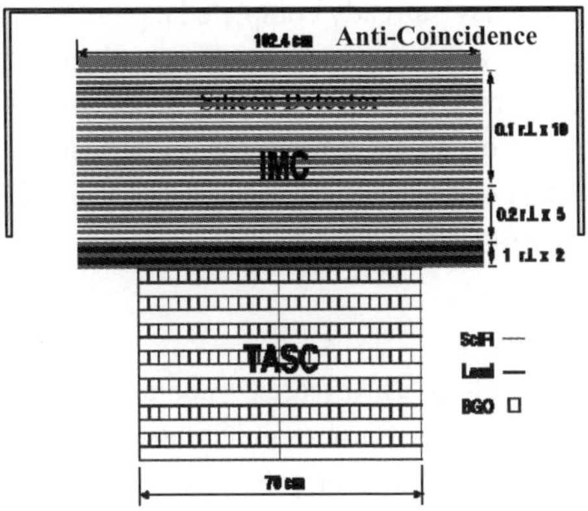

Figure 4. Schematic side view of CALET detector .

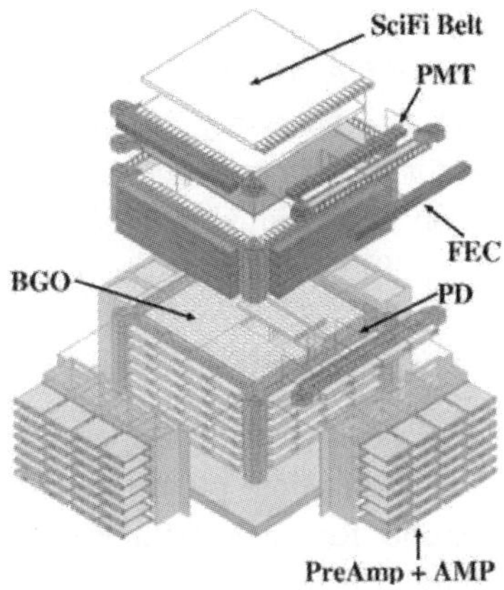

Figure 5. Conceptual structure of CALET.

3.2. *Accommodation Study for JEM*

The CALET will be launched by a Japanese carrier, HII Transfer Vehicle (HTV), and attached to the EFU #9, which is capable to maintain a heavy payload up to 2,500 kg in mass and has a wider field of view, 45 degrees. Figure 6 shows a schematic view of the CALET payload on ISS/JEM. The main structure of CALET is designed by adopting an interface structure of a usual exposed facility.

The structure, therefore, includes a pallet to sustain the detector, which is used both for launching by HTV and for attaching to JEM. The structure was optimized to meet the requirements from the ISS for both the vibration condition and the heat condition.

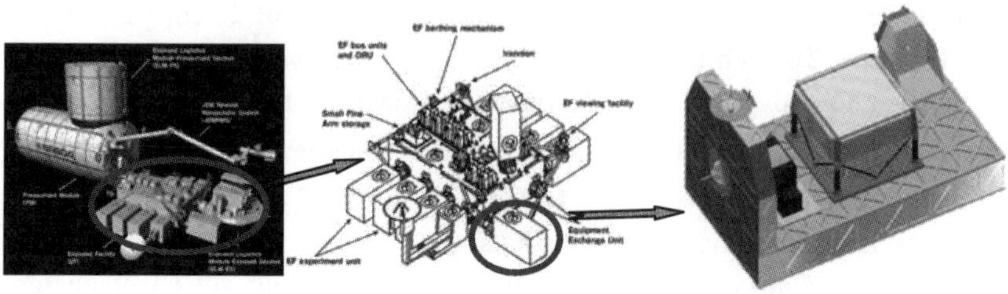

Figure 6. A schematic view of CALET on the JEM-EF platform.

4. Summary and Future Prospect

The CALET mission is proposed to perform a crucial observations of electrons and γ-rays at the high energy frontier. We have already completed a pre-phase A study for CALET within the last 3 years, and have successfully developed the electronics necessary to read-out the SciFi and the BGO. To confirm the detector performance for the experiment on space station, the flight test by a balloon is scheduled in 2006. We expect to begin operations on the ISS /JEM around 2012 and the mission life is supposed to be three-years.

Acknowledgments

This work is partially supported by Ground-based Research Announcement for Space Utilization promoted by Japan Space Forum andGrants-in-Aid for Scientific Research from the Ministry of Education , Science, Sports and Culture in Japan.

References

1. S.Torii, *et al., ApJ* , **559**, 973 (2001).
2. S.Torii, *et al., NIM* **A 452** 81 (2000).
3. K.Kasahara, *et al., Phys. Rev.D* **66,** 052004 (2002).
4. J.Nishimura, *et al., ApJ* , **238**, 394 (1980).
5. C.S.Shen and C.Y.Mao, *Astrophysical Letters*, **9**, 169 (1971).
6. T.Kobayashi, *et al., ApJ*, **601**, 340 (2004).
7. L.Bergstrom, *et al., Phys. Rev. D*, **63**, 083515 (2001)
8. H.C.Cheng, et al. , *Phys. Rev. Lett.* **89**, 211301 (2002).
9. M.Pohl and J.A.Esposito, *et al., Ap J,* **507**, 327 (1998)
10. L.Bergstrom, *et al., Phys. Rev. D* **63**, 083515 (2001).

INITIAL RESULTS FROM SUZAKU

TADAYUKI TAKAHASHI, KAZUHISA MITSUDA

Department of High Energy Astrophysics,
Institute of Space and Astronautical Science, JAXA,
3-1-1 Yoshinodai, Sagamihara, Kanagawa 229-8510, Japan
**E-mail: takahasi@astro.isas.jaxa.jp*
www.astro.isas.jaxa.jp

HIDEYO KUNIEDA

Department of Astrophysics, Nagoya University,
Furo-cho, Chikusa, Nagoya 464-8602 , Japan

ON BEHALF OF THE SUZAKU TEAM

Suzaku is the fifth in the series of Japanese astronomy satellites devoted to observations of celestial X–ray sources launched on a Japanese M-V rocket on July 10, 2005. *Suzaku* features the excellent X–ray sensitivity, with high throughput over a broad-band energy range of 0.2 to 600 keV. *Suzaku*'s broad bandpass, low background, and good CCD resolution makes it a unique tool capable of addressing a variety of outstanding problems in astrophysics.

Keywords: X-ray Astronomy, Hard X-ray Astronomy, AGN, Galactic Center

1. Introduction

Suzaku is the fifth in the series of Japanese astronomy satellites devoted to observations of celestial X–ray sources, following the highly successful Hakucho, Tenma, Ginga and ASCA satellites. Like ASCA, *Suzaku* is a joint Japanese-US mission, developed by the Institute of Space and Astronautical Science (part of the Japan Aerospace Exploration Agency, ISAS/JAXA) in collaboration with the National Aeronautics and Space Administration's Goddard Space Flight Center (NASA/GSFC) and many other institutions[1].

Suzaku was launched on a Japanese M-V rocket on July 10, 2005 from the JAXA Uchinoura Space Center. *Suzaku* is placed in a near-circular orbit with an apogee of 568 km, an inclination of 31.9 degrees, and an orbital period of about 96 minutes. With this constraint, most targets will be occulted by the Earth for about one third of each orbit, but some objects near the orbital poles can be observed nearly continuously. The current projection is that the observing efficiency of the satellite will be about 43%.

Despite initial success, on August 8, 2005 a thermal short between the helium and neon tanks resulted in the liquid helium coolant venting to space, leaving the X–Ray Spectrometer (XRS) inoperable[1,2]. However, the X–ray Imaging Spectrometer (XIS)[3] and Hard X-ray Detector (HXD)[4,5] are all working well. As a result, *Suzaku* retains its excellent X–ray sensitivity, with high throughput over a broad-band energy range of 0.2 to 600 keV. *Suzaku*'s broad bandpass, low background, and good CCD resolution makes it a unique tool capable of addressing a variety of outstanding problems in astrophysics.

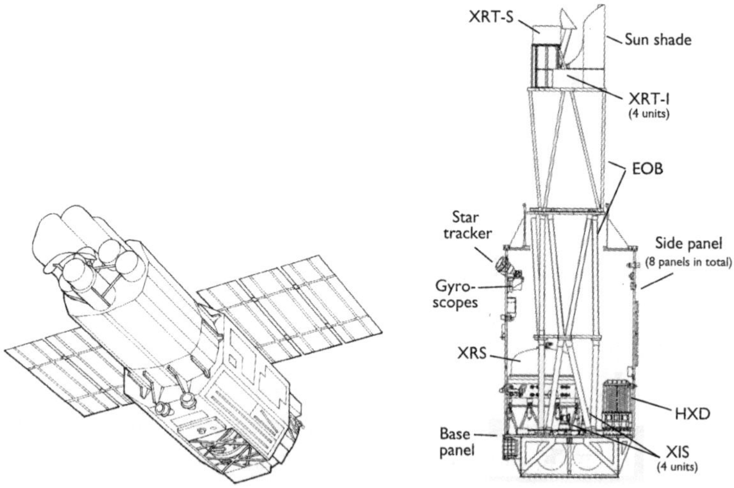

Figure 1. Schematic pictures of the the *Suzaku* satellite. The five sets of X-ray mirrors are mounted on top of the EOB and five focal plane detectors and a hard X-ray detector are mounted on the base panel of the spacecraft. The spececraft length is 6.5 m along the telescope axis after the deployment of the EOB.

2. Scientific Instrumentation of Suzaku

2.1. *XRT*

Suzaku has five light-weight thin-foil X–Ray Telescopes (XRTs)[6]. One is for the XRS and other four (XRT-Is) are for the XISs. The XRTs have been

S/C	Orbit Apogee	568 km
	Orbital Period	96 minutes
	Observing Efficiency	$\sim$ 45%
XRT	Focal length	4.75 m
	Field of View	17′ at 1.5 keV
		13′ at 8 keV
	Plate scale	0.724 arcmin/mm
	Effective Area	440 cm^2 at 1.5 keV
		250 cm^2 at 8 keV
	Angular Resolution	2′ (HPD)
XIS	Field of View	17.8′ × 17.8′
	Bandpass	0.2–12 keV
	Pixel grid	1024×1024
	Pixel size	24 μm× 24 μm
	Energy Resolution	$\sim$ 130 eV at 6 keV
	Effective Area	330 cm^2 (FI), 370 cm^2 (BI) at 1.5 keV
	(incl XRT-I)	160 cm^2 (FI), 110 cm^2 (BI) at 8 keV
	Time Resolution	8 s (Normal mode), 7.8 ms (P-Sum mode)
HXD	Field of View	4.5° × 4.5° ($\gtrsim$ 100 keV)
	Field of View	34′ × 34′ ($\lesssim$ 100 keV)
	Bandpass	10 – 600 keV
	– PIN	10 – 60 keV
	– GSO	30 – 600 keV
	Energy Resolution (PIN)	$\sim$ 3.0 keV (FWHM)
	Energy Resolution (GSO)	$7.6/\sqrt{E_{MeV}}$ % (FWHM)
	Effective area	$\sim$ 160 cm^2 at 20 keV, $\sim$ 260 cm^2 at 100 keV
	Time Resolution	61 μs
HXD-WAM	Field of View	2π (non-pointing)
	Bandpass	50 keV – 5 MeV
	Effective Area	800 cm^2 at 100 keV / 400 cm^2 at 1 MeV
	Time Resolution	31.25 ms for GRB, 1 s for All-Sky-Monitor

developed jointly by NASA/GSFC, Nagoya University, Tokyo Metropolitan University, and ISAS/JAXA. These are grazing-incidence reflective optics consisting of compactly nested, thin conical elements. Because of the reflectors' small thickness, they permit high density nesting and thus provide large collecting efficiency with a moderate imaging capability in the energy range of 0.2-12 keV, all accomplished in telescope units under 20 kg each. The new improvement from the ASCA telescope is the addition of a pre-collomater which reduce stray lights from nearby bright sources located outside the FOV of the mirrors.

The angular resolutions of the XRTs range from 1.8′ to 2.3′, expressed in terms of half-power diameter, which is the diameter within which half of the focused X–ray is enclosed. The angular resolution does not significantly depend on the energy of the incident X–ray in the energy range of *Suzaku*, 0.2-12 keV. The effective areas are typically 440 cm^2 at 1.5 keV and 250 cm^2 at 8 keV. The focal lengths are 4.75 m for the XRT-I. Individual XRT quadrants have their component focal lengths deviated from the design values by a few cm. The optical axes of the quadrants of each XRT are aligned within 2′ from the mechanical axis. The field of view for XRT-Is is about 17′ at 1.5 keV and 13′ at 8 keV.

2.2. *XRS*

In the XRS, the energy of an incoming X-ray photon (between 0.3–12 keV) is measured precisely by detecting the temperature rise resulting from the absorption of the photon in the micro-calorimeter. The XRS onboard Suzaku is the results of long-term efforts of scientists who joined the project. During the initial operation, 60 mK was achieved with the Solid Ne/Liq. He/ADR in the XRS system, which is the lowest temperature achieved in space. With the calibration source of Mn K, we successfully obtained energy resolution of 7 eV averaged over the 6x6 pixel array in orbit. The reason of the failure of the XRS, together with a performance we obtained in the orbit is described in Kelley et al. [2] . The excellent energy resolution of ∼6 eV of the X-ray micro-calorimeter array employed in the X-ray Spectrometer (XRS) of Suzaku was expected to give a significant contribution for understanding the high energy universe. Therefore, the loss of the XRS is a tragic event for all of us and, at the same time, it is of great importance to recover the XRS in very near future mission.

2.3. *XIS*

X-ray sensitive silicon charge-coupled devices (CCDs) are a key device for the X-ray astronomy. The four *Suzaku* XISs utilize X-ray CCDs [3] by following the great success of the *ASCA* satellite in which X-ray CCD is first introduced[7]. These detectors are named XIS-S0, S1, S2 and S3, each located in the focal plane of an X–ray Telescope. Each CCD camera has a single CCD chip with an array of 1024 × 1024 picture elements ("pixels"), and covers an 18′ × 18′ region on the sky. Each pixel is 24 μm square, and the size of the CCD is 25 mm × 25 mm. One of the XISs, XIS-S1, uses a back-side illuminated CCDs, while the other three use front-side illuminated CCDs. The XIS has been partially developed at MIT (CCD sensors, analog electronics, thermo-electric coolers, and temperature control electronics), while the digital electronics and a part of the sensor housing were developed in Japan, jointly by Kyoto University, Osaka University, Rikkyo University, Ehime University, and ISAS.

2.4. *HXD*

The Hard X–ray Detector is a non-imaging, collimated hard X–ray scintillating instrument sensitive in the ~ 10 keV to ~ 600 keV band[4,5]. It has been developed jointly by the University of Tokyo, Aoyama Gakuin University, Hiroshima University, ISAS/JAXA, Kanazawa University, Osaka University, Saitama University, SLAC, and RIKEN. Its main purpose is to extend the bandpass of the *Suzaku* observatory to the highest feasible energies, thus allowing broad-band studies of celestial objects. While the bandpass of previous Japanese X-ray satellites was typically below ~20 keV, where thermal emission predominates, the energy range of 10 - several 100 keV is where the radiation from the high-energy celestial sources is mainly non-thermal. In addition to the main detector parts which consists of 16 well-type units, tight active shielding of the HXD results in a large arrays of guard counters surrounding the main detector parts. These anticonicidence counters work as an excellent γ-ray burst monitor with limited angular resolution ($\sim 5^{\circ}$).

Since the background level sets the limit of the sensitivity in the hard X-ray region, the HXD has been designed to achieve an extremely low in-orbit background ($\sim 10^{-4}$ c s^{-1} cm^{-2} keV^{-1})[4]. A detailed pre-flight calibration confirm that the performance of the HXD meets the design goal of the experiment, including the threshold of ~ 10 keV of the PIN diode, and the very low on-ground background of 1 – 5 $\times 10^{-5}$ counts sec^{-1} keV $^{-1}$ cm $^{-2}$[5].

3. Scientific Capabilities and Initial Results

Suzaku was designed to be highly complementary to the two large missions which were already in orbit at launch, XMM-Newton [9] and Chandra [8]. The key feature of *Suzaku*, the high-sensitivity wide-band X-ray spectroscopy all in one observatory, was confirmed with the ~ 8 months of the observations in the performance verification (PV) phase. During the PV phase, we have observed more than 70 targets (see the list at http://www.astro.isas.ac.jp/suzaku/accept/swg/) according to the selection by Science Working Group. It is characterized by low background and good energy resolution, in particular a good line spread function in low energy range.

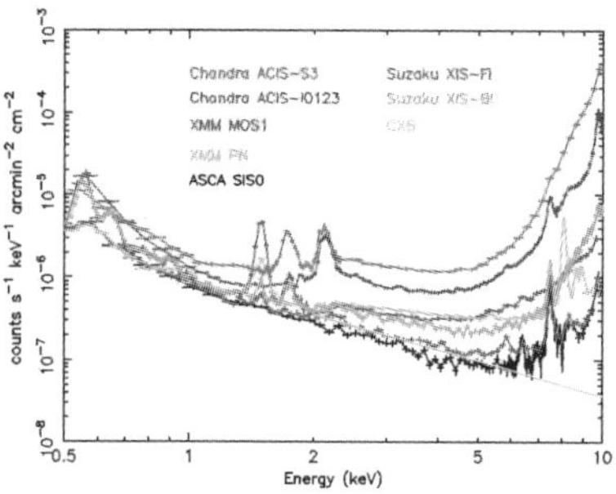

Figure 2. XIS background counting rate as a function of energy. The background rate was normalized with the effective area and the field of view, which is a good measure of sensitivity determined by the background for spatially extended sources. The background rate of *ASCA*, *Chandra*, and *XMM-Newton* are also shown for comparisons[1].

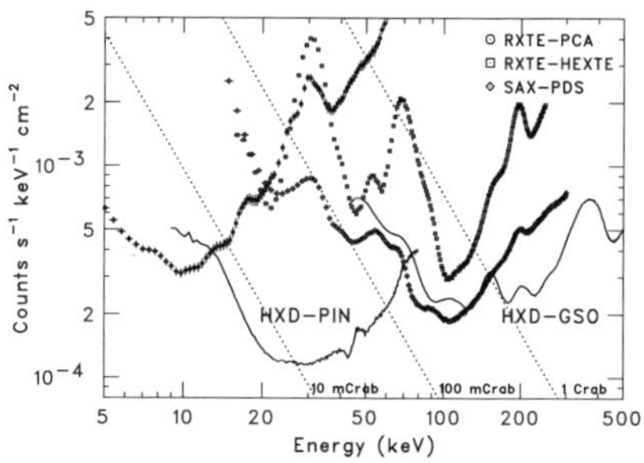

Figure 3. Background counting rate of HXD as a functions of energy. The background rate was normalized by the effective area. Background of *INTEGRAL*. *Beppo-SAX*, and *RXTE* are shown for comparison. The intensity of the Crab nebula is also shown[1].

In figure 2, we show the XIS background counting rate as a function of energy in a 0.5 to 10 keV range. Here the background is normalized by the effective area and the field of view. This is a reasonable measure of sensitivity determined by background for spatially extended sources. Among the instrument listed here the *ASCA* SIS has the lowest background, and *Suzaku* XIS (BI and FI CCD) has a low background comparable to *ASCA* SIS. Figure 3 shows the HXD background counting rate as a function of energy for 10 to 400 keV region. The sensitivity in this energy region is determined by the accurately of background estimation. The background rate of *Suzaku* is lowest among the existing missions at most of energies. At present we can reproduce the background spectrum with an accuracy of 5% of the background level. In a near future after we have accumulated more data, we expect to reach 1–3% level. Another large advantage of using Suzaku is the good energy response of the CCD's to low energy (<1

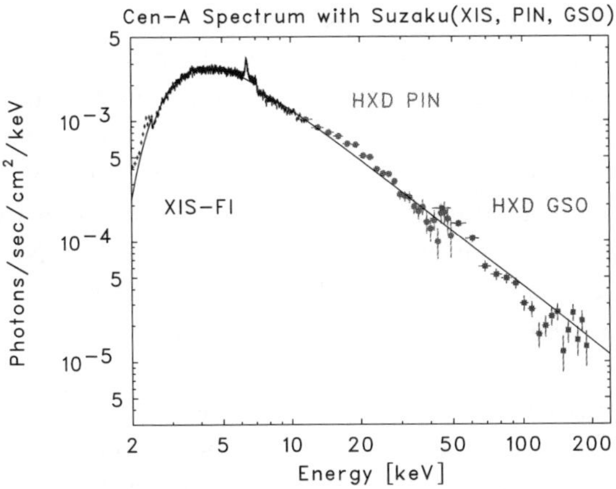

Figure 4. Energy spectrum of Cen A radio galaxy obtained with *Suzaku*. The spectrum was obtained with XIS FI CCDs (XIS-0, 2, 3) and HXD and deconvolved using an absorbed power-law model[1].

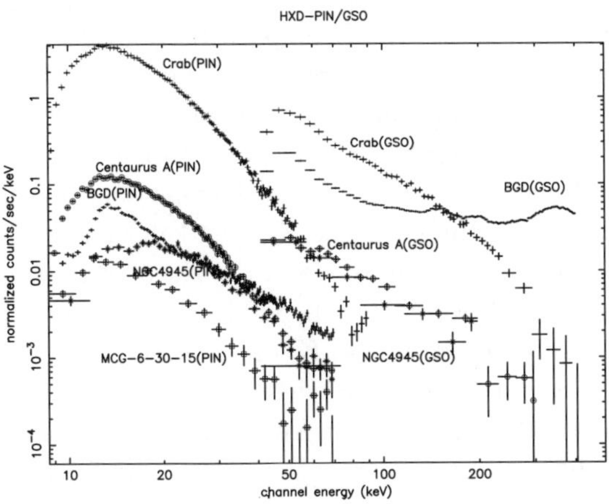

Figure 5. HXD PIN/GSO spectra of the Crab nebula and some AGNs obtained in the early test phase[15].

keV) X-rays. The line spread function of Suzaku CCD is very symmetric in shape even in the low energy range below 1 ked, in other words it has only a very low low-pulse height tail. This makes it possible to perform high signal to noise line spectroscopy in low energy bands.

The first publication from *Suzaku* was for the bright planetary nebula BD+30°3639. At the end of the life, light stars loose their outer shell through stellar winds to create hot plasma bubbles in the interstellar medium. Suzaku observed BD+30°3639, one of the brightest planetary nebulae in X-rays. Thanks to the large effective area in low energy X-ray band and high energy resolution of the XIS, the K- lines from C VI, O VII, and O VIII were resolved for the first time, and C/O, N/O, and Ne/O abundance ratios were determined. The C/O abundance ratio exceeds the solar value by several tens times, and that of Ne/O by at least a factor of 5[10].

The high sensitivity of the HXD has been demonstrated by the observation of the binary X-ray pulsar A0535+262. In spite of the very low source intensity (about 30 mCrab at 20 keV), its electron cyclotron resonance was detected clearly with the HXD, in absorption at about 45 keV. The resonance energy is found to be essentially the same as those measured when the source is almost two orders of magnitude more luminous [11].

Our galactic center region is opaque for visible light and soft X-ray due to dust and neutral hydrogen. Suzaku has spent hundreds of ksec for this region, since the sensitivity and resolution of imaging spectroscopy of Suzaku is suitable for the study of complex distribution of hot plasmas, reflection nebulae and other structure in this vicinity. The high energy resolution and the low background orbit provide excellent spectra of the Galactic Center diffuse X-rays (GCDX). The XIS found many emission loines in the GCDX near the energy of K-shell transitions of iron and nickel. The most pronounced features are the K_α (6.4 keV) line and K-edge (7.1 keV) of neutral or low ionization iron, and the K_α lines at 6.7 keV and 6.9 keV from He-like and hydrogenic iros. In addition, K_α lines of neutral or low inonization nickel and He-like nickel, K_β lines of Fe_{I}, Fe_{XXV} and Fe_{XXVI}, and K_γ lines of Fe_{XXV} and Fe_{XXVI} are detected for the first time. With the high energy resolution of XIS, the discussion on their spatial distribution is now available [12]. In addition to those new observations for line diagnostics, Suzaku has discovered new X-ray SNRs and XRNs from this region[13,14]

The broad band coverage of the Suzaku will play very important role to constrain the shape of the non-thermal spectrum from AGN. In figure 4, we show an example showing the power of *Suzaku*, a wide band spectrum of Cen A radio galaxy. Figure 5 shows the HXD spectra of several sources obtained during the test observation phase (August 17 – 23)[15]. The solid line at the bottom shows 3 % of background, which is the goal of background determination accuracy. One of the major discoveries with ASCA was the broad iron line feature from MCG-6-30-15 and some other AGN. However, it has been pointed out that some ambiguity of the continuum component exists, because ASCA covered only up to 10 keV. The Suzaku coverage up to 100-300 keV provides us with firm basis of the continuum components, such as reflection or (partially) absorbed ones. We spent a few hundreds of ksec for MCG-6-30-15 and confirmed clear and almost identical broad iron line feature to those observed by ASCA and Newton. Similar results are also obtained from MCG-5-23-16. Strong excess emission in the HXD above 10 keV is observed from these Seyfert galaxies, consistent with the presence of a reflection component from reprocessing in Compton-thick matter (e.g. the accretion disk)[16,17]. With its low background capability, Suzaku HXD is able to follow the spectral evolution during flares with a time scale of several 10 ks. This is of great importance when we study multi-band spectra from blazars through simultaneous observations with high energy observatories such as TeV Cherenkov telescopes and GLAST.

References

1. K. Mitsuda et al. *PASJ*, in press (2006).
2. R. Kelley et al. *PASJ*, in press (2006).
3. K. Koyama et al. *PASJ*, in press (2006).
4. T. Takahashi et al. *PASJ*, in press (2006).
5. M. Kokubun et al. *PASJ*, in press (2006).
6. P. Serlemitosos et al. *PASJ*, in press (2006).
7. Y. Tanaka, H. Inoue, & S.S. Holt, S.S. PASJ, **46**, L37 (1994).
8. M.C. Weisskopf, M. C. et al. *PASP*, **114**, 1 (2002).
9. F. Jansen et al. *A&A*, **365**, L1 (2001).
10. Murashima et al. *ApJL*, in press (2006).
11. Y. Terada et al. *ApJL*, in press (2006).
12. K. Koyama et al. *PASJ*, in press (2006).
13. K. Koyama et al. *PASJ*, in press (2006).
14. M. Tsujimoto et al. *PASJ*, in press (2006).
15. Y. Fukazawa et al. *SPIE*, in press (2006).
16. J. Reeves et al. *PASJ*, in press (2006).
17. G. Miniutti et al. *PASJ*, in press (2006).

X-RAY DIAGNOSTICS OF ACCELERATION PROCESS

A. BAMBA

RIKEN Cosmic Radiation Group,
2-1, Hirosawa, Wako-shi, Saitama, Japan
E-mail: bamba@crab.riken.jp

1. Introduction

Ever since the discovery of cosmic rays, the origin and the acceleration mechanism up to more than ~TeV have been long-standing problems. Koyama et al.(1995)[1] discovered synchrotron X-rays from shells of SN 1006, which is the first observational result indicating that SNRs accelerate electrons up to ~TeV. Now, several SNRs have recently been categorized into synchrotron X-ray emitters. These discoveries provide good evidence for the cosmic ray acceleration at the shocked shell of SNRs. The most plausible process of the cosmic ray acceleration is the diffusive shock acceleration (DSA)

Apart from the global success of DSA, there are still many remaining problems. We have not yet fully understood detailed but important information, such as the maximum energy of particles, the configuration of magnetic fields, the injection efficiency from thermal plasma to accelerated particles, and the acceleration and de-acceleration history of particles around the shock fronts, and so on. This partly comes from the fact that there is no information about the spatial distribution of accelerated electrons of practical objects, which strongly reflects the above uncertainties. Moreover, we have no information on accelerated protons, which are the main component of cosmic rays.

Accelerated electrons emit synchrotron radiation in the hard X-ray band, then hard X-ray observations are the strong tool to explore the acceleration sites. Bamba et al.(2003)[2] considered the spatial distribution of emission from accelerated particles as new information in order to understand the acceleration on the shock in SNRs; they observed the synchrotron X-ray emitting shell of SN 1006 with the excellent spatial resolution of the *Chandra* X-ray observatory and found that the emitting regions of synchrotron X-rays are incredibly thin ("filaments"). This result constrains the DSA theory and magnetic-field configurations; both an efficient acceleration scenario with strong magnetic field parallel to the shock normal and an inefficient one with a weak and perpendicular magnetic field are allowed[3,4].

Recently, TeV γ-ray Telescope HESS discovered several unknown TeV-emitters on the Galactic plane[5]. They have no counterparts in any other wavelength, although the follow-up observations are still insufficient. Their emission mechanism is still unknown; via inverse Compton emission from accelerated electrons, or via π^0 decay from accelerated protons and targets (molecular clouds and so on). Hard X-rays are the strong tool to distinguish the emission process, since accelerated electrons must emit synchrotron X-rays.

In this paper, we introduce how to use X-rays to make quantitative discussion on the acceleration of electrons and protons.

2. X-ray diagnostics of electron acceleration efficiency in Young SNRs

In order to explore the acceleration efficiency of electrons, we used *Chandra* archival data of the ACIS of five SNRs, Cas A, Kepler's remnant, Tycho's remnant, SN 1006, and RCW 86. The satellite and the instrument are described by Weisskopf et al. (2002)[6] and Garmire et al. (2000)[7], respectively. We know their precise age as shown in Table 1, since all the SNRs have historical records of their explosion.

Fig. 1 shows hard X-ray images of these SNRs taken by *Chandra*. In all SNRs, we can see clear filamentary structures ("filaments") on the outer rims of SNRs. We made profiles of these filaments and fitted with an exponential model[2]:

$$f(x) = \begin{cases} A\exp\left(-\left|\frac{x_0-x}{w_u}\right|\right) & \text{on the upstream side} \\ A\exp\left(-\left|\frac{x_0-x}{w_d}\right|\right) & \text{on the downstream side,} \end{cases} \qquad (1)$$

where A and x_0 are the flux and position at the emission peak, respectively. The quantities w_u and w_d represent the e-folding widths on the upstream and the downstream sides of the apparent emission peak, respectively (hereinafter, "u" and "d" represent upstream and downstream sides,

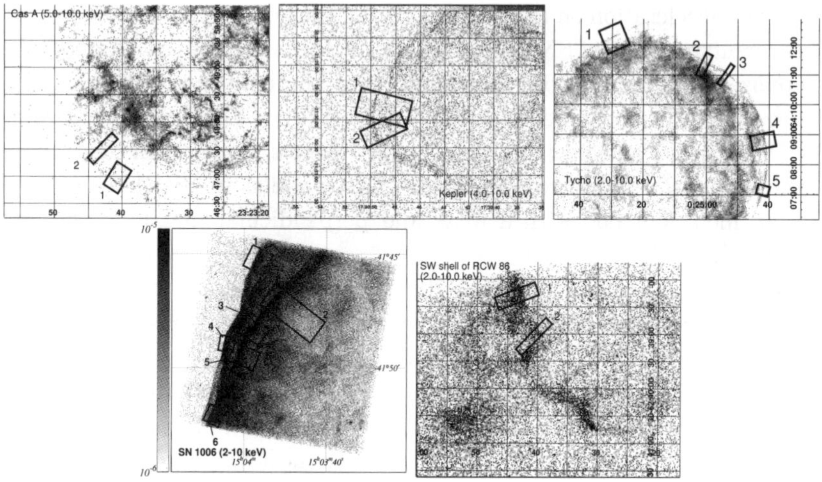

Figure 1. Close-up views of the SNRs in the hard X-ray band with *Chandra*.

respectively). As a result, we found that the value of w_u are as small as the point spread function of Chandra, whereas slightly larger on the downstream sides. w_d are listed in Table 1. We also made the their spectra, which are hard and have no line, in other words, nonthermal. This fact indicates that they are the acceleration sites of high energy electrons. They were well fitted with absorbed SRCUT model[8]. This model represents synchrotron emission from electrons with power-law distribution plus exponential cutoff. The model has a break called roll-off frequency (ν_{roll}),

$$\nu_{roll} = 1.6 \times 10^{16} \left(\frac{E_{max}}{10\ \mathrm{TeV}}\right)^2 \left(\frac{B}{10\ \mu\mathrm{G}}\right) , \tag{2}$$

where E_{max} and B are the maximum energy of accelerated electrons and the magnetic field. The best-fit roll-off frequencies for the five samples are also listed in Table 1.

Table 1. Spatial and spectral fitting results of the filaments

Parameters	Cas A	Kepler	Tycho	SN 1006	RCW 86
t_{age} [yrs]†	328	396	428	994	1,816
w_d [pc]	(<0.029)	$0.045^{+0.014}_{-0.010}$	$0.045^{+0.008}_{-0.005}$	$0.27^{0.05}_{0.04}$	$0.026^{+0.13}_{-0.06}$
ν_{roll} [10^{17} Hz]	24(> 1.6)	$3.6^{+4.3}_{-1.6}$	$0.83^{+3.1}_{-0.24}$	$2.6^{+0.7}_{-0.7}$	$2.1^{+3.8}_{-1.0}$

Table 1 represents that the scale widths on the downstream side (w_d) are very small compared with the radius of each SNR. Moreover, they may increase as the age of SNRs becomes larger. In fact, w_d is a function of t_{age} as follows;

$$w_d = 3.0^{+0.7}_{-0.6} \times 10^{-6} \left(\frac{t_{age}}{[1\ \mathrm{yr}]}\right)^{1.60^{+0.01}_{-0.30}} \quad [\mathrm{pc}] \ , \tag{3}$$

although the value of reduced χ^2 (9.68/3) did not accept the fitting.

To relate the width of the filament to the roll off frequency, we construct a function **B**, such that

$$\mathbf{B} = \nu_{roll} {w_d}^{-2} \quad [\mathrm{Hz\ pc}^{-2}] \ , \tag{4}$$

which shows monotonous decay as seen in Fig. 2. This function, **B**, depends on only magnetic field and shock velocity as written later, then we examined the time dependence of this function. A power-law fit was statistically acceptable with the best-fit value of $\mathbf{B} = C{t_{age}}^{\alpha}$,

$$C = 2.6^{+1.2}_{-1.4} \times 10^{27} \quad [\mathrm{Hz\ pc}^{-2}] \ , \tag{5}$$

$$\alpha = -2.96^{+0.11}_{-0.06} \ , \tag{6}$$

with a reduced χ^2 of 2.03/3, respectively. This result implies that the function **B** might include some physical quantities that evolve with time.

Here, we make a scenario in order to explain the time evolution of **B**. Our assumption is that the spatial profile found in our present analysis reflects that of accelerated electrons[2]. The maximum energy of accelerated electrons, E_{max}, is determined by the age of the SNR, or the synchrotron energy loss, such as $t_{acc} \sim min\{t_{age}, t_{loss}\}$, where t_{acc} and t_{loss} are the acceleration and the synchrotron loss time scale, respectively[3]. Then, we simply obtain the scale width of the accelerated electrons in the radial direction (w_d) as $w_d \sim v_d \times min\{t_{age}, t_{loss}\} \sim v_d t_{acc}$, where v_d is the downstream fluid velocity. The acceleration time t_{acc} is on the order of K/v_s^2, where v_s and K are the shock velocity and the diffusion coefficient, respectively. The quantity K is assumed to be proportional to the gyro radius of the accelerated electrons, $r_g = E_{max}/eB_d$, where B_d is the downstream magnetic field [3]. As a result, we find $w_d \sim K/v_s \propto {B_d}^{-1} E_{max} {v_s}^{-1}$. The value of ν_{roll} is proportional to $B_d {E_{max}}^{2}$[8]. Therefore, the quantity **B** is proportional to ${B_d}^3 {v_s}^2$. The time evolution of **B** may represent that of ${B_d}^3 {v_s}^2$.

Let us consider the time evolution of the magnetic field and the shock velocity as $B_d \propto {t_{age}}^{-\ell}$ and $v_s \propto {t_{age}}^{-m}$. The index, m, changes from 0

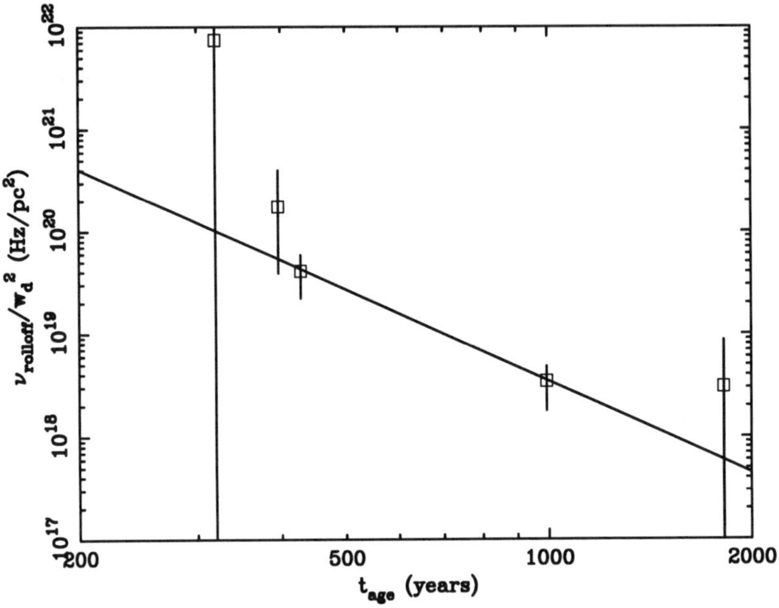

Figure 2. Time evolutions of $\mathbf{B} \equiv \nu_{\rm roll}/{w_d}^2$. The solid line represents the best-fit model.

to 3/5 as SNRs evolve from the free-expansion phase to the Sedov phase. Together with the observational fact of eq.(6), the index ℓ takes the value of 0.57–1.02. Our result suggests that the magnetic field decreases as the SNR ages. This is the first observational implication of the time evolution of the magnetic field in the SNR. The most interesting case is $\ell = 0.6$, which is allowed when SNRs are in the Sedov phase ($m = 0.6$). In this case, since the density of the thermal plasma is almost constant, the energy density of the magnetic field ($u_B \propto {B_d}^2$), and the kinetic and thermal energy densities of the shock ($u_{th} \propto$ shock temperature and $u_{kin} \propto {v_s}^2$), evolves according to the same time dependency (index $= -1.2$). The time evolution of the energy density of accelerated protons (u_p) may also be discussed from the relations by Lucek et al. (2000)[9] as

$$u_p = \frac{v_s}{v_A} u_B \propto B_d v_s \propto {t_{age}}^{-1.2} \quad , \tag{7}$$

where $v_A \equiv B/\sqrt{4\pi\rho}$ and ρ are the Alfvén velocity and the fluid density, suggesting that accelerated protons in an SNR evolve with the same energy density evolutions as that of the other energy carriers.

Since Bamba et al. (2003)[2] indicated that these energy densities are roughly in equipartition with each other, our result implies that they evolve while maintaining equipartition. Bell et al. (2001)[10] suggested theoretically that the magnetic field on the shock evolves according to the same time dependency as the shock velocity, which is consistent with our result.

3. Observations of Proton Accelerators with X-rays ?

HESS unID sources[5] are located on the Galactic plane and extended, indicating that they might be supernova remnants. We make a possible scenario to illustrate their nature. Young SNRs are bright in thermal and nonthermal X-rays, and radio band. As they ages, electrons decelerate and heated plasmas cool down, then becomes dim in any wavelength. However, protons never decelrates in such a short time scale. Once shock fronts of SNRs hit molecular clouds, accelerated protons make shower and emits γ-rays via π^0 decay. In such cases, we cannot see anything in the X-ray band. Therefore, strong upper limit in the X-ray band for such sources must be the first strong observational evidence of the proton acceleration. X-ray Imaging Spectrometer (XIS) onboard *Suzaku* has advantages to distinguish the origin of TeV γ-ray emission with low background and large effective area. With *Suzaku*, we might be able to "see" proton accelerators for the first time. Detailed discussions are shown in Yamazaki et al.(2006)[11].

References

1. Koyama, K., Petre, R., Gotthelf, E.V., Hwang, U., Matsura, M., Ozaki, M., and Holt S. S. *Nature*. **378**, 255 (1995)
2. Bamba, A., Yamazaki, R., Ueno, M., & Koyama, K. *ApJ* **589**, 827 (2003)
3. Yamazaki, R., Yoshida, T., Terasawa, T., Bamba, A., & Koyama, K. *A&A*, **416**, 595 (2004)
4. Berezhko, E. G., Ksenofontov, L. T., & Völk, H. J. A&A, **412**, L11 (2003)
5. Aharonian, F. et al. *Science*, **307**, 1938 (2005)
6. Weisskopf, M. C., Brinkman, B., Canizares, C., Garmire, G., Murray, S., & Van Speybroeck, L. P. *PASP*, **114**, 1 (2002)
7. Garmire, G., Feigelson, E. D., Broos, P., Hillenbrand, L. A., Pravdo, S. H., Townsley, L.,& Tsuboi, Y. *AJ*, **120**, 1426 (2000)
8. Reynolds, S. P., & Keohane, J. W. *ApJ*, **525**, 368 (1999)
9. Lucek, S. G. & Bell, A. R. *MNRAS*, **314**, 65 (2000)
10. Bell, A. R. & Lucek, S. G. *MNRAS*, **321**, 433 (2001)
11. Yamazaki, R., Kohri, K., Bamba, A., Yoshida, T., Tsuribe, T., & Takahara, F. *MNRAS* submitted (astro-ph/0601704)

SUPERNOVAE IN THE UNIVERSE

SHOICHI YAMADA
Department of Physics, School of Sceince & Engineering, Waseda University,
3-4-1 Okubo, Shinjuku, Tokyo 169-8555, Japan
**E-mail: shoichi@waseda.jp*
www.heap.phys.waseda.ac.jp

In this article, I review the current understanding of the mechanism of core-collapse supernovae, one of the most energetic events in the present universe. I will not only summarize the neutrino-heating mechanism, the standard paradigm at present, but also pay special attention to the recent topics such as the standing accretion shock instability and the acoustic revival scenario proposed by Burrows very recently.

Keywords: Supernovae; Neutrinos; Gravitational Collapse; Instabilities.

1. Introduction

The Core-collapse supernova occupies an important position in the high energy universe. One of the reasons is that it emits a large amount of energy, $\sim 10^{53}$erg, in various forms, trigering other processes subsequently. Neutrinos take the largest share, carrying most, $\sim 99\%$, of the energy. The rest of the energy goes to the kinetic energy, $\sim 10^{51}$erg, as well as to the radiations, $\sim 10^{49}$erg. A strong shock wave generated in the supernova explosion is, on the other hand, supposed to produce high energy, non-thermal cosmic rays. And last but not the least, the supernova is a promising source of the gravitational radiation, since the collapse of massive stars is in general non-spherical as observed for SN1987A. Aside from the great energy budget, the core-collapse supernova is also an important contributor to the chemical evolutions of universe. It distributes heavy elements which are synthesized not only in the hydrostatic phase prior to the explosion but also during the explosion itself, thus increasing the metalicity of galaxies. The collapse-driven supernova is also expected to be a promising r-process site, since ejecta are rich in general neutron. In these days, its association with gamma ray bursts, another big player in the high energy universe, is yet another reason the core-collapse supernova is attracting many researchers' interest.

The mechanism of core-collapse supernova has eluded our understanding for more than 40 years.[1] It is an explosive phenomenon that is supposed to occur at the end of the evolution of massive stars ($\gtrsim 8M_\odot$), triggered by the gravitational collapse of a central core after it exceeds a certain critical density ($\sim 10^{9.5}$g/cm^3) or temperature ($\sim 10^{9.7}$K). The event is followed by the formation of a neutron star or a black hole. The energy source of explosion is the gravitational binding energy liberated by the formation of a compact object, typically a neutron star. The difficulty in the modelling of core-collapse supernova arises from the fact that it involves a rich variety of physics. Microphysics such as weak interactions and nuclear physics is supposed to dictate the macroscopic dynamics in a critical way while the latter must set an appropriate stage for the former. We do not know at the moment which is more crucial in the explosion mechanism and, in fact, the focus of the supernova society has been swinging between them for decades.

In this article I will review our current understanding on the supernova mechanism, starting with the neutrino-heating scenario, supposedly the most promising mechanism at present, and then discussing another new scenario, the so-called acoustic revival mechanism advovated very recently by Burrows[2].

2. Neutrino-heating Scenario

It should be no surprise from the energetics point of view that the theoretical research of the core-collapse supernova has been done mainly in the context of the so-called neutrino-heating mechanism. It is currently taken for granted that a shock wave generated at the core bounce is not energetic enough to push through the stellar core and becomes an accretion shock somewhere in the middle. It is supposed in the neutrino-heating scenario that the stalled shock wave is re-energized by absorbing neutrinos copiously emitted by the proto neutron star. It is noted again that the gravitational energy of the proto neutron star, $E_{PNS} \sim 10^{53}$erg, that is liberated by the collapse of a stellar core, is mainly transported by these neutrinos. The problem in this scenario, however, comes from the fact that neutrinos interact with matter only through the weak interactions and the deposition of their enormous amount of energy to matter is inefficient. I will explain the difficulty more in detail in the followings.

2.1. *Critical Neutrino Luminosity*

Although the eventual success or failure of the shock revival has to be judged by quantitative multi-dimensional radiation-hydrodynamical simulations, the qualitative condition for the shock revival has been clarified by various means over the years.

It is now well known that there is a critical neutrino luminosity for a given accretion rate, above which the standing accretion shock will be moved outwards.[3,4,5] Hence the issue is whether this citical value is reached in the actual evolution or not. For example, Fig. 1 shows our recent results.[6] We solved the fully general relativistic hydrodynamical equations and Boltzmann equations for neutrinos simultaneously. The evolutions were followed for more than a second after the core bounce for two realistic nuclear EOS's. The left panel shows the mass trajectories for EOS by Shen et al. (SH-EOS). As is evident, there is no hint of shock revival. In fact, the shock stalls at around 200km from the center and becomes an accretion shock and then starts to recede onto the proto neutron star. The evolution is not qualitatively different for the other EOS by Lattimer & Swesty (LS-EOS) as shown in the right panel, where the evolutions of shock wave positions for the two models are given. Although the employed numerical techniques are different among them, the results obtained by other groups[7,8,9,6] are also consistently negative. These results suggest that the neutrino luminosity does not reach the critical value in the spherical symmetric collapse.

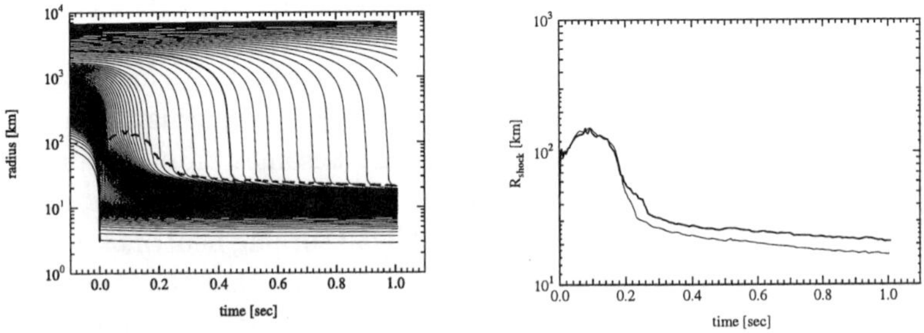

Figure 1. The radial trajectories of mass elements for SH-EOS (left panel) and the radial positions of the shock wave (right panel) as a function of time after bounce for 15 $M_\odot$ model. In the left panel, the dahed line indicates the location of the shock wave. In the right panel, the thick and thin lines represent the results for SH-EOS and LS-EOS, respectively.

This is more clearly demonstrated by simplified models, where we employ the quasi-steady approximation to describe the long-term post-bounce evolutions.[10] After the shock is stagnated, the accretion flows through the standing shock wave onto proto neutron star are quasi-steady. Variations of controlling parameters, namely, the accretion rate and neutrino luminosity, are much slower than the advection and sound-crossing over the region. Hence the accretion flows are approximately expressed by a series of steady flows with different accretion rates and neutrino luminosities. The accretion rate is determined by supersonic infall of the envelope, which we solve numerically for the realistic progenitor model. The neutrino luminosities are obtained from the quasi-static cooling of the proto neutron star. We have combined these approximate solutions for three different regions by imposing appropriate conditions at each boundary to find the evolutions of the whole system, We have taken into account the accretion onto the proto neutron star when the cooling is computed. The contraction of the proto neutron star radius is considered in solving the steady accretion flows. The initial condition is an outcome of the dynamical simulation at $\sim$ 300ms after bounce. We have found a reasonable agreement between the results from the approximate treatment and the dynamical simulation up to 1s after bounce. The quasi-steady approach not only allows us to follow the evolution of system much longer ($\sim$ 10s) than dynamical computatioins, but also facilitates a comparison of the actual neutrino luminosities with the critical values as demonstrated in Fig. 2. We find that the typical luminosity is smaller than the critical value by a factor of $\sim$ 2.

2.2. *Standing Accretion Shock Instability*

So far we have tacitly assumd spherical symmetry of the dynamics. However, several observations suggest that the supernova explosions are intrinsically asymmetric in general. For example, Leonard et al.[11] and Wang et al.[12] observed a few percent of linear polarizations for photons from core-collapse supernovae and argued that the stellar envelopes are globally asymmetric, oblate or prolate with an aspect ration of $\sim$ 2. It is widely believed that some convective motions occur in the supernova core. We also know that supernova remnants are asymmetric in general. Hence we have to consider the supernova mechanism bearing in mind the intrinsic multi-dimensionality of dynamics.

It is Blondin et al.[13] that first pointed out the instability, which they referred to as the standing accretion shock instability (SASI), using 2-

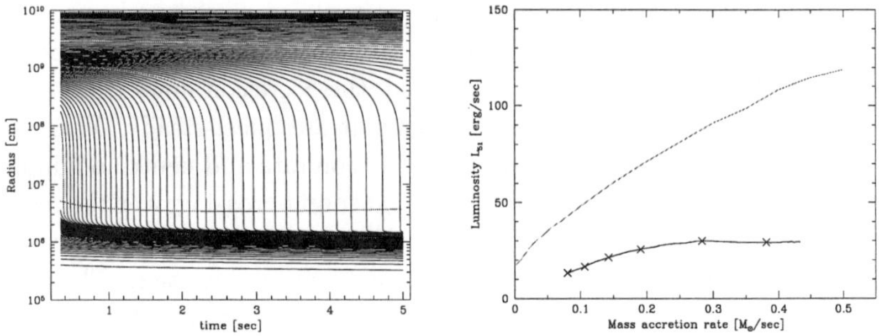

Figure 2. The radial trajectories of mass elements (left panel) and the evolution of the luminosity (right panel) as a function of time after bounce in the quasi-steady approximation. The horizontal line in the left panel shows the radial position of the shock wave. The upper and lower lines in the right panel represent the critical luminosity and the actual luminosity, respectively. The crosses on the line correspond to the following times from right to left: 500, 1000, 2000, 3000, 4000, 5000 ms after bounce.

dimensional hydrodynamical simulations for adiabatic accretions through a standing shock onto a proto neutron star. Figure 3 shows our 2-dimensional numerical experiments on SASI.[14] The distributions of entropy (the left half of the panel) and density (the right half) in the meridian section are displayed for the post-stagnation accretion models with $L_\nu = 5.5, 6.0 \cdot 10^{52}$ erg/s after 1% of the $\ell = 1$ single-mode velocity perturbation is added. The accretion rate and luminosity are fixed at constant values. For both models, we observe the growth of the perturbations. In the case of $L_\nu = 5.5 \cdot 10^{52}$ erg/s, the shock surface is deformed at first by the increasing amplitude of the non-radial mode and then begins to oscillate with a large amplitude. In the case of $L_\nu = 6.0 \cdot 10^{52}$ erg/s (right panels), on the other hand, in addition to the oscillations of the shock surface, we observe the substantial increase of the average shock radius as the time passes. In fact, after $t = 400$ ms, the shock radius continues to increase and appears to produce an explosion. Since the model is stable against radial perturbations, the non-radial instability and the neutrino heating therein are responsible for the explosion. This is a reconfirmation of the claim that the instability, whatever the origin, behind the shock is helpful for the shock revival.

For the random multi-mode velocity perturbations, we find that the modes with small $\ell's$, especially those with $\ell = 1, 2$, grow rapidly in the linear regime. This is particularly the case for the model without a negative entropy-gradient and the growths of the modes with $\ell > 10$ are negligibly

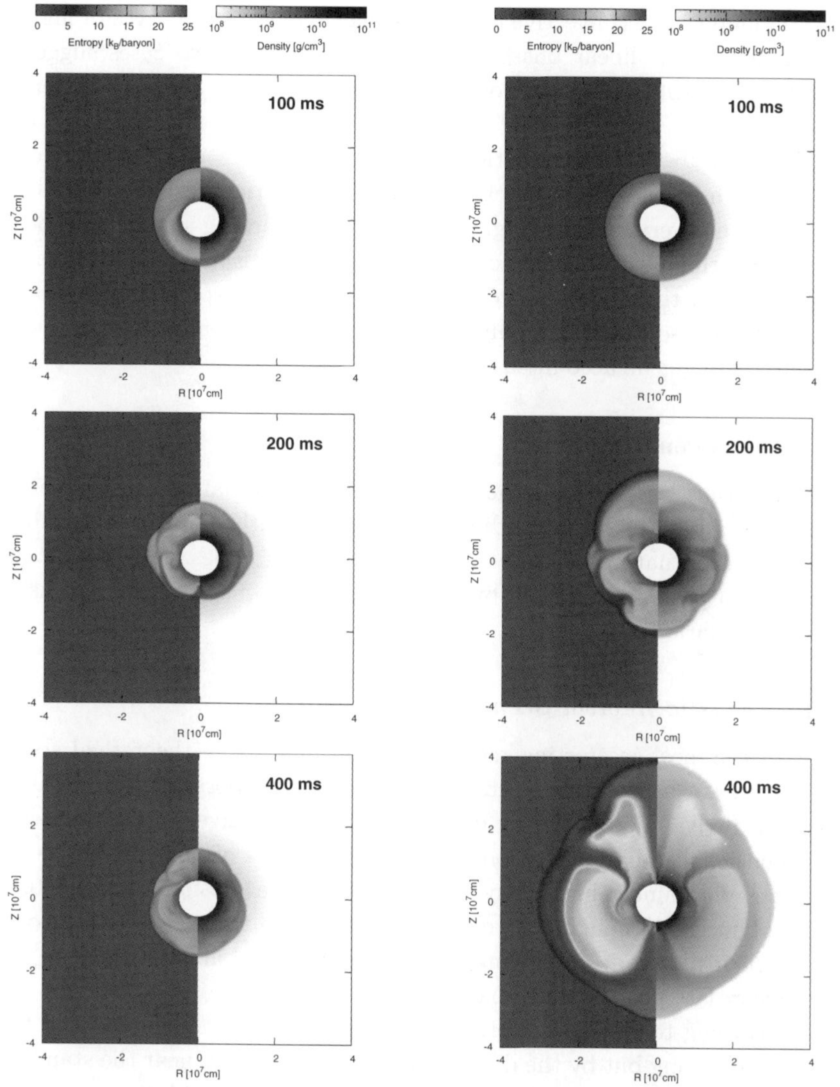

Figure 3. Entropy- (the left half of each panel) and density- (the right half) distributions in the meridian section for 1% of the $\ell = 1$ single-mode velocity perturbation. $L_\nu = 5.5 \cdot 10^{52}$ erg/s is assumed for the left panels and $L_\nu = 6.0 \cdot 10^{52}$ erg/s is for the right panels.

small. With a negative entropy-gradient, the broadening of the spectral ℓ-distribution is observed although the dominance of smaller ℓ modes can be still found. The convective instability may enhance the growth of higher harmonics in the linear phase. The similarity of the two cases suggests again that SASI is dominant over the convection even when the latter is operating.

We have to wait for realistic simulations before we judge if SASI can give enough boost for the shock revival. The numerical results[15] obtained so far are not very encouraging. We are thus still lacking a sufficient boost if the neutrino-heating were to be the true mechanism. It is needless to say, however, that SASI and convections should be explored in detail in 3-dimensional settings. Not only the linear proper motion but also the spin of pulsars might be accounted for by them.[16]

3. Other Scenarios

We have thus far seen that the neutrino-heating mechanism has not been very successful even for multi-dimensional models although it is still viable. It is, however, natural then to consider other possibilities. Here we pic up a new idea proposed recently by the Arizona group,[2] that is, the acoustic revival scenario.

3.1. *Acoustic Revival Scenario*

One of the most interesting recent developments in the theoretical study on supernova mechanism is, in my opinion, the proposal of the so-called acoustic revival mechanism by Burrows et al.[2] They have done a long-term 2-dimensional radiation-hydrodynamical simulations of non-rotating core-collapse and found a highly asymmetric one-sided explosion after $\sim$ 500ms after bounce.@Although the treatment of neutrino transfer is still incomplete, their new code employing an unstructured grid allowed them to explore very late post-bounce phase.

According to their descriptions, the explosion was not induced by the neutrino-heating but by the dissipation of acoustic waves near the standing shock wave. These acoustic waves are generated by oscillations of proto neutron star with predominantly g-mode feature, which the authors suppose are excited by turbulent motions that are induced initially by SASI. In fact the acoustic power dominates over the neutrino heating after $\sim$ 400ms after bounce. The advantages of this mechanism are the high efficiency of the energy transfer from the acoustic waves to the stalled shock and the self-regulation in the sense that the acoustic power is accumulated so that the explosion would eventually occur. Since the explosion is somewhat delayed, the problem of small neutron star mass that is often encountered in the models with artificially enhanced neutrino luminosities may be alleviated.

There are, of course, many questions to be answered. What is the excitation mechanism of the g-modes in the proto neutron star? How efficiently do the g-modes emit acoustic waves? What is the final explosion energy? What is the dependence on the progenitor, EOS and so on? Other groups should certainly be able to reproduce the results. Note that the reason why the mechanism had eluded the discovery by other groups is supposed that earlier simulations were not long enough or had put some restrictions to motions that would suppress the production of sound waves, such as putting an artificial inner boundary or assuming equatorial symmetry.

Motivated by the paper, we have calculated the eigen values and eigen functions of g-modes for the proto neutron star.[17] We have taken the density and entropy distributions from the paper by Dessart et al.[18] for the background model. We have also computed the excitation of these modes by SASI. The pressure perturbations have been calculated at the surface of the proto neutron star for the SASI model we presented in section 2.2 and have employed them as an external source for the excitation of the g-modes. The left panel shows the eigen functions for the $\ell = 1$ modes. g_1- and g_2 modes have a single- and double radial nodes, respectively. Since there is a convection region inside the proto neutron star, the g_1 mode has most of the power outside of the convection region while g_2 mode has more power inside the convection region.

The right panel shows the power spectrum of the external forces with the $\ell = 1$ angular dependence exerted on the proto neutron star surface by the non-linear SASI. Although the non-linear coupling broadens the spectrum, the mismatch of the frequencies between the g-mode and SASI is apprent. As a result, the excitation is rather inefficient with the growth time of 177ms for the g_1 mode and longer for other modes, at least for this particular model. Since the eigen functions depend rather sensitively on the background model, we are currently studying the growth rates etc. for other background models and will publish them elsewhere soon. It should be noted that there remains to be clarified how efficiently the acoustic waves are generated by the g-modes and are dissipated near the shock wave subsequently.

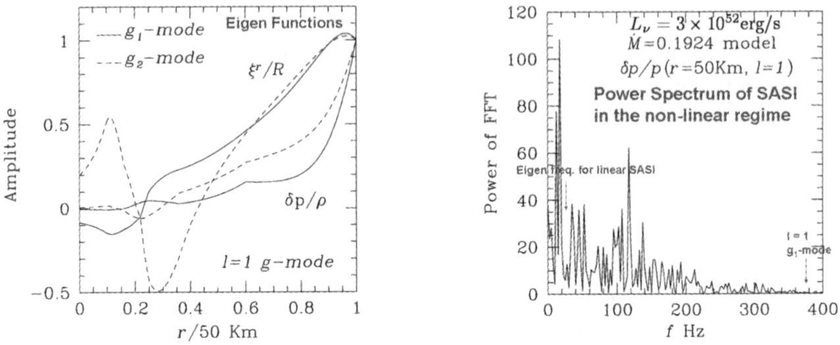

Figure 4. The eigen functions of g-modes for the $\ell = 1$ angular dependence (left) and the eigen frequency of g_1 mode and power spectrum of the $\ell = 1$ non-linear SASI (right). g_1- and g_2 modes have a single- and double radial nodes, respectively.

4. Conclusion

The core-collapse supernova is a very important player in discussing the energy budget of the high energy universe. After decades of intensive and extensive theoretical investigations, however, we are unfortunately still unable to pin down the most essential ingredient for the successful supernova explosion. Although the neutrino-heating mechanism has been supposed to be the most promising, it has been so far not very successful. Various hydrodynamical instabilities, particularly SASI, should be explored more in detail, since it is not only helpful for the revival of stalled shock wave but also may account for the observed asymmetry of explosion and the proper motions together with the spins of pulsars.

On the other hand, we had better look for other possibilities as well. The MHD mechanisms, particularly the magnetorotational instability (MRI), are promising candidates and certainly shold be investigated further. The acoustic revival scenario is quite new and should be confirmed by others and each step in the scenario should be scrutinized both by analytic and numerical methods. It is no doubt that this mechanism will be the focus of the society in the years to come.

Finally, I would like to touch another potentially important core-collapse event. It is natural that if the progenitor mass exceeds a certain threshold, the outcome will be a black hole. This black-hole-forming core-collapse is no less quiet in neutrino emissions. Aside from the gamma ray bursts and, probably hypernovae, which are supposed to be induced by the rapidly rotational collapse of very massive stars, the slowly rotating collapse will provide us with an opportunity to probe the properties of hot and dense hadronic matter, since the neutrino signals from them are different from those from the ordinary supernovae and, moreover, are very sensitive to the stiffness of the equation of state.[19] The events will be observed by neutrinos and possibly gravitational radiatons, but not by photons. It is emphasized that they are also an important player in the high energy universe.

Acknowledgments

This paper is based on the results obtained in an extended collaboration. I would like to thank K. Sumiyoshi, K. Kotake, H. Sawai, K. Nakazato, M. Watanabe, T. Yamasaki, N. Ohnishi and S. Yoshida for doing numerical computations. The numerical calculations were partially done on the supercomputers in RIKEN and KEK (KEK supercomputer Projects No.02-87 and No.03-92). This work is in part supported by Grants-in-Aid for the Scientific Research from the Ministry of Education, Science and Culture of Japan (No.S14102004, No.14079202, No.17540267), and Grant-in-Aid for the 21st century COE program "Holistic Research and Education Center for Physics of Self-organizing Systems"

References

1. K. Kotake, K. Sato, and K. Takahashi, submitted to *Rep. Prog. Phys.* (2005), astro-ph/0509456.
2. A. Burrows, E. Livne, L. Dessart, C. Ott, and J. Murphy, accepted to *Ap. J.* (2006), astro-ph/0510687.
3. A. Burrows and J. Goshy, *Ap. J. Lett.* **416**, L75 (1993).
4. H. -Th. Janka and E. Müller, *A.&Ap.* **306**, 167 (1996).
5. T. Yamasaki and S. Yamada, *Ap. J.* **623**, 1000 (2005).
6. K. Sumiyoshi, S. Yamada, H. Suzuki, H. Shen, S. Chiba, and H. Toki, *Ap. J.* **629**, 922 (2005).
7. M. Liebendöfer et al., *Nucl. Phys.* **A719**, 144 (2003).
8. M. Rampp and H. -Th. Janka, *Ap. J. Lett.* **539**, L33 (2000).
9. T. A. Thompson, A. Burrows, and P. A. Pinto, *Ap. J.* **592**, 434 (2003).
10. M. Watanabe and S. Yamada, (2006) in preparation.
11. D. C. Leonard, A. V. Filippenko, A. J. Barth, and T. Matheson, *Ap. J.* **536**, 239 (2000).
12. L. Wang et al., *Ap. J.* **579**, 671 (2002).
13. J. M. Blondin, A. Mezzacappa, and C. DeMarino, *Ap. J.* **584**, 971 (2003).
14. N. Ohnishi, K. Kotake, and S. Yamada, *Ap. J.* **641**, 1018 (2006).
15. H. -Th. Janka et al., *Nucl. Phys. A* **758**, 19 (2005).
16. A. Mezzacappa, in Proceedings of "Origin of Matter and Evolution of Galaxies (OMEG05) New Horizon of Nuclear Astrophysics and Cosmology", Tokyo, Nov. 8-11, edited by S. Kubono.

17. S. Yoshida, N. Ohnishi and S. Yamada, *Ap. J.* (2006) in preparation.
18. L. Dessart, A. Burrows, E. Livne, and C. D. Ott, accepted to *Ap. J.* (2006), astro-ph/0510229.
19. K. Sumiyoshi, S. Yamada, H. Suzuki, and S. Chiba, submitted to *PRL* (2006).

SEARCH FOR SUPERNOVA NEUTRINOS AT SUPER-KAMIOKANDE

MASAYUKI NAKAHATA

Kamioka Observatory, ICRR, University of Tokyo, Higashi-Mozumi, Kamioka, Gifu 506-1205, Japan

e-mail: nakahata@suketto.icrr.u-tokyo.ac.jp

Sensitivity of the Super-Kamiokande detector to supernova neutrinos and preliminary analyses of supernova neutrinos are presented.

1. Super-Kamiokande Detector

The first phase of Super-Kamiokande (SK-I) was started in April 1996 and about 5 years' data was taken with 11,146 PMTs until July 2001. After the accident in November 2001, SK was reconstructed using 5,182 PMTs with acrylic and FRP(Fiber Reinforced Plastic) cases to prevent shock wave production and the second phase data (SK-II) was taken from December 2002 through October 2005. New PMTs which will recover the original PMT coverage of SK have been produced from 2003 to 2005 and they were mounted from October 2005 to April 2006. The third phase of SK (SK-III) will start in Summer 2006 after the filling of pure water is finished. The fraction of photocathod coverage is 40%(19%) in SK-I and SK-III(in SK-II) and Cherenkov light yield is about 6 p.e./MeV (2.8p.e./MeV). The detector trigger threshold was about 4.1MeV at the end of SK-I and it was 5.5MeV in SK-II.

2. Sensitivity to Supernova Burst Neutrinos

Because of the large photo-sensitive volume of the SK detector(the volume inside the inner detector is 32 ktons), we expect large number of neutrino events for a galactic supernova burst. Figure 1 shows the expected number of supernova events for each interaction in the SK detector. In this figure and following figures in this section, the energy spectrum and the time profile simulated by Livermore group[1] is used. The main reaction is $\bar{\nu}_e + p \rightarrow e^+ + n$ and the number of expected events is about 7,300 for a supernova

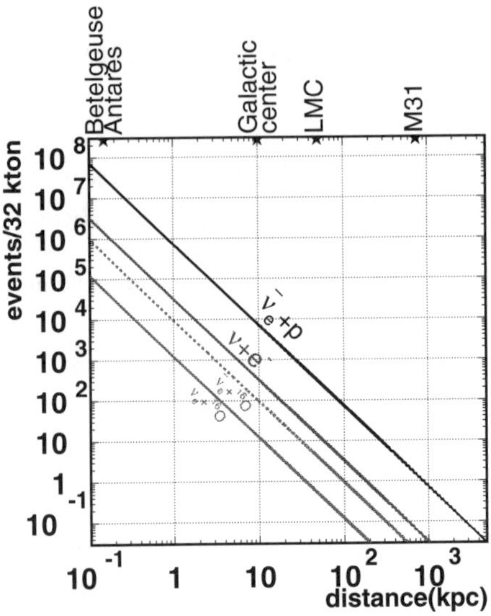

Figure 1. Expected number of events as a function of distance to supernova.

at 10kpc. The number of expected $\nu + e$ scattering events is about 300 and that of ν_e(or $\bar{\nu}_e$)+^{16}O processes is about 100. Neutrino oscillations are not taken into account here. If we take into account neutrino oscillations, the number of $\bar{\nu}_e + p$ interactions and $\nu + e$ scattering events increase by 10-30% because energies of ν_μ and ν_τ are generally higher than ν_e and $\bar{\nu}_e$. The number of $\nu + e$ scattering events at neutronization burst is about 5 events in 10 msec. If neutrino mass hierarchy is normal and crossing probability at H resonance(P_H) is near zero (i.e. adiabatic case), the expected number of neutronization events is about 0.7. If neutrino mass hierarchy is either inverted or normal with P_H=1(non-adiabatic case), the expected number of events is about two.

Figure 2 shows the visible energy spectrum of each interaction. The energy of electrons/positrons from $\bar{\nu}_e + p$, $\bar{\nu}_e +^{16} O$ and $\nu_e +^{16} O$ interactions preserve the energies of the original neutrinos. Total energy of a produced positron is the neutrino energy (E_ν) minus 1.3 MeV for $\bar{\nu}_e + p$ interaction. That for $\bar{\nu}_e +^{16} O$ ($\nu_e +^{16} O$) interaction is approximately $E_\nu - 10.5MeV$ ($E_\nu - 14.5MeV$). On the other hand, the recoil electron energy spectrum of $\nu + e$ scattering extends to lower energies and the total cross section of $\nu + e$ scattering increases almost linearly with energy while cross sections

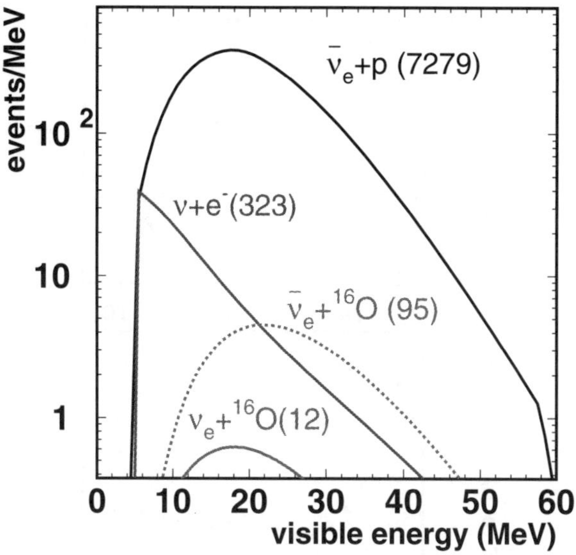

Figure 2. Expected visible energy spectrum assuming a supernova at 10 kpc. The horizontal axis is total energy of electrons/positrons produced by neutrino interactions. The number in parentheses is the total number of expected events integrated over time. The energy spectrum and time profile of neutrinos are from Livermore simulation[1].

of other processes increase quadratically. Because of those effects, the ratio of the number of $\nu + e$ scattering events to $\bar{\nu}_e + p$ events is higher at lower energies.

The large number of $\bar{\nu}_e + p$ events for a galactic supernova enable us to study the time profile of the neutrino burst very precisely. Figure 3 shows the time variation of the mean energy of positrons as a function of time. The error bar in the upper figure and the value in the lower figure show the statistical error of the mean energy obtained by events in each logarithmic time bin. The error is about 0.5MeV after 100 msec. So, if the increase of mean energy is a function of time, which is expected assuming the Livermore simulation is correct, we can easily see the effect as shown in Fig.3.

Figure 4 shows a simulation of angular distributions to the supernova. The peak in the direction from the supernova is due to $\nu + e$ scattering events and a nearly flat distribution at lower energies and the uprising distribution at higher energies are due to $\bar{\nu}_e + p$ interactions. Direction to

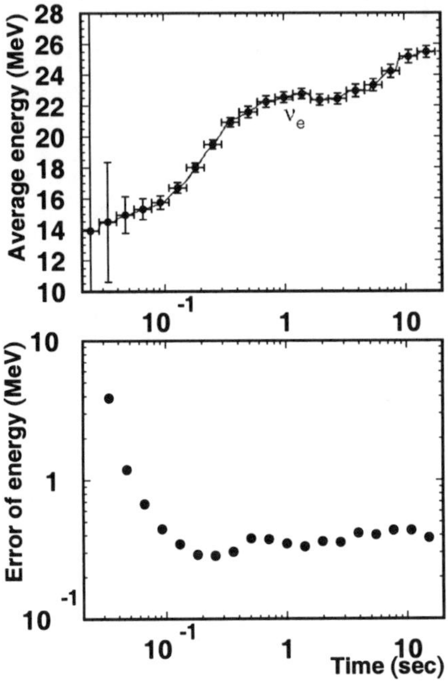

Figure 3. Simulated time variation of the mean energy of positrons(upper) and accuracy of mean energy measurement(lower). The distance to the supernova is assumed to be 10kpc.

the supernova can be easily obtained with an accuracy of a few degrees using events at lower energies. By statistically subtracting the nearly flat $\bar{\nu}_e + p$ component from the angular distribution, an energy spectrum of scattered electrons can be extracted using a technique developed for solar neutrino analysis. The recoil electron energy spectrum thus obtained can give additional information to compare data with burst models.

3. SK Data Analysis

In order to detect the supernova promptly, an online supernova alarm system has been running at the detector site. As soon as the online data acquisition system of SK creates a raw data file every several minutes, the file is sent to the online supernova alarm system. The signature of the supernova signal is time-clustered events whose vertex positions are randomly distributed over the fiducial volume of the detector. Selection criteria of the online alarm system are (1)$\geq$25 events within 10 sec and (2) R_{mean}

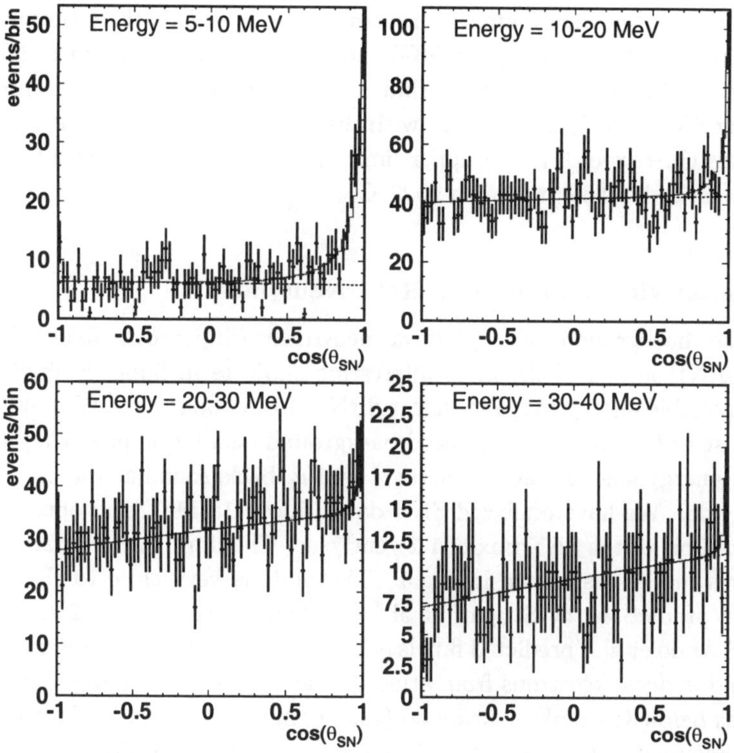

Figure 4. Simulation of angular distribution to a supernova for each energy range. The distance to the supernova is assumed to be 10kpc.

(avaraged distance between vertex positions) $\geq$ 7.5m, where events which fire a hardware trigger with a threshold of ~5.7MeV (~8MeV) for SK-I(SK-II) and have vertex within the fiducial voume are used. Most of the candidates are flashing PMTs and spallation products induced by energetic cosmic ray muons. Those candidates are rejected by clustering of vertex positions. Candidate event clusters generate an alarm to shift people who then visually inspect relevant data. During the SK-I and SK-II data taking periods, no galactic supernova was observed. An off-line supernova search was also performed in the SK-I data (1496 days; from May 31, 1996 to July 15, 2001) with an energy threshold of 4.5 MeV and SK-II data(622 days: from December 22, 2002 to March 19, 2005) with an energy threshold of 7 MeV. Selection criteria of the offline analysis were (1) at least 3, 4 or 8 events observed within 0.5, 2, or 10 seconds, respectively, (2) $R_{mean} \geq 10$

m. There was no candidate which satisfied the criteria on the time windows of 2 and 10 seconds during the SK-I and SK-II data taking periods. For the criterion of at least 3 events within 0.5 seconds, 8 clusters were observed during SK-I, which is consistent with the expected number of chance coincidence clusters of 5.7. An upper limit of the galactic supernova rate was obtained to be 0.39 yr^{-1} with 90 % CL.

4. Sensitivity to Supernova Relic Neutrinos

Diffuse background of supernova neutrinos (Supernova Relic Neutrinos(SRN)) is one of the most interesting subjects in Super-Kamiokande. The possible energy range to detect SRN is from about 15 MeV to 40 MeV because of the cosmic ray induced background (spallation products) at the lower energy side and atmospheric ν_e and $\bar{\nu}_e$ background at the higher energy side. We have analyzed SK-I data[2] above 18 MeV and obtained an upper limit of the SRN flux of 1.2 /cm^2/sec. The limit was compared with various models as shown in Fig.5. The limit is very close to the recent model predictions by Strigari et al.[3] (predicted flux is 0.3~1.2 /cm^2/sec) and S.Ando et al.[4](predicted flux is ~1.1 /cm^2/sec). The most serious background is decay electrons from atmospheric ν_μ interactions which produce muons below Cherenkov threshold (called "invisible muons"). Possible improvements of the analysis are (1) lower energy threshold down to 15 MeV by improving the spallation cut, (2) enlarge fiducial volume by 0.5~1m to increase statistics by about 20%, and (3) reject events which have preactivity induced by nuclear gamma emissions. They are under study now.

Figure 6(left) shows an expected energy spectrum assuming the SRN flux prediction by Ando et al.[4] and the background rate of invisible muons measured in SK-I. The expected number of events from 15 MeV to 30 MeV is 22.7 events assuming detection efficiency of 80 % for 10 years' livetime. The number of invisible background is about 115 events/10years in this energy range. So, the statistical significance of the excess due to SRN is expected to be only around the 1.9σ level. In order to reduce background of invisible muons, we are thinking about possibilities of the identification of $\bar{\nu}_e + p \rightarrow e^+ + n$ interactions through neutron tagging. The first possibility is using a 2.2MeV gamma-ray produced by $n + p \rightarrow D + \gamma$ capture reaction. The number of PMT hits by a 2.2MeV gamma ray is estimated to be about 6. So, it is difficult to trigger a 2.2MeV gamma with high efficiency. But, if we collect information of all hit PMTs within several hundred microsec-

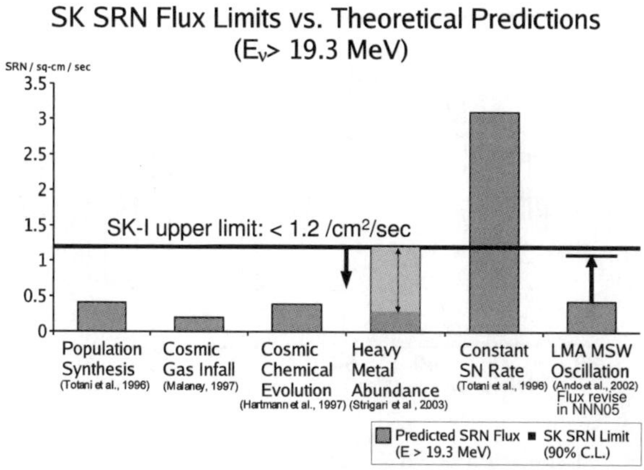

Figure 5. Flux upper limit of SK-I relic neutrino data (<1.2 /cm^2/sec) compared with various model predictions.

onds after candidate events, we may detect 2.2MeV gamma rays associated with SRN signals. The distance between the vertex positions of the prompt positron event and the 2.2 MeV gamma ray is less than about 50cm. So, the PMT hits induced by the 2.2MeV gamma ray could be extracted even though vertex position cannot be reconstructed. Another possibility is to detect gamma rays from a neutron capture reaction on gadolinium[5]. Because of the large gadolinium neutron capture cross section, 0.2% solution of $GdCl_3$ is enough to capture neutrons with >90% efficiency. The total energy of gamma rays is about 8 MeV for a Gd(n,γ)Gd reaction and multiple gamma rays are emitted. Supposing we can reduce invisible muon background by a factor of ten, the expected prompt energy spectrum should look like Fig.6(right). As seen in the figure, a clear excess due to SRN should be visible. The statistical significance of the excess between 15 MeV and 30 MeV would be 3.8 σ using 10 years' SK data (signal: $\sim$23 events, background: $\sim$13 events).

5. Conclusion

A large number of events is expected for a galactic supernova (>8,000 at 10kpc) and we can study details of the supernova explosion, e.g. time variation of the mean energy of $\bar{\nu}_e$. We can also discuss the energy spectrum of other neutrinos by extracting $\nu + e$ scattering events using the angular

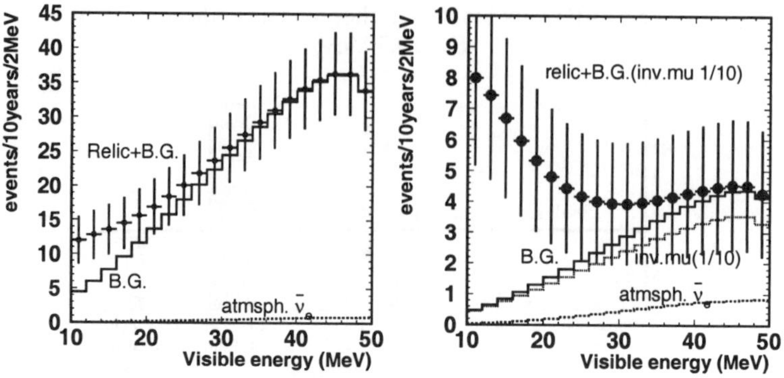

Figure 6. Expected spectrum of SRN signal with invisible muon background(left) and assuming that the invisible background can be reduced by a factor of 10. 10 years' SK data is assumed with the detection efficiency of 80%. The SRN flux calculation of [4] is used.

distribution to the supernova. Supernova relic neutrinos could be observed with about 10 years of SK data, if $\bar{\nu}_e + p \rightarrow e^+ + n$ interactions can be tagged by neutrons.

References

1. T. Totani, K. Sato, H. E. Dalhed and J. R. Wilson, ApJ.496, 216(1998).
2. Super-Kamiokande Collaboration (M. Malek et al.). , Phys.Rev.Lett.90,061101(2003).
3. L. E. Strigari, M. Kaplinghat, G. Steigman, and T. P. Walker,JCAP 0403,007(2004).
4. S. Ando, K. Sato and T. Totani, Astropart. Phys. 18, 307(2003); SRN flux expectation was updated in NNN05.
5. J. F. Beacom and M. R. Vagins,Phys.Rev.Lett.93, 171101(2004).

ASPECTS OF NEUTRINO PRODUCTION IN SUPERNOVAE

TODD A. THOMPSON*

Department of Astrophysical Sciences
Princeton University
Peyton Hall — Ivy Lane
Princeton, New Jersey, 08544 USA
E-mail: thomp@astro.princeton.edu

I discuss neutrino production in supernovae (SNe) and the detection of both Galactic core collapse events and the diffuse extra-galactic MeV neutrino background expected from the integrated history of star formation. In particular, I consider what processes might affect our expectations for both. I focus on "rapid" rotation, defined as leading to millisecond initial neutron star spin periods. Rotation affects the neutrino luminosity, the average neutrino energy, the duration of the Kelvin-Helmholtz cooling epoch, and the ratios of luminosities and average energies between neutrino species; it can strongly suppresses the $\bar{\nu}_e$ as well as ν_μ, $\bar{\nu}_\mu$, ν_τ, and $\bar{\nu}_\tau$ fluxes relative to ν_e. As a result, depending on the prevalence of rapid rotation in SN progenitors through cosmic time, this may affect predictions for the MeV neutrino background and the history of nucleosynthetic enrichment. I emphasize connections between the MeV neutrino background and tracers of the star formation rate density at high redshift in other neutrino and photon wavebands.

1. Introduction

When the iron core of a massive star collapses, the implosion is reversed at nuclear densities when nuclei dissociate into free nucleons. The equation of state stiffens dramatically, driving a bounce shockwave into the supersonic infalling outer core. The bounce shock stalls almost immediately as a result of neutrino losses, the ram pressure of the infalling material, and the dissociation of nuclei into free nucleons across the shock. After the shock stalls, a characteristic post-bounce accretion structure obtains that is quasi-steady-state. The hot (~10 MeV) newly-born "proto"-neutron star (PNS) has a

*Lyman Spitzer Jr. Fellow

neutrinosphere radius of ~80 km.[a] Overlying the PNS is a subsonic accretion flow, bounded by the stalled stand-off shockwave at ~200 km. Just above the PNS is a cooling layer and beyond that a region of net neutrino heating (the "gain" region) provided by both accretion luminosity as the matter falling through the bounce shock is incorporated into the PNS and by core neutrino luminosity as the PNS cools and deleptonizes.

The revival of the shock to an energy of order 10^{51} ergs is the focus of supernova theory. Any mechanism for this revival must fundamentally rely on the transfer of gravitational binding energy to the post-shock mantle.

The "neutrino mechanism"[1,2] employs neutrino interactions — primarily $\nu_e n \leftrightarrow pe^-$ and $\bar{\nu}_e p \leftrightarrow ne^+$ — to transfer binding energy to the shocked matter in the post-bounce epoch. With the standard suite of microphysics and a solution to the Boltzmann equation for all neutrino species, the neutrino mechanism fails in one spatial dimension; the stalled bounce shock remains trapped forever, accreting the overlying stellar progenitor.[3–9] Although some models show that multi-dimensional effects might be necessary for success of the neutrino mechanism,[10–13] recent calculations employing more sophisticated neutrino transport fail to explode[14] — albeit marginally. In some cases, successes are obtained.[15] However, the recent calculation of Ref. 16 suggests that energy deposition via neutrino interactions may be sub-dominant at late times with respect to acoustic heating generated by oscillations of the PNS generated by anisotropic accretion; the systematics of this new "acoustic" mechanism have yet to be elaborated. Many recent complimentary works focus on the stability of the shockwave and the post-shock material, and its importance for the mechanism.[17–19]

Quite apart from the details of the explosion mechanism, the total energy budget dictating the character of neutron star birth is set by the gravitational binding energy of the neutron star:

$$E_{\rm bind} \approx 3 \times 10^{53} M_{1.4}^2 R_{10}^{-1} \text{ ergs}, \tag{1}$$

where $M_{1.4} = M/1.4\,{\rm M}_\odot$ and $R_{10} = R/10\,{\rm km}$ are the neutron star mass and radius, respectively. Theoretical models suggest[20,21] and detection of neutrinos from SN 1987A confirm that a fraction of order unity of this energy is radiated in ~ 10 MeV neutrinos on the Kelvin-Helmholtz timescale $\tau_{\rm KH} \sim 10-100$ s, long with respect to the collapse and explosion timescales.

[a]Where the optical depth to ν_e neutrinos ~2/3, an energy- and time-dependent quantity.

Using these results, a number of studies have been made of the MeV neutrino background expected from SNe.[22–24] Indeed, the constraints on the background from SuperK[25] have become tight enough that, for example, a total energy radiated in $\bar{\nu}_e$ neutrinos of $E_{\rm bind}/3$ per SN with $\langle\varepsilon_{\bar{\nu}_e}\rangle > 20\,{\rm MeV}$ is excluded.[24] Estimates of the background must assume the total neutrino energy radiated per neutrino species per SN, the average neutrino energy per species, and neutrino oscillation parameters. For light water detectors like SuperK the dominant detection mode is single-species ($\bar{\nu}_e p \rightarrow n e^+$) and because the cross section for this interaction is proportional to the neutrino energy squared the partitioning of energy, mixing, and the spectral shape are essential. The typical assumption is that the energy is partitioned equally between species, that the spectrum is thermal, and that the average energy is $\sim 10 - 20\,{\rm MeV}$.

Although the total energy budget (in all channels) per SN is dictated by equation (1), it is interesting to consider what processes might affect the best estimates of the diffuse neutrino background and the expected neutrino signal from the next Galactic SN at order unity. Potential processes are both microphysical and macrophysical/astrophysical. As an example of the former, it is possible that a change to the neutrino opacities or the equation of state for dense matter may alter the neutrino luminosity (L_ν) and average energy ($\langle\varepsilon_\nu\rangle$) during the cooling epoch, while the total energy radiated per SN is unaltered.[21] Alternatively, as an example of astrophysical uncertainty, one may include the assumption of a universal IMF with constant (neutron-star-producing) SN rate per solar mass per year of star formation as a function of redshift, environment, and metallicity. In addition, there are a number of potential macrophysical effects that might modify our expectations at first order, including rapid rotation of the massive progenitor's iron core just before collapse.[32]

Millisecond rotation of the PNS is interesting for several reasons. First, it changes L_ν and $\langle\varepsilon_\nu\rangle$ and their ratios between species. This strongly effects the neutrino signature of core collapse and it may affect nucleosynthesis in the inner SN ejecta by altering the ratio of the ν_e to $\bar{\nu}_e$ fluxes during explosion.[26,27] Second, rapid rotation may be accompanied by more significant gravitational wave emission than non-rotating progenitors, thereby opening up a new channel of emission for a fraction of $E_{\rm bind}$. Third, rapid rotation may affect the morphology and nucleosynthesis of the remnant.[28,29] Finally, rapid rotation at birth has been theoretically motivated by the existence of magnetars. Ref. 30 argue that millisecond rotation may be neceesary for production of the very high magnetic field strengths ($\sim 10^{15}$G) inferred for the magnetar neutron star subclass. Although uncertain, simple estimates imply that $> 10\%$ of all SNe produce magnetars.[31]

In §2, I summarize results[8] from a set of 1D simulations of core collapse, including rotation in a parameterized way, and I briefly discuss the implications of these models for neutrino detection. In §3 I discuss the neutrino background and its connection to other observable backgrounds.

2. Neutrinos from Rotating Core Collapse

Details of the models presented here are given in Ref. 8. A rotating $15\,M_\odot$ progenitor (model "E15A") from Ref. 32 and a set of $11\,M_\odot$ progenitors from Ref. 33 with imposed initial rotation profiles are calculated. The imposed profile is taken to be $\Omega(r) = (2\pi/P_0)[1 + (r/R_\Omega)]^{-2}$, where $R_\Omega = 1000\,\mathrm{km}$; thus, the iron core is in roughly solid-body rotation out to R_Ω. The $11\,M_\odot$ models span spin periods of $1 \lesssim P_0 \lesssim 10\,\mathrm{s}$ — the former corresponding to near-, but sub-Keplerian rotation after collapse and the latter yielding results virtually indistinguishable from non-rotating models. The $\Omega(r)$ profile for the $15\,M_\odot$ model most resembles the $P_0 = 2\,\mathrm{s}$ $11\,M_\odot$ progenitor (see Fig. 1 of Ref. 8).

Figure 1 shows L_ν and $\langle\varepsilon_\nu\rangle$ as a function of time after bounce for each of our rotating models. None of the models reach a centrifugal barrier and none explode. Higher initial rotation rates yield lower core temperatures after bounce for the PNS. On average, after ν_e breakout, this effect produces lower core L_ν and $\langle\varepsilon_\nu\rangle$ for shorter P_0. The fractional differences in $\langle\varepsilon_\nu\rangle$ between the model with $P_0 = 1.25$ s and a non-rotating model 200 ms after bounce are approximately 15%, 17%, and 30% for ν_μ, $\bar{\nu}_e$, and ν_e, respectively. The same comparison for L_ν yields fractional differences of 75%, 63%, and 33% for L_{ν_μ}, $L_{\bar{\nu}_e}$, and L_{ν_e}, respectively. The difference in L_{ν_μ} at 100 ms after bounce between our slowest and fastest rotators is a factor of ~ 6. Despite the fact that model E15A has rapid rotation, it has a much larger L_{ν_e} after breakout than, for example, the $P_0 = 2$ s $11\,M_\odot$ model, even though they have similar initial $\Omega(r)$ profiles. This is due to the extended density profile of model E15A relative to the $11\,M_\odot$ model and the associated larger accretion luminosity after bounce.

Figure 2 shows the ratio of L_{ν_e} and $\langle\varepsilon_{\nu_e}\rangle$ to L_ν and $\langle\varepsilon_\nu\rangle$ for each species as a function of time in each rotating model. The total luminosity in

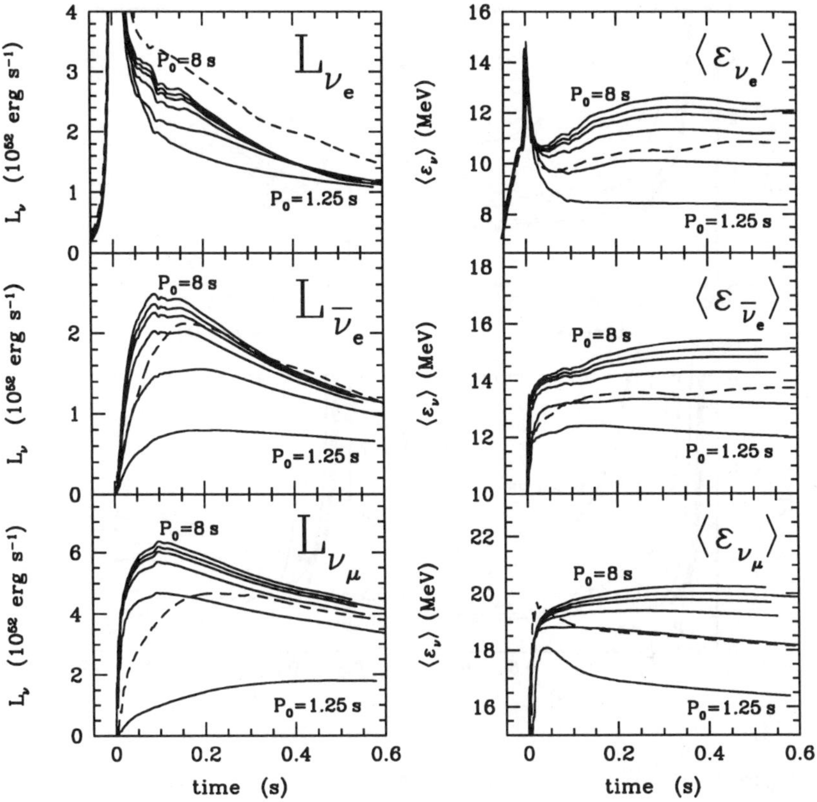

Figure 1. L_ν (*left column*) and $\langle\varepsilon_\nu\rangle$ (*right column*) as a function of time after bounce for the 11 $M_\odot$ progenitor with initial core spin periods of $P_0 = 1.25$, 2, 3, 4, 5, and 8 s (*solid lines*) and for the model E15A (*dashed lines*). 200 ms after bounce the lowest L_ν and $\langle\varepsilon_\nu\rangle$ correspond to the shortest initial spin period. This hierarchy is preserved at all times and in all panels except in the case of L_{ν_e} 400 ms after bounce and at breakout. The ν_e breakout pulse is not shown in the upper-left panel. It is largest for the fastest rotator, with peak L_{ν_e} ~30% larger than that for the slowest rotator ($\approx 2.4 \times 10^{53}$ ergs s^{-1}).

$\mu-$ and $\tau-$type neutrinos, L_{ν_μ}, is suppressed by rapid rotation: for slow rotation ($P_0 = 8$ s) the ratio $L_{\nu_e}/L_{\nu_\mu} \approx 0.3$ at 0.5s after bounce, whereas for $P_0 = 1.25$ s, $L_{\nu_e}/L_{\nu_\mu} \approx 0.8$. A similar enhancement of L_{ν_e} relative to $L_{\bar{\nu}_e}$ is also seen in the upper left panel. Although Figure 1 shows that $\langle\varepsilon_{\nu_e}\rangle$ is decreased on average by rotation, the right panels here show that both $\langle\varepsilon_{\bar{\nu}_e}\rangle$ and $\langle\varepsilon_{\nu_\mu}\rangle$ are decreased yet more. The two upper panels may

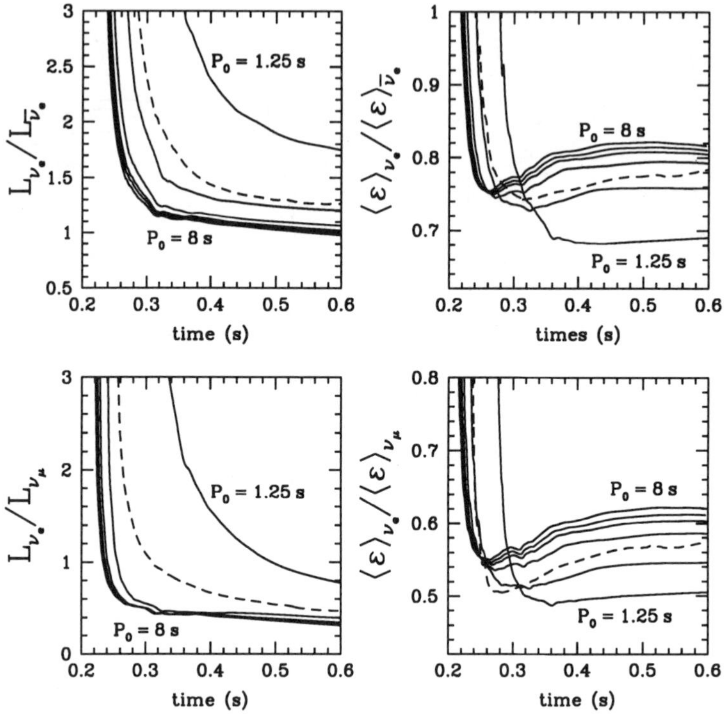

Figure 2. Ratios of L_νs (*left column*) and $\langle\varepsilon_\nu\rangle$s (*right column*) from Fig. 1.

be particularly important for the nucleosynthesis of inner ejecta in the gain region, should an explosion occur. Several studies address how the electron fraction and nucleosynthesis are determined by $L_{\nu_e}/L_{\bar{\nu}_e}$ and $\langle\varepsilon_{\nu_e}\rangle/\langle\varepsilon_{\bar{\nu}_e}\rangle$ as the explosion commences.[26,27] A general study with these ratios motivated by Figure 2 may constrain the frequency of SN events with rapid rotation.

Figure 3 shows L_ν and $\langle\varepsilon_\nu\rangle$ at 0.5 s after bounce for the 11 $M_\odot$ progenitor as a function of P_0. In the effectively non-rotating case $P_0 = 8$ s, $L_{\nu_e} : L_{\bar{\nu}_e} : L_{\nu_\mu}/4 :: 1 : 1.1 : 0.95$. As Figure 3 makes clear, this equality between neutrino species does not persist when rotation is rapid; for $P_0 = 1.25$ s, $L_{\nu_e} : L_{\bar{\nu}_e} : L_{\nu_\mu}/4 :: 1 : 0.6 : 0.4$. Although not explored systematically as a function of P_0, multi-D simulations of rotating collapse also exhibit suppression of $\bar{\nu}_e$ and ν_μ at the level described here.[34,35]

The strong suppression of both the $\bar{\nu}_e$ and ν_μ luminosity and average

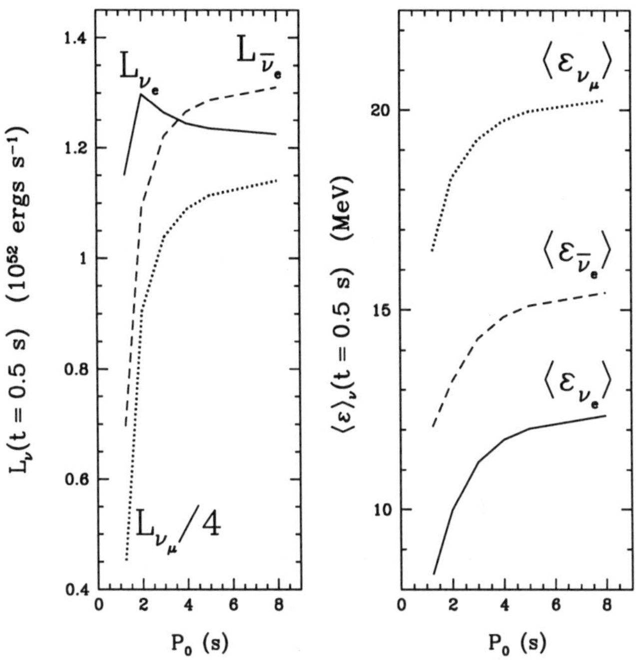

Figure 3. L_ν and $\langle\varepsilon_\nu\rangle$, at $t = 0.5$ s after bounce, as a function of P_0 (see Figs. 1 & 2). Although L_{ν_e} is fairly constant, $L_{\bar{\nu}_e}$ and L_{ν_μ} decrease sharply with more rapid initial core rotation.

energy is important for the detection of neutrinos from the next Galactic SN. Figure 4 shows the integrated number of events expected in SuperK from the 11 $M_\odot$ progenitors with $P_0 = 8$ s and 1.25 s at a distance of 10 kpc from the interactions $\bar{\nu}_e p \rightarrow ne^+$ and $\nu_e e^- \rightarrow \nu_e' e^{-\prime}$. Neutrino oscillations within the SN progenitor envelope, while the neutrinos are in transit, or within the Earth are neglected. The number rate of neutrinos detected is $\dot{N} \propto [L_\nu/\langle\varepsilon_\nu\rangle]\sigma$, where σ is the cross section. Because $\sigma \propto \langle\varepsilon_{\nu_e}\rangle$ for inelastic $\nu_e - e^-$ scattering and because L_{ν_e} is roughly independent of P_0 (Fig. 3), the total number of $\nu_e - e^-$ scattering events in SuperK is approximately independent of P_0, although the ν_e breakout burst is somewhat larger in the rapidly rotating case, $P_0 = 1.25$ s. However, the dominant charged-current interaction $\bar{\nu}_e p \rightarrow ne^+$ has $\sigma \propto \langle\varepsilon^2_{\bar{\nu}_e}\rangle$ so that for this process $\dot{N} \propto L_{\bar{\nu}_e}\langle\varepsilon_{\bar{\nu}_e}\rangle$. Thus, the strong decrease in both $L_{\bar{\nu}_e}$ and $\langle\varepsilon_{\bar{\nu}_e}\rangle$ with P_0 (Fig. 3) leads to a very large decrease in $N(< t)$: 0.5 s after bounce the ratio between the total number of neutrino events for the two models is $\approx 1190/350 \approx 3.4$.

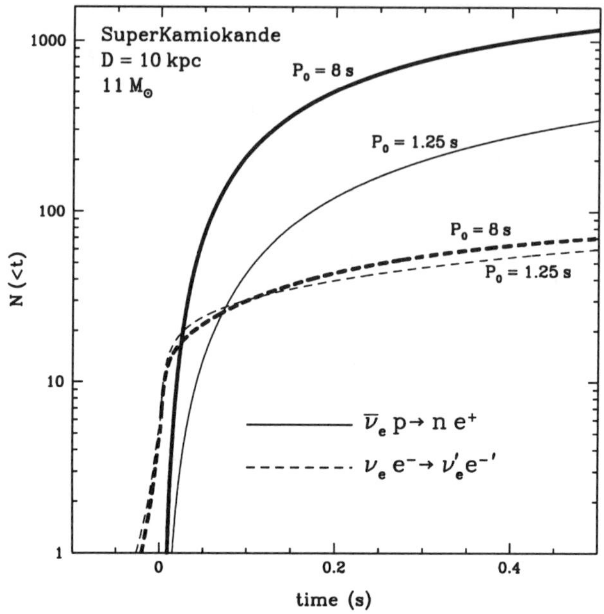

Figure 4. Integrated number of events in SuperK $N(<t)$ for a SN at 10 kpc as a function of time after bounce. Only the $P_0 = 8$s (*heavy lines*) and $P_0 = 1.25$s (*thin lines*) models are shown. Event rates for the reactions $\nu_e e^- \rightarrow \nu_e' e^{-\prime}$ (*dashed lines*) and $\bar{\nu}_e p \rightarrow pe^+$ (*solid lines*) are shown separately. In the rapidly rotating model the $\bar{\nu}_e$ signal is strongly suppressed and the prompt ν_e signal persists for ≈ 0.05s before being swamped by the $\bar{\nu}_e$s, much longer than for the $P_0 = 8$s model.

3. The Energy Budget of the Universe

A simple estimate for the contribution of SNe to the MeV neutrino background can be made by relating the total energy expected in neutrinos from each SN to the star formation rate, $\dot{M}_\star$, which is related to the total IR luminosity ($L_{\rm TIR}$ [8 − 1000]μm) by $L_{\rm TIR} = \epsilon \dot{M}_\star c^2$, where ϵ is an IMF-dependent constant. Assuming that the SN rate per unit star formation $\Gamma_{\rm SN}$ is a constant fraction of $\dot{M}_\star$ and that the total energy radiated is $E_\nu^{\rm tot} = 10^{53.5}$ ergs per SN, one finds that (averaging over a suitably large number of systems or time)

$$L_\nu \approx 3 L_{\rm TIR} E_{53.5} \beta_{17}, \tag{2}$$

where L_ν is the total neutrino luminosity and $\beta_{17} = (\Gamma_{\rm SN}/\epsilon)/17\,{\rm M}_\odot$ has only a weak dependence on the assumed IMF because massive stars dominate photon and neutrino production. The total specific intensity in MeV SN neutrinos expected from the star formation history of the universe can then be scaled from the integral over the comoving star formation rate density as a function of redshift, $\dot{\rho}_\star(z)$. Taking $\Omega_m = 0.3$ and $\Omega_\Lambda = 0.7$ and using "RSF2" for $\dot{\rho}_\star(z)$ from Ref. 36, one finds that $\zeta = \int dz[\dot{\rho}_\star(z)/\dot{\rho}_\star(z=0)]/\{(1+z)^2[\Omega_m(1+z)^3+\Omega_\Lambda]^{1/2}\} \approx 2$. Thus,

$$F_\nu^{\rm tot} \sim 4\,(\zeta/2) \quad 10\,{\rm MeV\ cm^{-2}\ s^{-1}\ sr^{-1}}, \tag{3}$$

where $H_0 = 71\,{\rm km\,s^{-1}\,Mpc^{-1}}$ has been assumed and the dependences on $E_\nu^{\rm tot}$ and β have been dropped. The dominant reaction for detection of relic neutrinos in SuperK is $\bar{\nu}_e p \rightarrow ne^+$. Assuming that $F_{\bar{\nu}_e} \approx F_\nu^{\rm tot}/6$, $\langle\varepsilon_{\bar{\nu}_e}\rangle \approx 15\,{\rm MeV}$, and accounting for the fact that SuperK's sensitivity to the diffuse background is maximized for $\varepsilon_{\bar{\nu}_e} \sim 20\,{\rm MeV}$, considerably larger than $\sim$15/(1 + z) MeV for $z \approx 1$, one finds an event rate in SuperK of $\sim$1 yr^{-1}, in accord with more complete estimates.[22–24] As Figures 3 & 4 make clear, if a large fraction of all SNe are born rapidly rotating, then the background estimate is significantly decreased because of the strong decrease in $\langle\varepsilon_{\bar{\nu}_e}\rangle$ and $\langle\varepsilon_{\bar{\nu}_\mu}\rangle$ with decreasing P_0.

As implied by equations (2) and (3), the massive stars that generate MeV neutrinos also produce a corresponding total IR background of $F_{\rm TIR}^{\rm tot} \approx 2\times10^{-5}$ ergs cm^{-2} s^{-1} sr^{-1} $\approx$ 20 nW m^{-2} sr^{-1}. The SNe that accompany this star formation also accelerate cosmic ray electrons and protons to very high energies. Inelastic collisions between cosmic ray protons and gas in the ISM of star-forming galaxies produce $\pi^{+,-}$ and π^0, which subsequently decay to $e^{-,+}$ and high-energy neutrinos, and γ-rays, respectively. The primary electrons and secondary electrons/positrons suffer synchrotron losses in the host galactic magnetic field. Inverse Compton and bremsstrahlung losses likely also contribute significantly to cooling.[37] The observed tight linear correlation between the IR and radio luminosity of star-forming and starburst galaxies implies that the contribution to the IR background from star formation comes together with radio emission at the level νI_ν(radio) $\approx 3\times10^{-11}$ ergs cm^{-2} s^{-1} sr^{-1}, assuming a flat cosmic ray spectrum — consistent with $\sim$5% of the 10^{51} ergs of asymptotic kinetic energy of each SN going into cosmic rays.[38] Recent work suggests that inelastic $p-p$ collisions in the dense ISM of starburst galaxies may contribute significantly to the diffuse GeV γ-ray and GeV$-$TeV ν_μ-neutrino background at the level of $\sim10^{-7}$ GeV cm^{-2} s^{-1} sr^{-1}.[38,39] Thus, stars that produce SNe may dominate — or contributor importantly to — the neutrino (MeV & GeV$-$TeV), γ-ray, IR, and radio backgrounds.

Acknowledgments

I thank Adam Burrows, Eliot Quataert, and Eli Waxman for collaboration and many stimulating conversations.

References

1. Bethe, H. & Wilson, J. R. 1985, ApJ, 295, 14
2. Janka, H.-Th. 2001, A&A, 368, 527
3. Rampp, M., & Janka, H.-T. 2000, ApJL, 539, L33
4. Rampp, M., & Janka, H.-T. 2002, A&A, 396, 361
5. Mezzacappa, A., et al. 2001, PRL, 86, 1935
6. Liebendörfer, M., at al. 2001, PRD, 63, 103004
7. Thompson, T. A., Burrows, A., & Pinto, P. A. 2003, ApJ, 592, 434
8. Thompson, T. A., Quataert, E., & Burrows, A. 2005, ApJ, 620, 861
9. Sumiyoshi, K., et al. 2005, ApJ, 629, 922
10. Herant, M., et al. 1994, ApJ, 435, 339
11. Burrows, A., Hayes, J., & Fryxell, B. A. 1995, ApJ, 450, 830
12. Janka, H.-T., & Mueller, E. 1996, A&A, 306, 167
13. Fryer, C. L. 1999, ApJ, 522, 413
14. Buras, R., et al. 2003, PRL, 90, 241101
15. Buras, R., et al. 2006, A&A, 447, 1049
16. Burrows, A., et al. 2006, ApJ, 640, 878
17. Blondin, J. M., Mezzacappa, A., & DeMarino, C. 2003, ApJ, 584, 971
18. Ohnishi, N., Kotake, K., & Yamada, S. 2006, ApJ, 641, 1018
19. Foglizzo, T., et al. 2006, arXiv:astro-ph/0606640
20. Burrows, A. & Lattimer, J.M. 1986, ApJ, 307, 178
21. Pons, J. A., et al. 1999, ApJ, 513, 780
22. Ando, S. & Sato, K. 2002, Prog. Theoretical Phys., 107, 957
23. Strigari, L. E., et al. 2005, JCAP, 4, 17
24. Yuksel, H., Ando, S., & Beacom, J. 2005, arXiv:astro-ph/0509297
25. Malek, M., et al. 2003, PRL, 90, 061101
26. Pruet, J., et al. 2005, ApJ, 623, 325
27. Fröhlich, C., et al. 2006, ApJ, 637, 415
28. Thompson, T. A., Chang, P., & Quataert, E. 2004, ApJ, 611, 380
29. Bucciantini, N., et al. 2006, MNRAS, 368, 1717
30. Duncan, R. C., & Thompson, C. 1992, ApJL, 392, L9
31. Woods, P. M., & Thompson, C. 2004, arXiv:astro-ph/0406133
32. Heger, A., Langer, N., & Woosley, S. E. 2000, ApJ, 528, 368
33. Woosley, S. E. & Weaver, T. A. 1995, ApJS, 101, 181
34. Fryer, C. L., & Heger, A. 2000, ApJ, 541, 1033
35. Dessart, L., et al. 2006, ApJ, 644, 1063
36. Porciani, C., & Madau, P. 2001, ApJ, 548, 522
37. Thompson, T. A., et al. 2006, ApJ, 645, 186
38. Thompson, T. A., Quataert, E., & Waxman, E. 2006, arXiv:astro-ph/0606665
39. Loeb, A., & Waxman, E. 2006, JCAP, 5, 3

INTEGRAL

R. SUNYAEV*, E. CHURAZOV, M. REVNIVTSEV and S. SAZONOV

Max-Planck Institute for Astrophysics,
85741 Garching bei München, Germany
and
Space Research Institute, Russian Academy of Sciences,
117997 Moscow, Russia
**E-mail: sunyaev@mpa-garching.mpg.de*

INTEGRAL is an orbital observatory covering a broad energy range from keVs to MeVs. Its strongest features are sensitive imaging in hard X-rays (15–100 keV) and ultra-fine spectroscopy of gamma-ray lines. We present selected results of INTEGRAL observations in 2003–2006 on such subjects as positron annihilation in the Milky Way, activity of the central Galactic black hole in the recent past, Galactic absorbed X-ray sources, statistics of nearby AGN and the cosmic X-ray background.

Keywords: Hard X-ray Surveys; Galactic Center; Cosmic X-ray Background

1. Introduction

The **Inte**rnational **G**amma-**R**ay **A**strophysics **L**aboratory[1] is the ESA's mission with the participation of Russia and the USA, dedicated to fine spectroscopy and imaging of celestial gamma-ray sources in the energy range from 15 keV to 10 MeV with simultaneous monitoring in the X-ray (3–35 keV; JEM-X) and optical (V-band, 550 nm; OMC) energy ranges. INTEGRAL was launched by a PROTON launcher from Baikonour/Kazakhstan on October 17, 2002. The satellite operates in a 72-hour orbit with an apogee of 153,600 km and an inclination of 52.5 deg. Below we summarize some of the most interesting results obtained with INTEGRAL so far.

2. Electron–positron annihilation in the Milky Way

The electron–positron annihilation line at 511 keV is the brightest gamma-ray line of the Galaxy. First observed with a NaI scintillator as a ~ 476 keV line coming from the Galactic Center (GC) region,[2] it was subsequently identified with a narrow (FWHM < 3.2 keV) e^-–e^+ annihilation line using germanium detectors.[3]

The SPI instrument[4] on board INTEGRAL is a coded-mask germanium spectrometer, well suited for 511 keV studies. The instrument consists of 19 Ge detectors, has a field of view of $\sim 16°$ (fully-coded), effective area ~ 70 cm^2 and energy resolution ~ 2 keV at 511 keV. SPI studies of the 511 keV emission are reported in several recent papers.[5–8] Here we present updated results of our analysis in Ref. 7, using data covering a period from the early 2003 till May 2006.

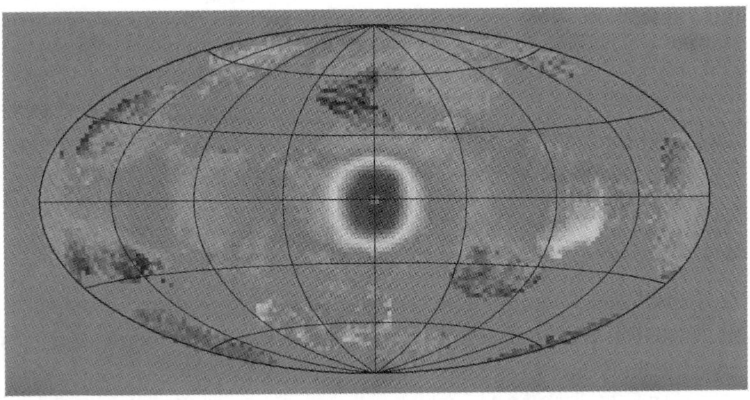

Fig. 1. Heavily smoothed image of the Milky Way in the electron–positron annihilation line. The central bright spot is centered on the Galactic Center.

Figure 1 shows the map of the Galaxy in the narrow energy band 508–514 keV, containing the 511 keV line. This image was constructed by averaging the detected signal over the SPI field of view, thus effectively convolving the true sky image with a $\sim 30°$-wide filter. One can see that the GC region, rather than the Galactic plane, is the most prominent source of annihilation emission. Detailed analysis[6,7] shows that the flux is mostly coming from a $\sim 6°$-wide Gaussian centered at the GC. Such spatial distribution contrasts with that of the 1.8 MeV line of Al^{26} (Fig. 2), implying that most of the positrons are not produced by decay of Al^{26}. Furthermore, the INTEGRAL data witness against a positron origin associated with Type 2 supernovae or massive stars, since these classes of object are concentrated toward the Galactic disk rather than the bulge. The data strongly favor bulge-dominated populations of positron sources, such as Type 1a supernovae, low-mass X-ray binaries and dark matter.

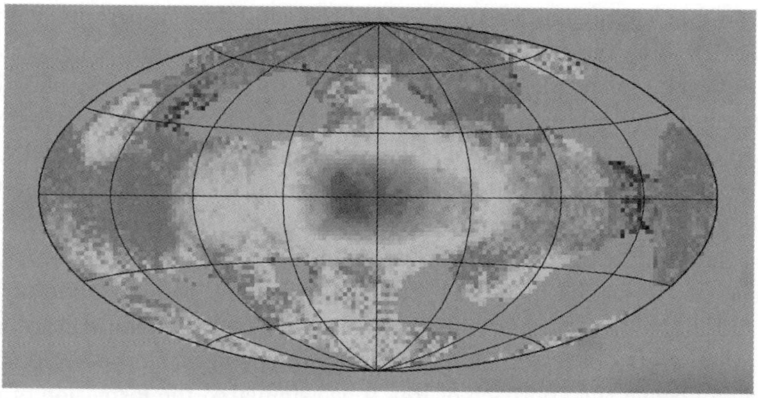

Fig. 2. Heavily smoothed image of the Milky Way emission in the 1809 keV Al^{26} line. The flux distribution is more extended along the disk than concentrated in the bulge.

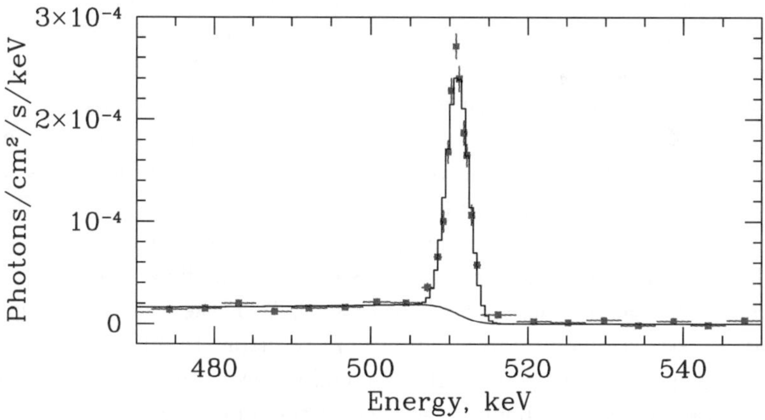

Fig. 3. Spectrum of the e^-–e^+ annihilation radiation observed by SPI from the GC region vs. the model consisting of a Gaussian line and an ortho-positronium continuum.

Figure 3 shows the spectrum of annihilation emission together with a model consisting of a Gaussian line and an ortho-positronium continuum. The flux in the narrow line is 0.7×10^{-3} phot s^{-1} cm^{-2}. The position of the line centroid agrees very well with the laboratory value: $E/m_ec^2 = 1.00002 \pm 0.00007$. This limits the bulk motion of the annihilating positrons to be less than 20–30 km s^{-1} relative to the Earth. The observed flux implies that $\sim 2 \times 10^{43}$ positrons are annihilating in the inner Galaxy every second.

To a first approximation the shape of the annihilation spectrum can be characterized by the effective width of the 511 keV line and the strength of the continuum below 511 keV associated with three-photon decay of ortho-positronium. All other channels, including the decay of para-positronium, lead to the generation of two photons and contribute only to the 511 keV line. Since ortho- and para-positronium are produced in 3:1 proportion, the continuum/line flux ratio determines the fraction of positrons $F_{\rm PS}$ annihilating through the formation of positronium: $F_{\rm PS} = 2/(1.5 + 2.25 F_{2\gamma}/F_{3\gamma})$.

The two measured quantities, the line width and the flux ratio of the line and 3-photon continuum, impose constraints on the temperature and ionization degree of the annihilation medium. Using a Monte-Carlo code largely following the approach of Ref. 9, we simulated the formation of the annihilation spectrum in a pure hydrogen, dust-free plasma. The predicted annihilation line has a non-Gaussian shape and contains a broad and a narrow components. Interestingly, there is indeed an indication of non-Gaussianity in the observed spectrum (Fig. 3).

In Fig. 4 the effective line width and positronium fraction are shown as functions of temperature and ionization degree. These theoretical curves are compared with the constraints provided by the INTEGRAL data. We assume that positrons are born with energies exceeding 100 keV and decelerate in the gas and plasma due to different collisional processes. There are two families of curves. One regime pertains to gas temperatures below ~6,000 K. In a cold and neutral medium about 94% of positrons form positronium in flight. The remaining 6% decrease their energy below the threshold for positronium formation (6.8 eV) and annihilate with bound electrons. The net line has a width of $\sim$ 4.6 keV. If the ionization degree exceeds $\sim 10^{-3}$, Coulomb losses cause more positrons to fall below 6.8 eV to subsequently experience radiative recombination with free electrons or annihilation with free or bound electrons. If the ionization degree exceeds several per cent, only radiative recombination and annihilation with free electrons are important, and both the positronium fraction and the line width approach the values expected for a fully ionized plasma.

The second family of curves corresponds to temperatures above ~6,000 K. In this regime thermalized positrons can form positronium by charge exchange with hydrogen atoms. This process dominates over radiative recombination and direct annihilation if the plasma is not strongly ionized. The positronium fraction approaches unity. For a significantly ionized plasma (~6–10%) at $T \geq 8,000$ K annihilation with free electrons becomes important and the positronium fraction decreases with increasing

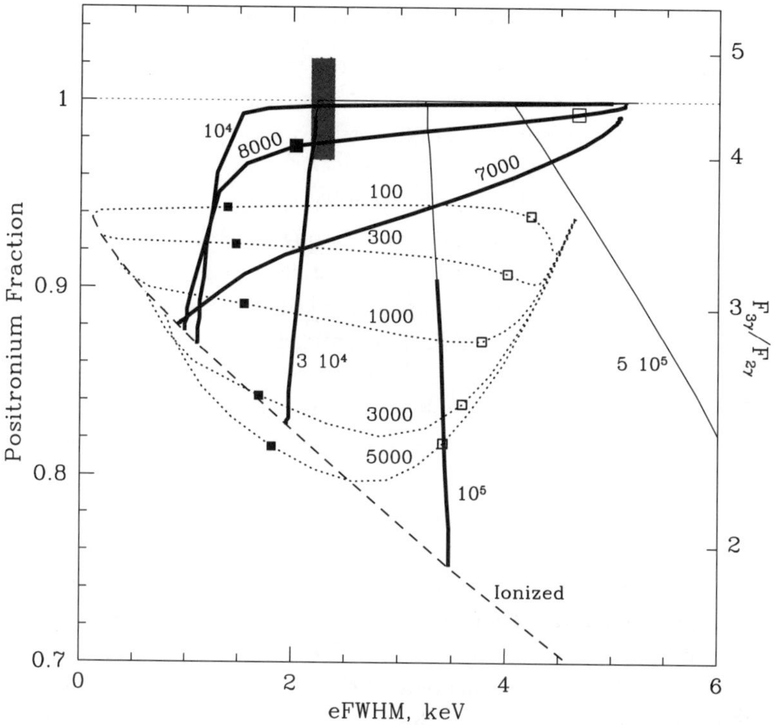

Fig. 4. Positronium fraction vs. effective width of the 511 keV line for different temperatures and ionization degrees of the medium. There are two groups of theoretical curves: low-temperature ($T \leq 5,000$ K, dotted lines), and high-temperature ($T \geq 7,000$ K, solid lines). The ionization degree varies from 0 to 1 along the curves. For the low-temperature curves and for the 8,000 K one, an ionization degree of 0.01 is indicated by empty squares, and 0.1 by dark squares. Each high-temperature curve has two regimes: thin (thick) lines correspond to the ionization degrees lower (higher) than expected for collision dominated plasma. The dashed line shows the prediction for a fully ionized plasma. The rectangle represents the range of parameter values allowed by INTEGRAL data.

ionization degree.

As follows from Fig. 4, the observed line width and strength of the ortho-positronium continuum can be explained by the positron annihilation taking place in a 8,000–10,000 K, ~10% ionized medium, which closely corresponds to the standard warm ISM phase.[10] However, certain combinations of several ISM phases are also allowed by the data.

3. Past activity of the central Galactic black hole

Our Galactic Center harbors a black hole of mass $\sim 3 \times 10^6 M_\odot$.[11] It has been a puzzle why the source Sgr A* associated with the black hole is faint despite the presence of significant amounts of ambient gas capable of fuelling it.[12]

Among many complex structures near the GC, X-ray telescopes have detected 8–20 keV continuum[13] and strong (1–2 keV equivalent width) 6.4 keV line diffuse emission[14–18] associated with giant molecular clouds, in particular Sgr B2 located at a projected distance $\sim$ 100 pc from Sgr A*. This was suggested to be radiation emitted by Sgr A* several hundred years ago, Compton scattered and reprocessed by the cloud's gas.[13,14] One could then expect Sgr B2 to be also a strong source of hard X-rays. As discussed below, such a hard X-ray source has indeed been discovered by INTEGRAL.[19]

INTEGRAL has extensively observed the central $20° \times 20°$ area of the Galaxy.[20] Figure 5 shows a hard X-ray image of the innermost $3.5° \times 2.5°$ region obtained from observations with a total exposure of 2.3 Ms. Thanks to the good angular resolution (12′) of the IBIS telescope,[21] practically all bright sources in this area are resolved. The newly discovered source IGR J17475−2822[20] is coincident with the Sgr B2 cloud. The observed 20–200 keV flux is 2.5 ± 0.1 mCrab, which corresponds to a luminosity of 6×10^{35} erg s^{-1} at a distance of 8.5 kpc. The fact that INTEGRAL sees X-ray emission above 20 keV from the zone of 6.4 keV emission in Sgr B2 provides strong support to the Sgr A* scenario.

Analysis of the broad-band spectrum of Sgr B2 further supports this scenario. As shown in Fig. 6, the spectral energy distribution of Sgr B2 measured with INTEGRAL at 20–200 keV matches the 3–20 keV spectrum measured with ASCA and GRANAT/ART-P. The combined spectrum at 3–200 keV can be well fit by a model in which X-rays from Sgr A* are scattered and reprocessed in a homogeneous spherical cloud of cold gas. The spectrum emerging from Sgr B2 depends on the slope (Γ) of the incident spectrum, the cloud radial optical depth to Thomson scattering [$\tau = \sigma_T(2n_{H_2})r$], the Fe abundance relative to solar (A), and the scattering angle (θ) for photons travelling from Sgr A* to Sgr B2 and then to us, and the ISM column density toward Sgr B2 (N_H). Using Monte Carlo simulations we found the following best-fit values: $\Gamma = 1.8 \pm 0.2$, $\tau = 0.4 \pm 0.1$, $A = 1.9 \pm 0.2$, $\theta = 80° \pm 10°$, and $N_H = (8 \pm 2) \times 10^{22}$ cm^{-2}. The best-fit model is shown in Fig. 6.

Based on the measured optical depth we can estimate the mass of the scattering gas in Sgr B2 as $M_{H_2} = (4\pi/3)(m_p/\sigma_T)\tau r^2 \approx 2 \times$

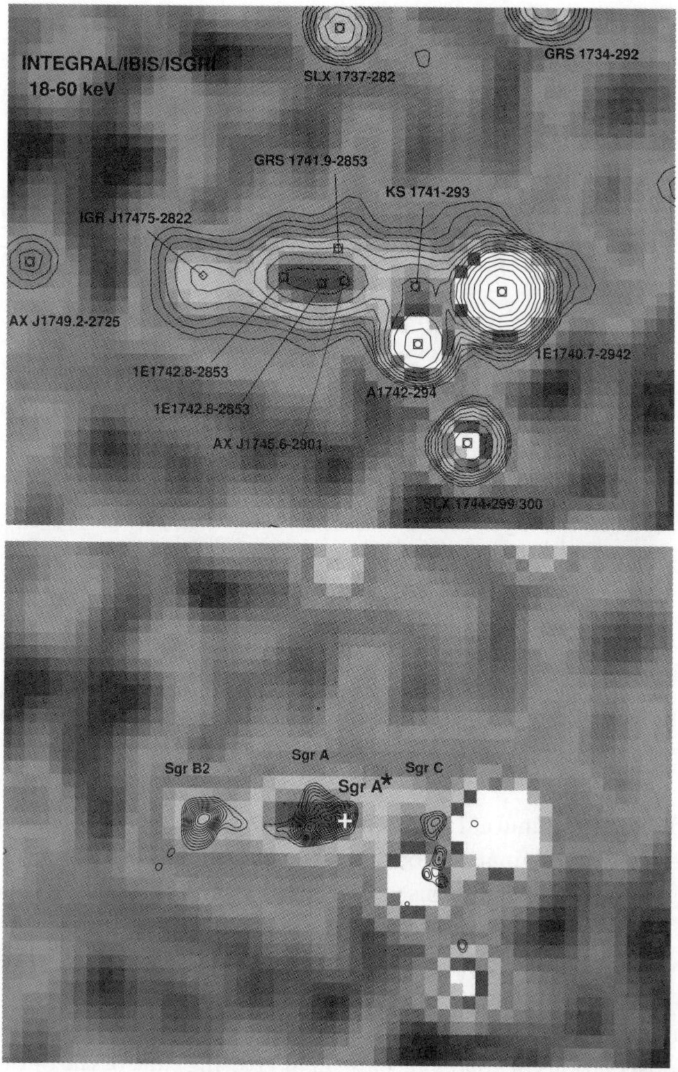

Fig. 5. *Top:* 3.5° × 2.5° 18–60 keV image of the GC region obtained with INTEGRAL/IBIS. *Bottom:* The same IBIS color image, with overplotted ASCA/GIS contours of brightness distribution in the 6.4 keV line. Largest molecular clouds are indicated and the position of Sgr A* is marked with a cross.

$10^6 M_\odot (r/10\ \mathrm{pc})^2$. The high-energy rollover tentatively seen in the INTEGRAL spectrum can be explained by the Compton recoil of hard X-ray photons in the cloud and is sensitive to θ. We can then place an upper limit

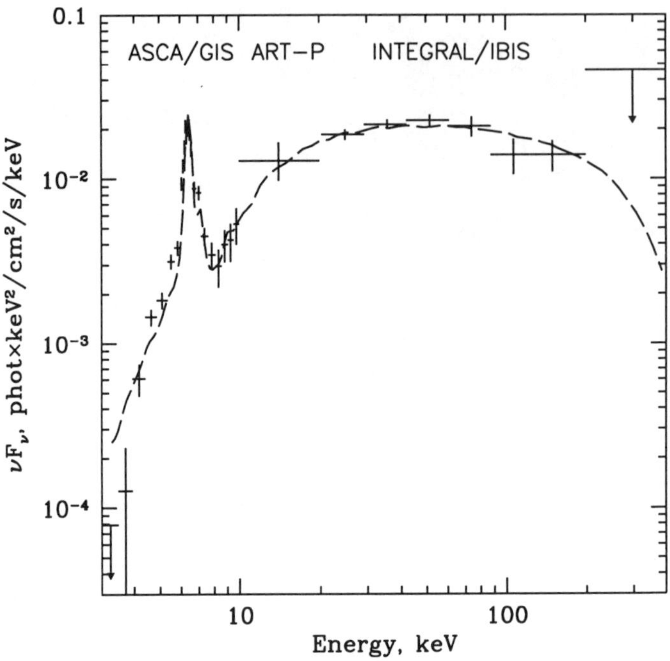

Fig. 6. Broad-band X-ray spectrum of the source IGR J17475−2822 associated with the Sgr B2 cloud. Data of ASCA/GIS (3–10 keV), GRANAT/ART-P (10–20 keV) and INTEGRAL/IBIS (20–400 keV) are presented. The dashed line is the best-fit.

of $\theta < 135°$ on the mutual position of Sgr A* and the Sgr B2 cloud.

Alternative explainations of the X-ray emission of Sgr B2 meet serious difficulties.[19] In particular, instead of attributing the primary emission to Sgr A* one could hypothesize that a transient source inside the Sgr B2 cloud was irradiating the molecular gas. The large equivalent width of the 6.4 keV line implies that we are seeing pure reprocessed emission but not the primary source. The source therefore should have faded away before the ASCA observations of 1993. Since the light crossing time of the Sgr B2 cloud is ~ 30 years, one would expect to see a decline of the 6.4 keV line flux by a factor of 2 from 1993 till now.[22] Using archival data of ASCA, BeppoSAX, Chandra and XMM observatories we found no significant variability of the line flux during the period 1993–2001. INTEGRAL observations similarly indicate that the continuum 18–60 keV flux was constant within 25% during 2003–2004. For the Sgr A* model, the constancy of the line flux merely means that the luminosity of Sgr A* remained approximately constant for more than 10 years a few hundred years ago.

The bombardment of molecular gas by cosmic-ray electrons was also put forward[23,24] to explain the 6.4 keV emission. In this model, the lack of a strong cutoff below 200 keV in the spectrum of Sgr B2 implies that electrons with energies higher than a few hundred keV should be present, while the slope of the observed spectrum ($\Gamma \approx 2$) constrains the distribution of electrons in energy. Given these constraints, only $(1\text{–}3)\times10^{-5}$ of the cosmic-ray electrons' energy can go into hard X-ray radiation around 50 keV. Thus, to produce the observed luminosity at 50 keV at least $(1\text{–}3)\times10^{40}$ erg s^{-1} of energy in cosmic-ray electrons ought to be dumped into the cloud. This power is comparable to the bolometric luminosity of Sgr B2[25] which is thought to be mostly due to hot stars. Since Sgr B2 is an almost perfect calorimeter, little room is left for additional energy in low-energy cosmic-ray protons. Furthemore, the observed ~ 2 keV equivalent width requires a factor of ~ 5 overabundance of iron in Sgr B2. The hypothesis of production of the 6.4 keV line by cosmic-ray ions[26,27] encounters similar problems.

Thus although the cosmic-ray scenario cannot be ruled out,[24] most likely Sgr B2 cloud is sending us an X-ray echo of violent activity of the GC black hole some 300 years ago, which lasted at least 10 years. The luminosity of Sgr A* at that time was $\sim 5\times10^{38}$ erg s^{-1} in the 2–10 keV band, i.e. a few $\times10^{5}$ times higher than it is now.[12] It is actually not surprising that our GC was so active in the recent past, as AGN with luminosities higher than 10^{38} erg s^{-1} (2–10 keV) are found in $\sim 50\%$ of galaxies morphologically similar to the Milky Way.[28] There is a significant probability that Sgr A* will become bright again in the foreseeable future. That would provide unique information about the duty cycle of activity in galactic nuclei.

4. Population of absorbed X-ray sources in the Galaxy

The INTEGRAL observatory is particularly well suited for searching strongly absorbed sources. While obscured AGN are immediately coming to mind as a principle target of such searches, INTEGRAL has also discovered a large population of strongly absorbed sources in the Milky Way. The discovery[29] of the first such source, IGR J16318−4848, with a very strong low-energy cutoff (Fig. 7), was followed by the discovery of more than a dozen similar objects. Based on the photometric observations of IGR J16318−4848 it was suggested[30] that it is a binary with a giant or supergiant secondary having a strong wind. This hypothesis was soon confirmed via spectroscopic observations of the optical counterpart.[31]

These strongly absorbed sources are strongly clustered on the sky.[32] A large group of sources was discovered in a relatively small region of the

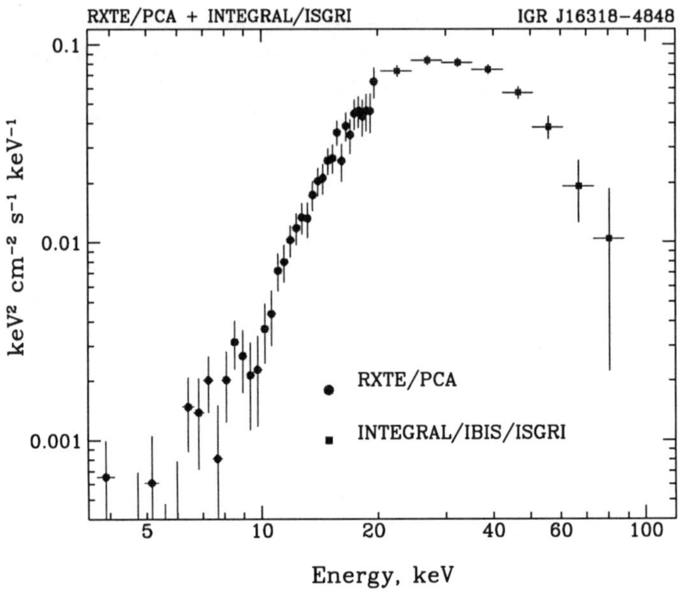

Fig. 7. Broad-band X-ray spectrum of the prototype of the family of "absorbed" Galactic sources discovered by INTEGRAL — IGR J16318−4848.

Galactic plane, close to the tangent to the Norma spiral arm (Fig. 8). Some of the discovered sources proved to be X-ray pulsars. A similar cluster of accreting X-ray pulsars was previously seen with GINGA in the direction of the Scutum spiral arm tangent.[33] Subsequent ASCA observations showed that a large fraction of them are intrinsically absorbed.[34]

Such clustering around spiral arms suggests that we are dealing with high-mass X-ray binaries (HMXBs) associated with regions of enhanced star formation. Previous studies of Galactic X-ray sources[33,35] suggest that sources with significant intrinsic X-ray absorption can be reliably classified as HMXBs, while the absence of significant absorption does not enable reliable classification. Such a difference between high- and low-mass X-ray binaries (LMXBs) can be understood if we recall that in most HMXBs the compact object accretes matter from the stellar wind which can provide significant absorbing column density. The best studied HMXBs such as Cyg X-3, Vela X-1, GX 301-2 and 4U 1700−37 are all characterized by strong X-ray absorption. On the contrary, in LMXBs accretion occurs mostly through the inner Lagrangian point and such systems usually lack strong intrinsic absorption.

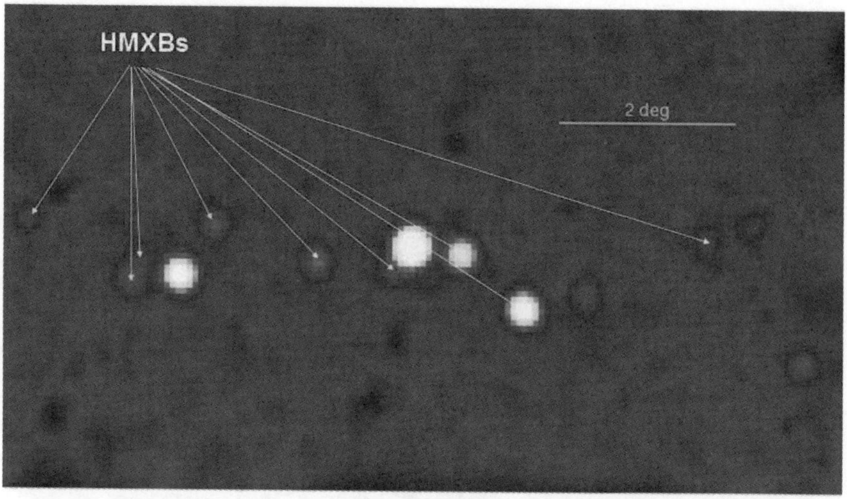

Fig. 8. 20–60 keV image of the Galactic plane near the tangent to the Norma spiral arm. The hard X-ray emission is dominated by strongly absorbed HMXBs.

One can therefore expect any soft X-ray survey to be biased against finding HMXBs with high intrinsic X-ray absorption and hard X-ray surveys are needed to provide a census of such objects. By now sensitive (~1–2 mCrab) surveys have been carried out by INTEGRAL for much of the Galactic plane[20,36,37] adding 18 objects to the list of Galactic HMXBs, 15 of which are significantly absorbed.

5. Cosmic X-ray background and nearby AGN

Supermassive black holes (SMBHs) apparently reside in the nuclei of most galaxies and each SMBH likely had a period of rapid growth in its history when a vast amount of radiation was emitted. Such objects are observed as Seyfert galaxies in our vicinity and as powerful quasars in the distant Universe. All together the growing SMBHs are believed to be making a dominant contribution to the cosmic X-ray background (CXB).

Understanding the nature of the CXB means that we know its flux and that we can account for all of it in terms of known populations of objects. At energies below several keV deep surveys with the X-ray telescopes Chandra and XMM-Newton have already resolved $\sim 90\%$ of the CXB, directly counting $\sim 10^3$ sources per square degree of the sky.[38] At higher energies such an exercise is far beyond the capabilities of existing instruments so other techniques have been employed by INTEGRAL to study the CXB above 20 keV and its composition, as described below.

5.1. *Using Earth for measuring the CXB*

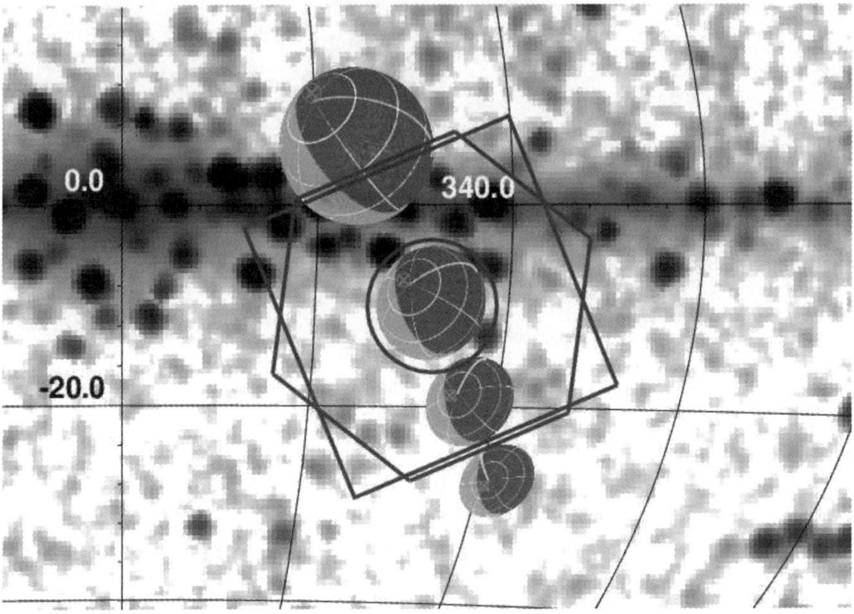

Fig. 9. Illustration of INTEGRAL observations of the Earth. The FoVs of JEM-X, IBIS and SPI are shown with a circle, box and hexagon, respectively, superposed on the RXTE 3–20 keV map of the sky. In this map apart from many point sources the extended X-ray emission of the Galactic Ridge is visible. During the observation the Earth crosses the instruments' FoVs. The Earth's day side is indicated by the lighter shade of gray.

Determination of the CXB spectrum is a difficult task because hard X-ray measurements are usually dominated by the internal detector background. To decompose the total background into particle-induced and CXB components one needs either an accurate model of the internal background or two observations with different relative contributions of these two components. The latter approach was behind recent INTEGRAL observations in which the Earth disk was used as a screen modulating the CXB.[41]

During these observations, carried out in four $\sim$ 30 ks periods in January–February 2006, the Earth was drifting through the field of view of INTEGRAL instruments (Fig. 9) producing flux modulation $\sim$ 200 mCrab at 30 keV. The Earth distance varied from $\sim$40,000 to $\sim$100,000 km, and the maximal solid angle subtended by the Earth was $\sim$ 100 sq. deg.

Extracting information about the CXB from the light curve and spectra measured by INTEGRAL is complicated by the fact that the Earth's atmosphere emits hard X-ray radiation induced by cosmic rays and also reflects a significant fraction of the CXB in all directions. Furthermore, several Galactic X-ray sources were occulted by the Earth together with the CXB during the INTEGRAL observations. We took all these complications into account in analyzing the data.

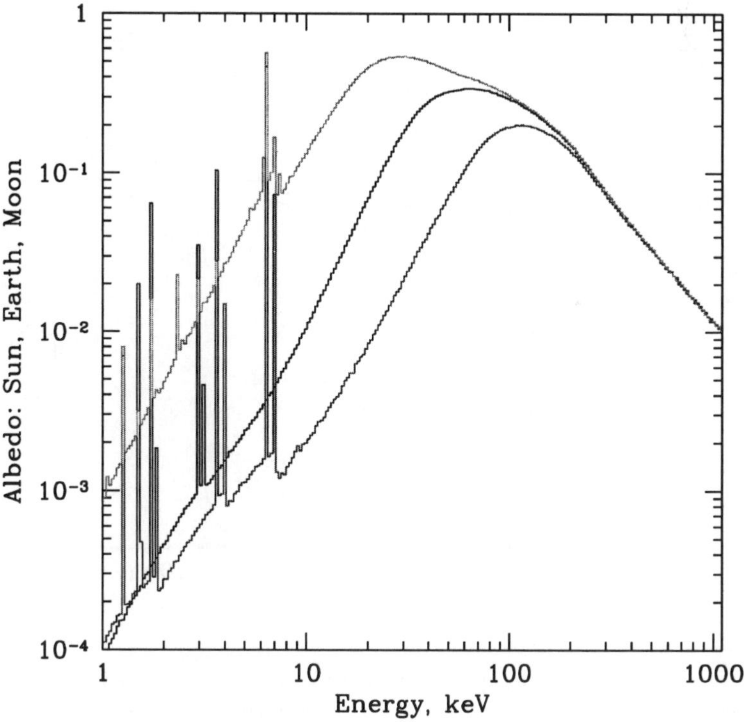

Fig. 10. Albedo with respect to the CXB spectrum calculated for: Solar atmosphere (upper curve), Earth's atmosphere (middle curve) and Moon's crust (bottom curve).

The problem of reflection of X-rays by the upper layers of the Earth's atmosphere closely resembles the well-studied case of the X-ray reflection from a stellar surface[39] or an accretion disk[40] except for the different chemical composition of the reflecting medium. The Earth X-ray albedo $A(E)$ peaks at $\sim$ 60 keV. To accurately evaluate it we performed Monte-Carlo simulations[42] (see Fig. 10) taking into account all relevant physical processes (photoabsorption, Compton scattering and Rayleigh scattering). The reflected CXB component was then included in our spectral model as $S_{\rm CXB}(E)A(E)$, where $S_{\rm CXB}(E)$ is the CXB spectrum.

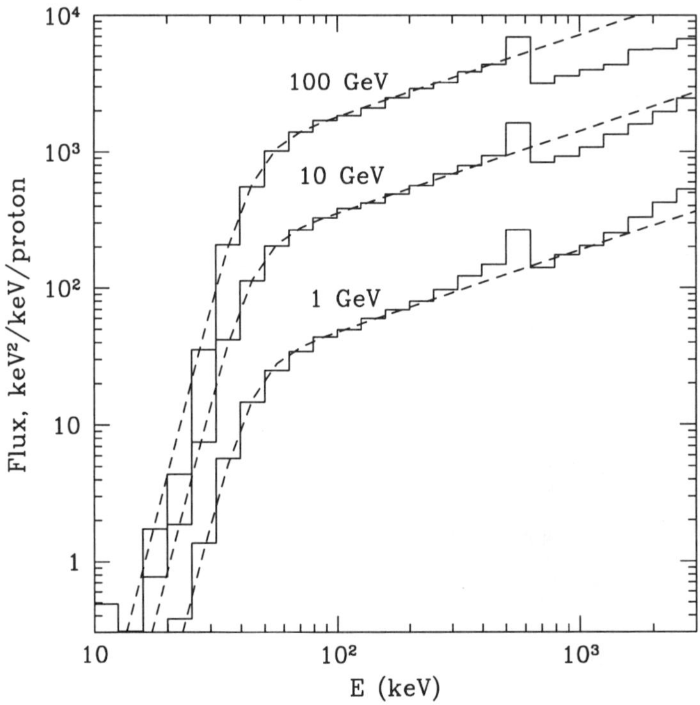

Fig. 11. Examples of simulated spectra (solid lines) of atmospheric emission produced by cosmic protons of given energy: $E_p = 1$, 10 and 100 GeV. It can be seen that in the photon energy range 25-300 keV the shape of the emergent spectrum is almost invariant and well fitted by Eq. (1) (the dashed lines).

Another major component in the observed spectra is produced by cosmic rays bombarding the Earth's atmosphere. Using realistic Monte-Carlo simulations based on the toolkit Geant 4, we found[43] that as a result of multiple Compton scatterings of photons the hard X-ray spectrum emerging from the atmosphere should be barely sensitive to the energy or type of the incident cosmic-ray particle and in the energy range 25–300 keV can be well fit (see Fig. 11) by the formula

$$S_{\mathrm{ATM}}(E) = \frac{C}{(E/44)^{-5} + (E/44)^{1.4}}. \tag{1}$$

The photon spectrum peaks around 44 keV; at lower energies the spectrum shows a rapid decline ($\propto E^{-5}$) due to photoabsorption.

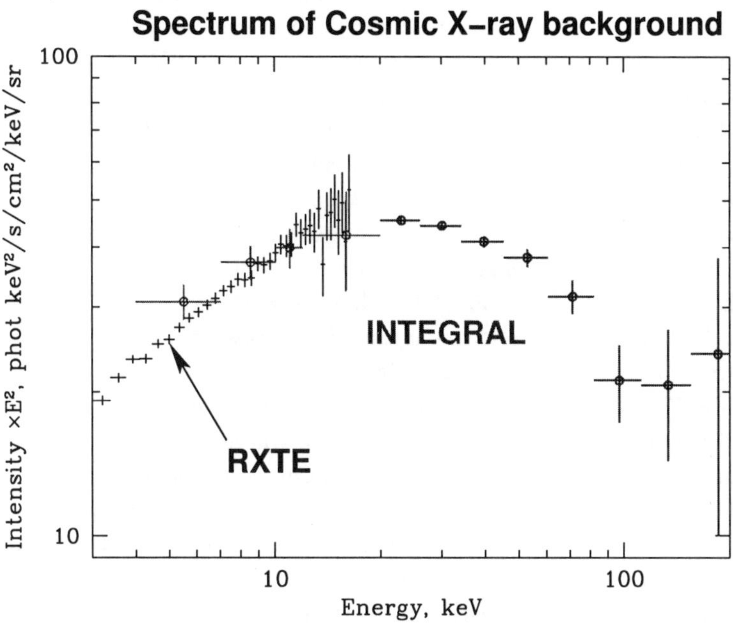

Fig. 12. Comparison of the CXB spectrum measured by INTEGRAL with the 3–20 keV CXB spectrum determined from RXTE data.[45] The IBIS/ISGRI and SPI data points were averaged in this plot.

After subtraction of the (constant) internal detector background the observed spectrum was fitted with a model consisting of three components: CXB decrement, reflected radiation and atmospheric emission. We adopted the Gruber et al. fitting formula of the HEAO-1 spectrum[44] and Eq. (1) for the spectral shapes of the CXB and atmospheric component, respectively, and considered the amplitudes of these components free parameters.

The inferred CXB spectrum (Fig. 12 lies $\sim 10\%$ higher than the previous measurement by HEAO-1.[44] The measured CXB flux near its peak at 29 keV is 46.9 keV^2 cm^{-2} s^{-1} keV^{-1} sr^{-1}. The INTEGRAL result is in good agreement with the recent CXB measurement at 3–20 keV by RXTE.[45]

5.2. *A census of nearby AGN*

Deep extragalactic X-ray surveys[38] revealed the phenomenon of AGN "downsizing" — transition from powerful quasars at high redshift ($z \gtrsim 2$) to less luminous Seyfert galaxies at $z \lesssim 1$. They also discovered numerous obscured AGN. Despite this success, the AGN census may still be incomplete because surveys performed at energies below $\sim$ 10 keV are biased against heavily obscured sources. Also, deep surveys do not probe the bright end of the AGN luminosity function at low redshift ($z \lesssim 0.3$). In studying heavily obscured and luminous nearby AGN deep X-ray surveys could be complemented by all-sky hard X-ray surveys. The INTEGRAL/IBIS telescope provides such a possibility owing to its good sensitivity above 20 keV, large field of view and good angular resolution.

By the spring of 2005 INTEGRAL observations had covered more than half of the sky, and we initiated a series of observations to cover the rest of the sky. This campaign has now been completed. Approximately 75% (50%) of the sky have been covered down to a limiting flux (for 5σ source detection) of 5 (3) mCrab, where 1 mCrab corresponds to $\sim 1.4 \times 10^{-11}$ erg s^{-1} cm^{-2} (17–60 keV). Our source identification program includes X-ray observations with Chandra[46] and spectroscopic observations on the Russian–Turkish 1.5-meter Telescope.[47] Many identifications have been done by other teams.[48,49] Based on the compiled source catalog we have studied such key statistical properties of local AGN as their hard X-ray luminosity function and absorption distribution.[50]

Our all-sky catalog comprises 127 identified AGN, 33 of which were discovered by INTEGRAL.[50] This includes 91 sources (82 Seyfert galaxies and 9 blazars) detected with more than 5σ significance on the average 17–60 keV IBIS/ISGRI map and 36 sources detected only during single observations; we excluded the latter from the analysis. All but one of the emission-line AGN are located at $z < 0.1$, while there are several blazars as distant as $z \sim$1–2.5, with isotropical luminosities up to $\sim 10^{48}$ erg s^{-1}.

Our all-sky AGN sample can be incomplete because of the presence of 24 unidentified sources. Since most of them are located near the Galactic plane, we restricted our statistical study to the region $|b| > 5°$ and also excluded blazars from consideration. The results presented below are thus based on 66 emission-line AGN located at $|b| > 5°$.

The measured hard X-ray (17–60 keV) luminosity function of nearby

($z \lesssim 0.1$) emission-line AGN (Fig. 13) is well fit by a broken power law,

$$\frac{dN_{\rm AGN}}{d\log L_{\rm hx}} = \frac{A}{(L_{\rm hx}/L_*)^{\gamma_1} + (L_{\rm hx}/L_*)^{\gamma_2}}, \tag{2}$$

the parameters of which are given in Table 1. Also quoted are the inferred number density and luminosity density of AGN with $\log L_{\rm hx} > 40$.

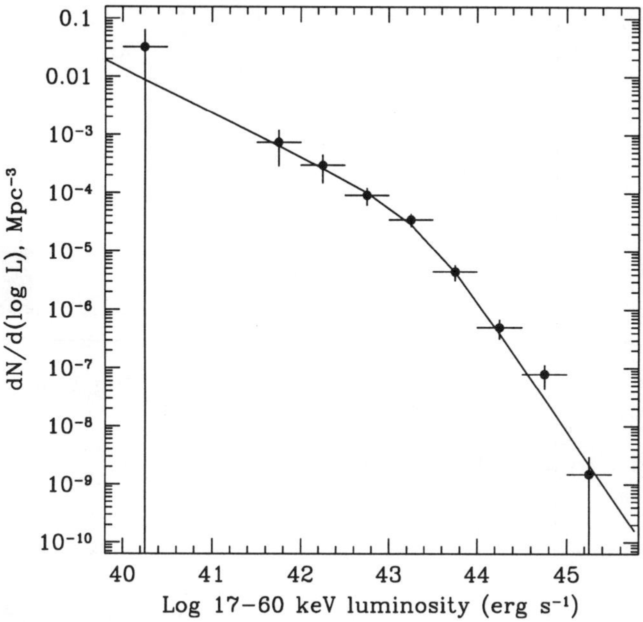

Fig. 13. Hard X-ray luminosity function of local emission-line AGN obtained with INTEGRAL. The line is the analytic approximation given by Eq. (2) and Table 1.

Parameter	Value and 1σ range
$\log L_*$	43.40 ($43.12 \div 43.68$)
γ_1	0.76 ($0.56 \div 0.94$)
γ_2	2.28 ($2.06 \div 2.56$)
A ($\rm Mpc^{-3}$)	3.55×10^{-5}
$n_{17-60\ \rm keV}(>40)$ (10^{-3} $\rm Mpc^{-3}$)	9 ($4 \div 18$)
$\rho_{17-60\ \rm keV}(>40)$ (10^{38} erg $\rm s^{-1}$ $\rm Mpc^{-3}$)	14.1 ($11.8 \div 17.1$)

One of the interesting findings of the recent RXTE 3–20 keV slew survey[51] was that the fraction of X-ray absorbed ($\log N_H > 22$) AGN drops with increasing luminosity.[52] To check this we divided our INTEGRAL sample into two parts: AGN with $\log L_{hx} < 43.6$ and those with $\log L_{hx} > 43.6$, corresponding to the faint and bright end of the luminosity function. The source absorption columns were either adopted from the literature or determined by analyzing spectral data from different X-ray astronomy missions.

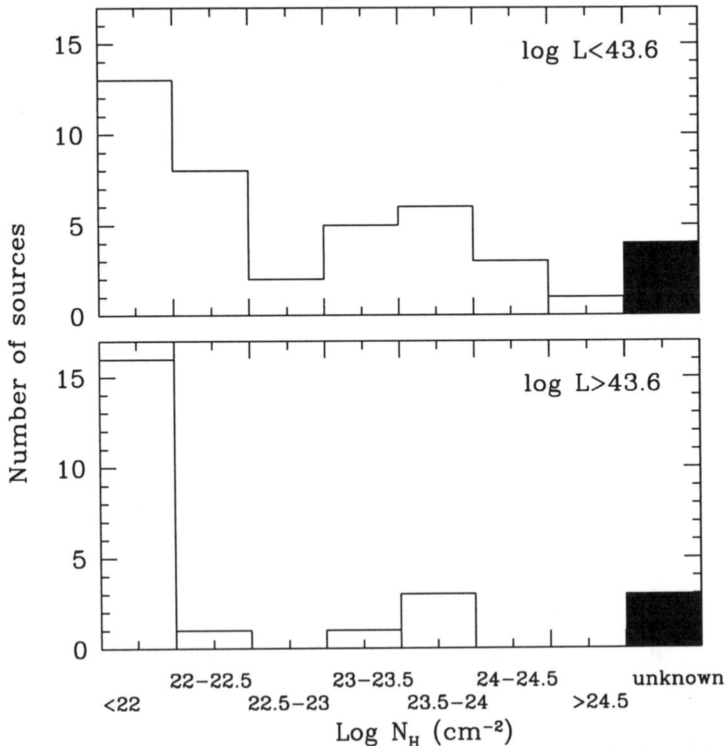

Fig. 14. Observed X-ray absorption distribution of AGN in the faint (top) and bright (bottom) end of the luminosity function.

The derived N_H distributions (Fig. 14) confirm and strengthen the RXTE result that obscured AGN are more frequent ($\sim 66\%$) in the faint end of the luminosity function than in its bright end ($\sim 24\%$). Another important result is that the local fraction of Compton thick AGN ($\log N_H \gtrsim 24$) is less than 20%, at least over the luminosity range ($\log L_{hx} \gtrsim 41$) effectively probed by INTEGRAL.

We can now ask the following question: is the distribution of X-ray absorption columns measured with INTEGRAL in the local AGN consistent with that required for distant quasars to explain the CXB spectrum? Suppose that all AGN have the same intrinsic spectrum, consisting of a cutoff power law and a Compton reflection component with relative amplitude R:

$$\frac{dN_\gamma}{dE} = AE^{-\Gamma}\exp(-E/E_{\rm f}) + Rf(E). \qquad (3)$$

We adopt $\Gamma = 1.8$, $E_{\rm f} = 200$ keV and $R = 0.5$. By propagating this fiducial intrinsic spectrum through different line-of-sight absorption columns and summing up the resulting spectra with weights inferred from the observed $N_{\rm H}$ distribution (Fig. 14), a composite spectrum $F(E)$ of local AGN can be constructed, normalized to the local AGN luminosity density measured with INTEGRAL, $\rho_{17-60\ {\rm keV}}(> 40)$.

Now, if this composite spectrum is universal and AGN experience pure luminosity evolution with redshift, as appears to be the case at $z \lesssim 1.5$ (where the bulk of the CXB is produced),[53] then the cumulative AGN spectrum observed at $z = 0$ will be (for a flat cosmology)

$$I(E) = \frac{c}{4\pi H_0}\int_0^\infty \frac{\epsilon(z)F((1+z)E)}{(1+z)\left[\Omega_{\rm m}(1+z)^3 + \Omega_\Lambda\right]^{1/2}}dz, \qquad (4)$$

where $\epsilon(z)$ describes the evolution of the AGN luminosity density. According to deep X-ray surveys,[53] $e(z) \propto z^\alpha$ with $\alpha \sim 3$ at $z \lesssim 1$ while at higher redshifts $e(z)$ is roughly bound between *const* and $\propto 1/z$.

Figure 15 shows the allowed range of CXB spectra arising in this scenario in comparison with the CXB spectrum actually measured by INTEGRAL. The model is in good agreement with the data. This suggests the possibility that (at least since $z \sim 1.5$) AGN have been evolving mainly in luminosity and much less in other properties such as the $N_{\rm H}$ distributions in the faint and bright ends of the luminosity function.

6. Acknowledgments

Part of this work has been supported by the DFG-Schwerpunktprogramme (SPP 1177).

References

1. C. Winkler et al., *A&A* **411**, L1 (2003)
2. W.N. Johnson, F.R. Harnden and R.C. Haymes, *ApJ* **172**, L1 (1972)
3. M. Leventhal, C.J. MacCallum and P.D. Stang, *ApJ* 225, L11 (1978)
4. G. Vedrenne G. et al., *A&A* **411**, L63 (2003)
5. B.J. Teegarden et al., *ApJ* **621**, 296 (2005)
6. J. Knödlseder et al., *A&A* **441**, 513 (2005)

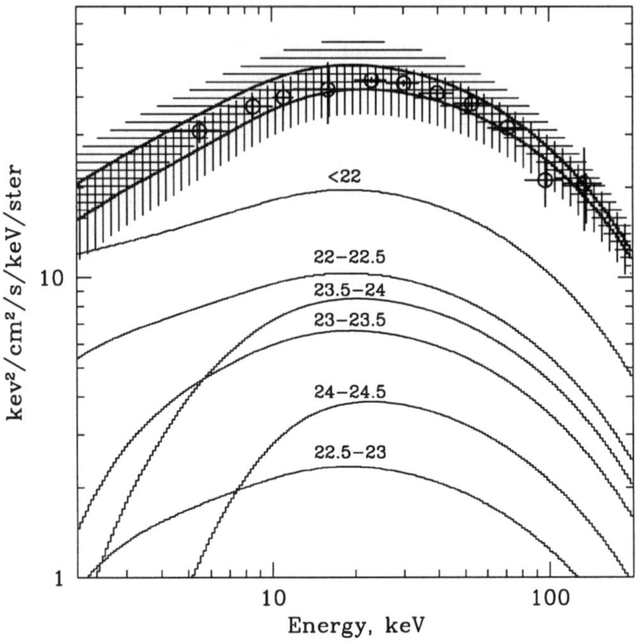

Fig. 15. Predicted CXB spectrum based on the N_H distribution measured in local AGN by INTEGRAL and redshift evolution of AGN measured with Chandra. The upper thick line corresponds to the scenario of flat luminosity density evolution at $z > 1$, the horizontally shaded region indicating the corresponding uncertainty. The lower thick line and the vertically shaded region correspond to the scenario of $\propto 1/z$ evolution at $z > 1$. Also shown are the predicted contributions to the CXB of AGN with different $\log N_H$ (the labels near the curves). The points with error bars show the CXB spectrum measured by INTEGRAL.[41]

7. E. Churazov et al., *MNRAS* **357**, 1377 (2005)
8. P. Jean et al., *A&A* **445**, 579 (2006)
9. R.W. Bussard et al., *ApJ* **228**, 928 (1979)
10. C.F. McKee and J.P. Ostriker, *ApJ* **218**, 148 (1977)
11. R. Schödel et al., *ApJ* **596**, 1015 (2003)
12. F.K. Baganoff et al., *ApJ* **591**, 891 (2003)
13. R.A. Sunyaev, M. Markevitch and M. Pavlinsky, *ApJ* **407**, 606 (1993)
14. K. Koyama et al., *PASJ* **48**, 249 (1996)
15. L. Sidoli et al., *A&A* **372**, 651 (2001)
16. H. Murakami, K. Koyama and Y. Maeda, *ApJ* **558**, 687 (2001)
17. P. Predehl et al., *Astron. Nachr.* **324**, 73 (2003)
18. S. Park et al., *ApJ* **603**, 548 (2004)
19. M.G. Revnivtsev et al., *A&A* **425**, L49 (2004)
20. M.G. Revnivtsev et al., *Astron. Lett.* **30**, 382 (2004)
21. P. Ubertini et al., *A&A* **411**, L131 (2003)

22. R. Sunyaev and E. Churazov, *MNRAS* **297**, 1279 (1998)
23. A. Valinia et al., *ApJ* **543**, 733 (2000)
24. F. Yusef-Zadeh et al., astro-ph/0608710 (2006)
25. M.A. Gordon et al., *A&A* **280**, 208 (1993)
26. V.A. Dogiel et al., *PASJ* **50**, 567 (1998)
27. V. Tatischeff, R. Ramaty and B. Kozlovsky, *ApJ* **504**, 874 (1998)
28. L.C. Ho et al., *ApJ* **549**, 51 (2001)
29. R. Walter et al., *A&A* **411**, L427 (2003)
30. M. Revnivtsev et al., *Astron. Lett.* **29**, 587 (2003)
31. P. Filliatre and S. Chaty, *ApJ* **616**, 469 (2004)
32. A. Lutovinov et al., *A&A* **444**, 821 (2005)
33. K. Koyama et al. *Nature* **343**, 148 (1990)
34. M. Sugizaki, *ApJS* **134**, 77 (2001)
35. I. Negueruela, Proceedings of JENAM-2004, astro-ph/0411759) (2004)
36. A.J. Bird et al., *ApJ* **607**, L33 (2004)
37. S.V. Molkov et al., *Astron. Lett.* **30**, 534 (2004)
38. W.N. Brandt and G. Hasinger, *ARA&A* **43**, 827 (2005)
39. M.M. Basko, R.A. Sunyaev and L.G. Titarchuk, *A&A* **31**, 249 (1974)
40. I.M. George and A.C. Fabian, *MNRAS* **249**, 352 (1991)
41. E. Churazov et al., astro-ph/0608250 (2006)
42. E. Churazov et al., astro-ph/0608252 (2006)
43. S. Sazonov et al., astro-ph/0608253
44. D.E. Gruber et al., *ApJ*, **520**, 124 (1999)
45. M. Revnivtsev et al., *A&A* **411**, 329 (2003)
46. S. Sazonov et al., *A&A* **444**, L37 (2005)
47. I.F. Bikmaev et al., *Astron. Lett.* **32**, 221 (2006)
48. N. Masetti et al., *A&A* **426**, L41 (2004)
49. L. Bassani et al., *ApJ* **636**, L65 (2006)
50. S. Sazonov et al. *A&A in press*, astro-ph/0608418 (2006)
51. M. Revnivtsev et al., *A&A* **418**, 927 (2004)
52. S. Sazonov and M. Revnivtsev, *A&A* **423**, 469 (2004)
53. A.J. Barger et al., *AJ* **129**, 578 (2005)

REVEALING THE DARK TEV SKY: THE ATMOSPHERIC CHERENKOV IMAGING TECHNIQUE FOR VERY HIGH ENERGY GAMMA-RAY ASTRONOMY *

TREVOR C. WEEKES

Whipple Observatory,
Harvard-Smithsonian Center for Astrophysics,
P.O. Box 6369, Amado, Arizona 85645-0097, U.S.A.
e-mail:tweekes@cfa.harvard.edu

The Atmospheric Cherenkov Imaging Technique has opened up the gamma-ray spectrum from 100 GeV to 50 TeV to astrophysical exploration. The development of the technique (with emphasis on the early days) is described as are the basic principles underlying its application to gamma-ray astronomy. The current generation of arrays of telescopes, in particular, VERITAS is briefly described.

1. Introduction

One of the last frontiers of the gamma-ray sky is that characterized by the distribution of TeV photons. These photons can be detected relatively easily with ground-based detectors (constituting a TeV "window" in the atmosphere); thus the detection of TeV gamma-ray sources did not have to await the availability of space platforms. In practice although the technology was available at an early date, it required the impetus of gamma-ray space astronomy to justify a major effort in a new discipline. Since it concerns the highest energy photons with which it is yet feasible to map the sky, it is of particular interest to high energy astrophysicists. Any source of TeV photons must be associated with a cosmic particle accelerator and of inherent interest to high energy particle physicists as well as students of the cosmic radiation.

To date almost all the observational results in the energy interval 100 GeV - 100 TeV have come from observations using the so-called "Atmospheric Cherenkov Imaging Technique (ACIT)." Although considerable ef-

*This work is respectfully dedicated to the memory of Neil A. Porter (1930-2006), one of the Founding Fathers of Very High Energy Gamma-ray Astronomy.

fort has been applied to the development of alternative techniques, they are more specialized and will not be considered here.

In this historical review of the ACIT, emphasis will be on the early days in which the technique was established; a brief outline of the general principles underlying atmospheric Cherenkov telescopes (ACT) will be given and a description, albeit incomplete, of the ACIT as currently used and the present generation of instruments will be described. More complete accounts can be found elsewhere[1,2,3,4,5,6].

2. Early History of the Atmospheric Cherenkov Technique

2.1. *Discovery of the Phenomenon*

In the Ph.D. dissertations of students studying the atmospheric Cherenkov phenomenon, the first reference is usually to the 1948 note in the Royal Society report on the study of night-sky light and aurora by the British Nobel Laureate, P.M.S. Blackett[7]; in that note he points out that perhaps 0.01% of the light in the dark night-sky must come from Cherenkov light emitted by cosmic rays and their secondary components as they traverse the atmosphere. Little attention was paid to this prediction (since it seemed unobservable) at the time. Fortunately five years later, when Blackett was visiting the Harwell Air Shower array, he brought his prediction to the attention of two Atomic Energy Research Establishment physicists, Bill Galbraith and John Jelley. After the visit, the idea occurred to them that, while the net flux of Cherenkov light would be impossible to measure, it might just be possible to detect a short light pulse from a cosmic ray air shower which involved some millions of charged particles (Figure 1).

Within a week Galbraith and Jelley had assembled the items necessary to test their hypothesis. A 5 cm diameter photomultiplier tube (PMT) was mounted in the focal plane of a 25 cm parabolic mirror (all housed in a standard-issue Harwell garbage can) and coupled to an amplifier with a state-of-the-art 5 MHz amplifier whose output was displayed on an oscilloscope. They observed oscilloscope triggers from light pulses that exceeded the average noise level of the night-sky background every two minutes. They noted that the pulses disappeared when the garbage can lid was put in place and a padding lamp was adjusted to give the same current in the PMT as was observed from the night-sky[8]. Jelley noted that if the rate had been any lower than that observed they would probably have given up and gone home early![9]. It is not often that a new phenomenon can be discovered with such simple equipment and in such a short time, but it may

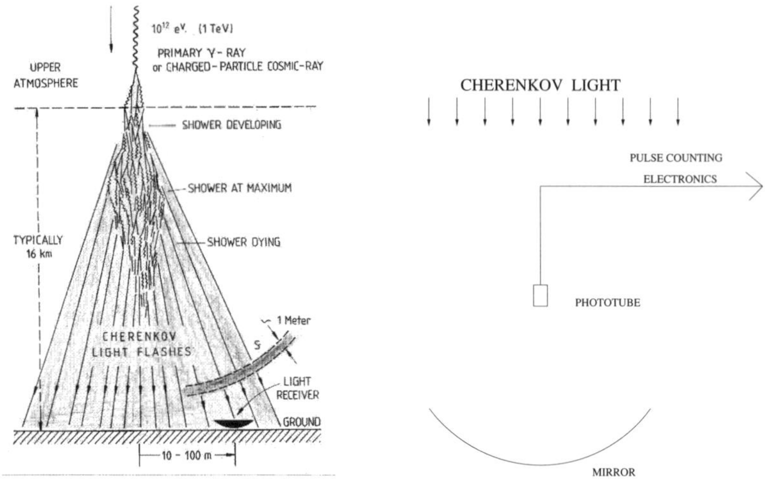

Figure 1. **Left:** Cartoon of the atmospheric Cherenkov shower phenomenon, as drawn by John Jelley in 1993. **Right:** The essential elements of an Atmospheric Cherenkov Detector

also be true that it is not often that one finds experimental physicists with this adventurous spirit! Whereas the modern physicist would not embark on a speculative venture of this nature without extensive simulations, John Jelley had a great suspicion of excessive computation and relied instead on his gut feelings for the inherent physics of the phenomenon; he was seldom wrong!

2.2. *The Power of the Technique*

With the Harwell air shower array (one of the largest such arrays then in existence) in close proximity, it was easy to show that the light pulses were indeed associated with air showers. In the years that followed, Galbraith and Jelley made a series of experiments in which they determined the basic parameters of the Cherenkov radiation from air showers. The account of these elegant experiments is a must-read for all newcomers to the field[10,11]. The basic detector elements of the ACT are extremely simple (Figure 1). It was realized at an early stage that the phenomenon offered the possibility of detecting point sources of cosmic ray air showers with high efficiency. Since charged primaries are rendered isotropic by the intervening interstellar magnetic fields, in practice this meant the detection of point sources of neutral quanta, i.e., gamma-ray photons or perhaps neutrons. The lateral spread of the Cherenkov light from the shower as it strikes the ground is $\approx$ 100-200 m so that even a simple light receiver of modest dimensions has an effective collection area of some tens of thousands of square meters. The fact that the light pulse preserves much of the original direction of the primary particle and that the intensity of light is proportional to the total number of secondary particles, and hence to the energy of the primary, makes the detection technique potentially very powerful.

The prediction by Cocconi[12] of a strong flux of TeV gamma rays from the Crab Nebula precipitated an experiment by the Lebedev Research Institute in the Crimea in 1960-64[13]. Supernova Remnants and Radio Galaxies had recently been identified as sources containing synchrotron-emitting electrons which suggested that they might be gamma-ray sources. A selection of these (including the Crab Nebula) were examined with a ACT system consisting of twelve 1.5 m aperture ex-World War II searchlight mirrors mounted on railway cars at a dark site near the Black Sea (Figure 2). This system did not attempt to discriminate between air showers initiated by gamma rays and those initiated by hadrons. No sources were found but the basic methodology involved in a search for point source anisotropies in the cosmic ray air shower distribution was defined. The technique was refined by John Jelley and Neil Porter in a pioneering British-Irish experiment in the Dublin Mountains in which the candidate source list was expanded to include the recently discovered quasars and magnetic variable stars (with null results [14]). This early experiment also used ex-World War II searchlight mirrors on a Bofors gun mounting (continuing the tradition of putting military hardware to good use) (Figure 3). The Smithsonian group led by Giovanni Fazio built the first large optical reflector for gamma-ray astronomy on Mount Hopkins in southern Arizona (Figure 4). This 10 m telescope is still in use after 38 years of service! This again was a first generation device in which the assumption was made that there was no easily measured differences in the light pulses from gamma-ray and hadronic primaries. The motivation for this large increase in mirror area (and decrease in energy threshold) was a refined prediction of a detectable flux of gamma rays from the Crab Nebula based on a Compton-synchrotron model[15].

Although these first generation detection systems were extremely simple and exploited the ease with which gamma rays could be *detected*, they did not provide the means of *identifying* gamma rays among the much more numerous cosmic ray background. Hence, until 1989 when the Crab Nebula was finally detected[16], there was no credible detection of a gamma-ray flux from any cosmic source.

Figure 2. The first ground-based experiment in TeV gamma-ray astronomy which was the Lebedev Institutes's of twelve 1.5 m searchlight mirrors in the Crimea; it had an energy threshold of 1.5 TeV

Figure 3. **Left:** Neil A. Porter (1930-2006) (Photo: D.J.Fegan) **Right:** The second ground-based gamma-ray telescope; the British-Irish experiment at Glencullen, Ireland c. 1964; the telescope consisted of two 90 cm searchlight mirrors on a Bofors gun mounting. The experiment was led by Jelley and Porter.

2.3. *Basic Principles*

The light signal (in photoelectrons) detected is given by:

$S = \int_{\lambda_2}^{\lambda_1} k\ E(\lambda)\ T(\lambda)\ \eta(\lambda)\ A\ d\lambda$

where C(λ) is the Cherenkov photon flux within the wavelength sensitivity bounds of the PMT, λ_1 and λ_2, E(λ) is the shower Cherenkov emission

Figure 4. The Whipple Observatory 10 m gamma-ray telescope was built in 1968; it is still in operation. It is composed of 250 glass facets, each of focal length 7.3 m.

spectrum (proportional to $1/\lambda^2$), $T(\lambda)$ is the atmospheric transmission and k is a constant which depends on the shower, and the geometry.

The signal must be detected above the fluctuations in the night-sky background during the integration time of the pulse counting system, τ.

The sky noise B is given by:
$B = \int_{\lambda_2}^{\lambda_1} B(\lambda)\ \eta(\lambda)\ \tau\ A\ \Omega\ d\lambda.$
Hence the signal-to-noise ratio is essentially
$S/N = S/B^{0.5} = \int_{\lambda_2}^{\lambda_1} C(\lambda)\ [\eta\ (\lambda)\ A\ /\Omega\ B(\lambda)\ \tau]^{1/2}\ d\lambda.$

The smallest detectable light pulse is inversely proportional to S/N; the minimum detectable gamma ray then has an energy threshold, E_T given by
$E_T \propto 1/C(\lambda)\ [B(\lambda)\ \Omega\ \tau/\eta(\lambda)\ A]^{1/2}$

If S = the number of gamma rays detected from a given source in a time, t, and A_γ is the collection area for gamma-ray detection, then $S = F_\gamma(E)\ A_\gamma\ t$. The telescope will register a background, B, given by:
$B = F_{cr}\ A_{cr}(E)\ \Omega\ t$, where $A_{cr}(E)$ is the collection area for the detection of cosmic rays of energy E. The cosmic ray background has a power law spectrum:
$F_{cr}(>E) \propto E^{-1.7}$ and if we assume the gamma-ray source has the form: $F_\gamma(>E_\gamma) \propto E_\gamma{}^{-a_\gamma}$.

Then the standard deviation, $\sigma \propto S/B^{1/2} \propto E^{1.7/2-a_\gamma}\,[A_\gamma/A_{cr}^{1/2}]\,t^{1/2}$.

The minimum number of standard deviations, σ, for a reliable source detection is generally taken as 5[6].

3. Early Development of the ACIT

3.1. *Discrimination Methods*

At an early stage it was realized that while the atmospheric Cherenkov technique provided a very easy way of *detecting* gamma rays with simple light detectors, it did not readily provide a method of discriminating the light pulse from gamma-ray air showers from the background of light pulses from the much more numerous cosmic ray showers; thus the *flux sensitivity* was severely limited. Although the hadron showers are isotropic, there is typically a ratio of 1,000-10,000 of cosmic rays to gamma rays recorded by the simple light detectors that were available in the two decades following the Harwell experiments. Once it was apparent that the early, very optimistic, predictions of the strength of the most obvious potential TeV sources were not to be realized, then attention turned to methods of improving the flux sensitivity of the technique. Although superficially very similar, Monte Carlo simulations of shower development and Cherenkov light emission suggested some differences that might be exploited to preferentially select gamma rays.

These differences are listed below:

- Lateral Spread at ground level: the light pool from gamma-ray showers is more uniform than that from cosmic ray showers. This feature is difficult to exploit since it requires numerous light detectors spread over relatively large areas; it has recently been used by the group at the Tata Institute at their Pachmari site[17]
- Time Structure: because the cosmic ray component contains penetrating particles (mostly muons) that survive to detector level, the duration of the light pulse can be longer. Many early versions of the ACT, particularly the Haleakala experiment[18], attempted to exploit this feature but it was not to prove very effective,
- Spectral Content: the penetrating component of cosmic ray showers is close to the light detector and its overall Cherenkov light at the detector is less attenuated in the ultraviolet; this feature was used as a discriminant in the early Whipple and Narrabri experiments of Grindlay and his collaborators[19] and in the Crimean

experiments[20]. It is mostly effective when combined with other discriminants.

- Angular Spread: the image of the light superimposed on the night-sky background has a more regular distribution from gamma-ray showers and is smaller and more uniform. This feature was recognized by Jelley and Porter[21] but not really exploited until some decades later. This was to prove the most powerful discriminant and to lead to the first successful credible detection of a TeV gamma-ray source[16].

The Cherenkov light image has a finite angular size which can, in principle, be used to refine the arrival direction, and perhaps even to distinguish it from the images of background cosmic rays[22]. However when a simple telescope with a single light detector (pixel) is used as a gamma-ray detector, this information is lost and the angular resolution is no better than the field of view of the telescope. Because the Cherenkov light images are faint and fast, it is not technically straight-forward to record them. Boley and his collaborators [23] had used an array of photomultipliers at Kitt Peak to study the longitudinal development of large air showers but these were from very energetic primaries. A pioneering effort by Hill and Porter[24], using a image intensifier system from a particle experiment, resulted in the first recorded images of Cherenkov light from air showers (Figure 5). These images, although relatively low resolution, demonstrated in a very vivid way the information contained in the Cherenkov image recorded at ground level. The potential advantages of using this detection technique as a means of separating out the gamma-ray component were recognized in a prophetic paper by John Jelley and Neil Porter[21]:

"For a long time it has been appreciated that the image intensifier offers potentialities in this field, and the photography of Cherenkov images against the night-sky is the first step in this direction. Temporarily postponing the technical problems, what are the advantages of this technique? First, with Schmidt optics, it is possible in principle to combine a wide field of view with a high resolution. Secondly, photographs already obtained of the Cherenkov images suggest that their shapes may be used to give detailed information both on the true direction of the shower and also the coordinates of its point of intersection with the ground, in relation to the position of the equipment. The third feature, and it is really the most important one for gamma-ray astronomy of 'point sources', is the high angular resolution which may be attained. Though the Cherenkov images are $\approx 2^\circ$ across, and are in general non-circular in shape, it should be possible to determine a shower direction to $\approx 0.2°$. Thus, we have, for a true point source, a discrimination (by solid angle) against showers from the general-field CR primaries, of $\approx$ 100 times better than that possible for drift-scans with a photomultiplier system. It might be added here that a stereoscopic technique, with two separated telescopes, would greatly enhance these potentialities."

However, because of the finite size of the photocathode on the image intensifiers then available, it was only possible to couple them to a relatively small mirrors which meant that only cosmic ray primaries above 100 TeV could be detected. Even then it was necessary to couple these state-of-the-art instruments to a phosphor with decay times of microseconds to allow the image intensifier to be gated and the image recorded photographically. Since this meant that the technique was limited to energies > 100 TeV where the attenuation of the gamma-ray flux by photon-photon pair production in intragalactic space was appreciable, this approach was not pursued at that time. A recent Japanese experiment has revived interest in this technique using the best modern image intensifiers (Sasaki, this workshop).

A novel approach to imaging was that pursued by Grindlay and his colleagues in the seventies[19] in which multiple light detectors separated by distances $\approx$ 100 m were used to detect the shower maximum associated with gamma-ray showers; this pinpointed the shower arrival direction. The penetrating, mostly muon, component from hadron showers was detected by a second detector and was used as a veto to preferentially select events that were initiated by gamma rays. This "Double Beam" technique was potentially powerful but was difficult to implement with the resources available at the time. Initially the detectors used were 1.5 m searchlight mirrors with single phototubes at their foci; later the 10 m reflector was incorporated into the system with two pixels. The technique received new life when the Narrabri Stellar Interferometer (in Australia) became available. With two large reflectors of 9 m aperture on a circular rail system, (Figure 6) the system, originally built to measure the diameters of bright stars using the intensity interferometer principle, was ideally suited for this technique. Although some detections were reported (the Crab pulsar, the Vela pulsar and Centaurus A)[25], they were not confirmed by later, more sensitive, observations. The Double-Beam technique, although ingenious, was not pursued after this although it can be seen as the stalking horse for imaging arrays (see below).

Activity in ground-based gamma-ray astronomy was at a low ebb in the

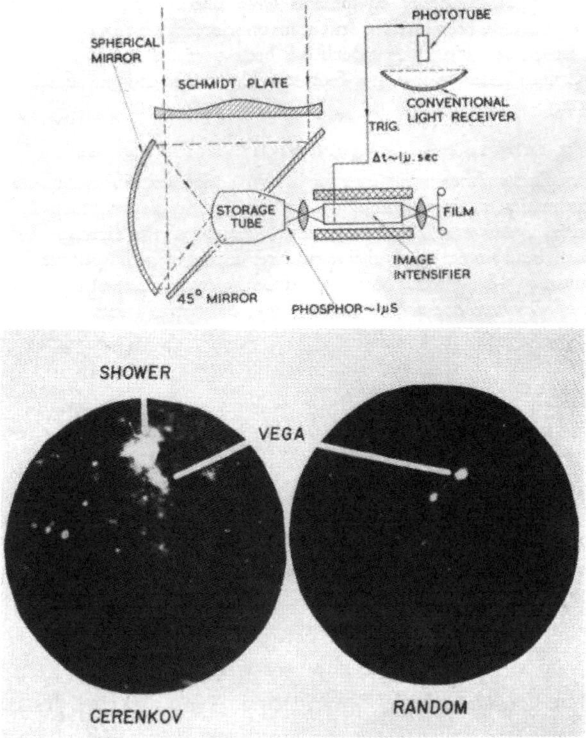

Figure 5. **Top:** Image Intensifier used by Hill and Porter to record the images of cosmic ray air showers [24]. **Bottom** Images of the night-sky triggered by an ACT (left) and triggered randomly (right). The field of view was $\pm 12.5°$.

seventies. Observations with the Whipple 10 m reflector had moved the energy threshold of the technique close to 100 GeV but this had only produced upper limits on the predicted sources. Smaller telescopes produced tentative detections of several binaries and pulsars but these were always on the edge of statistical credibility and were not subsequently verified (this controversial epoch of TeV gamma-ray astronomy has been reviewed elsewhere[26,27]).

3.2. *The Power of the Atmospheric Cherenkov Imaging Technique*

The concept of using electronic cameras consisting of matrices of phototubes in the focal plane of large reflectors to record the images of the

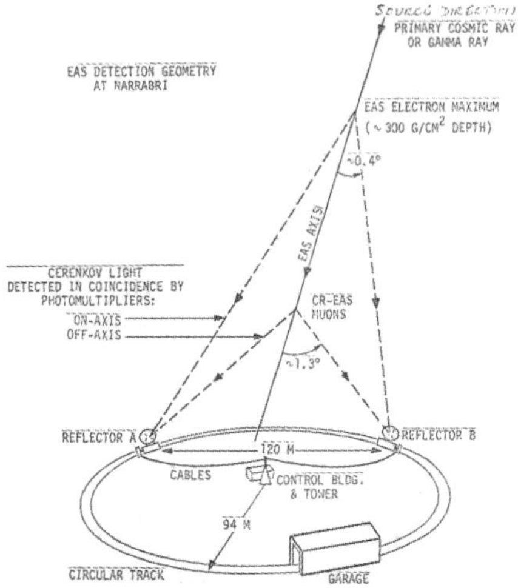

Figure 6. The Double Beam Technique developed by Grindlay in which two reflectors are used, each with two "pixels." The upper two define the shower maximum and the lower two define the penetrating component and act as a veto to reject hadronic showers.

Cherenkov light from small air showers was first suggested in a paper at a workshop in Frascati, Italy[28]. Entitled "Gamma-Ray Astronomy from 10-100 GeV: a New Approach" the emphasis was on lowering the energy threshold through the use of two large reflectors separated by 100 m, each equipped with arrays of phototubes in their focal plane. The motivation to go to lower energies came from the prediction from Monte Carlo simulations that the ratio of Cherenkov light from gamma-ray showers to cosmic ray showers of the same energy increases dramatically below 100 GeV. In this paper the physical explanation of this falloff was stated: "In a proton shower most of the Cherenkov light comes from the secondary electromagnetic cascades. Energy comes into these cascades via the production of pions by the primary and the subsequent nucleon cascade. Two thirds of the energy (approximately) goes to charged pions; they can decay to muons or undergo a collision.The latter process is a more efficient method of producing Cherenkov light; since the lifetime against decay is greater a higher energies, the chance of collisions is greater. At lower energies therefore, proportionally more energy comes off in muons whose energy may be below the Cherenkov threshold and hence the low energy showers are deficient in Cherenkov light". The idea of using an array of phototubes with limited resolution to image the Cherenkov light rather than the high resolution offered by image intensifiers was motivated by the experience of the author using CCD detectors in optical astronomy where the resolution achieved is significantly greater than the scale of the pixels. In the paper there was little emphasis on discrimination of the primaries based on the shapes of the images although it was claimed that there would be a significant improvement in angular resolution (to 0.25°). The use of two reflectors in coincidence was advocated to reduce the predicted muon background.

In this paper[28] the basic concept of the Cherenkov light imaging telescope was described; it consisted of an array of PMTs in the focal plane of a large reflector. Although the initial development centered on the use of a single large reflector (the Whipple 10 m reflector, Figure 4), the utility of an array with at least two such cameras was advocated. This has been the model for all subsequent telescopes using the ACIT. In general, in recording the Cherenkov light image from an air shower, the gamma-ray astronomer tries to characterize its nature (gamma-ray or hadron), determines its arrival direction, and gets some estimate of the primary that initiated the air shower. The factors that cause the observed shape and size of the image are many: the nature of the primary particle, its energy and trajectory, the physical processes in the particle cascade (principally pair production and bremsstrahlung in electromagnetic cascades with the addition of pion production in hadron initiated cascades), Coulomb scattering of shower electrons, the effect of geomagnetic deflections of the shower particles, the distance of the point of impact of the shower core from the optic axis, the Cherenkov angle of emission, and the effect of atmospheric absorption[4]. In addition the properties of the imaging system must be completely understood: the reflectivity of the mirrors, the quantum efficiency of the light detectors as a function of wavelength, the time response of the system, and the distortions introduced by the system's optics, cables, electronics and data readout.

Fortunately all of these factors are amenable to calculation or measurement. The physics of the various processes involved in the shower development are well known and Monte Carlo methods can be used to estimate the expected values from particular primaries. However since fluctuations play a major role in such development the expected values cover a range of possibilities and identification must always be a statistical process. It is relatively easy to predict the properties of the gamma-ray initiated showers; it is more difficult to predict the expected properties of the background which is mainly from charged cosmic rays. While every attempt is made to estimate both signal and background, it is usually found that the background contains some unpleasant surprises; hence although the gamma-ray detection rate can be reliably predicted, the efficiency of the identification of the gamma rays from the more numerous background requires the system to be actually operated in observations of a known source. Since the background is numerous and constant, its properties can be readily modeled from empirical databases of night-sky background events. There is an irreducible background from hadron showers which develop like electromagnetic cascades (most of the energy goes into a π^o in the first interaction) and from the electromagnetic cascades produced by cosmic electrons (whose fluxes in the range of interest are 0.1 - 0.01% of the hadron flux).

3.3. *The First Source*

When the imaging systems first went into operation it was not immediately obvious how the images should be characterized and discriminated from the background. There were no credible sources and Monte Carlo calculations were still being developed and were untested. The first such calculations available to the Whipple Collaboration indicated that fluctuations might effectively rule out any discrimination and did not encourage the development of sophisticated analysis techniques. The first Whipple camera had 37 pixels, each of 0.25° diameter[29]. A relatively simple image parameter, *Frac2*, defined as the ratio of the signal in the two brightest pixels to the total light in the image, was developed empirically and led to the first indication of a signal from the Crab Nebula[30,31]. This simple parameter picked out the compact images expected from electromagnetic cascades but did not provide any information on the arrival direction (other than that it was within the field of view of the detector). However the application of the same selection method on putative signals from the then popular sources, Cygnus X-3 and Hercules X-1, did not improve the detection credibility and initially cast doubt on the effectiveness of *Frac2* as a gamma-ray identifier.

Since the images were roughly elliptical in shape, an attempt was made to quantify the images in terms of their second and third moments[32]. However this was not applied to gamma-ray identification until Hillas undertook a new series of Monte Carlo calculations[33]. These calculations predicted that gamma-rays images could be distinguished from the background of isotropic hadronic images based on two criteria: the difference in the *physics* of the shower development, which led to smaller and better defined ellipses for gamma rays, and the difference in the *geometry* of image formation due to all images coming from a point source on axis having their major axes intersecting the center of the field of view. Fortunately the first property aids the definition of the second and provides potentially very good angular resolution. Hillas[33] defined a series of parameters which included the second moments (*Width* and *Length*), the parameter *Dist* which measures the distance of the centroid of the image from the optic axis, and *Azwidth* which measures the projected width of the image on the line joining the centroid to the center of the field of view. Later *Alpha*, the angle between this line and the major axis was added as was *Asymmetry*, the third moment. *Azwidth* was particularly simple; it is easy to use and proved to be very effective as it combined discrimination based on image size (physics) and arrival direction (geometry) and led to the first definite detection of a point source of TeV gamma-rays. In general multiple parameter selections were made. The parameters were first defined in Monte Carlo calculations but once the standard candle of the Crab Nebula was established[16], optimization was made on the strong and steady Crab signal to preferentially select gamma rays. This optimization led to an analysis package called Supercuts[34], which proved to be extraordinarily robust, and in various forms, was the basis of the data analysis used by the Whipple Collaboration to detect the first AGN[35,36,37,38,39]. Other groups have defined different parameters and analysis schemes but the basic methodology is the same.

4. ACT Observatories

4.1. *Third Generation Observatories*

By 1996 the ACIT was judged to have been very successful and a number of groups made plans for third generation ACTs. The limitation of a single telescope was easily seen from the results obtained using the Whipple telescope and camera[40]. At low trigger thresholds it was impossible to distinguish low energy gamma-ray events from the much more numerous background of partial muon rings (arcs). Despite intense efforts with sophisticated analysis methods, it was clear that the discrimination threshold was a factor of 2-3 above the trigger threshold. Hence although the fundamental threshold was $\approx$ 200 GeV, the effective gamma-ray threshold was $\approx$ 400 GeV. Since the muon Cherenkov emission is essentially a local phenomenon, this background is easily eliminated by demanding a coincidence with a second telescope separated from the first by a minimum distance of 50 m[28]. In fact the HEGRA experiment had already demonstrated[41] the power of an array of small imaging telescopes to improve the angular and energy resolution of the ACIT; at the threshold energies of these telescopes the muon background was not a problem.

Thus it was apparent that the next generation of the ACIT would involve arrays of reflectors with apertures in excess of 10 m, with better optics, with more sophisticated cameras, and with data acquisition systems capable of handling high rates. Such systems required an investment that was almost an order of magnitude greater than the previous generation of detectors (but the flux sensitivity would be improved by a similar factor). Of necessity the number of people involved in each experiment would be so large ($\approx$ 100) that the new collaborations would be more in line with the numbers of scientists found in particle physics experiments than in typical major astronomical projects.

4.2. *The Power of ACT Arrays*

ACTs arrays can be discussed under the headings of improvements offered in energy threshold, energy resolution, angular resolution and background discrimination. A comprehensive discussion can be found in[3]. A typical array provides multiple images of a single event as seen in Figure 7.

Energy Threshold: The basic quantities involved in determining the energy threshold of an ACT are given above in Section 2.3 and are fairly obvious: the mirror area should be as large as possible and the light detectors should have the highest possible quantum efficiency. To the first approximation (as demonstrated in[13]) it does not critically depend on how the mirror area is distributed, i.e., a cluster of small telescopes in close proximity operated in coincidence is the same as if their signals are added and is approximately the same as that of a single large telescope of the same total mirror area. Practical considerations tend to dominate: coincidence systems are more stable, the cost of telescopes scales as the Aperture$^{2.5}$, the relative cost of multiple cameras each on a small telescope versus the cost of a single camera on a large telescope, etc. However the simplest way to get the lowest energy threshold is to go for a single large telescope (although this may introduce other problems).

Angular Resolution: Angular resolution is important not only for reducing the background and identifying a potential source but also for mapping the distribution of gamma rays in the source. Stereoscopic imaging, the simplest form of "array" imaging, offers the immediate advantage of improving the angular resolution. This principle was established with the use of just two telescopes with a separation of $\approx$ 100 m, i.e., with the two

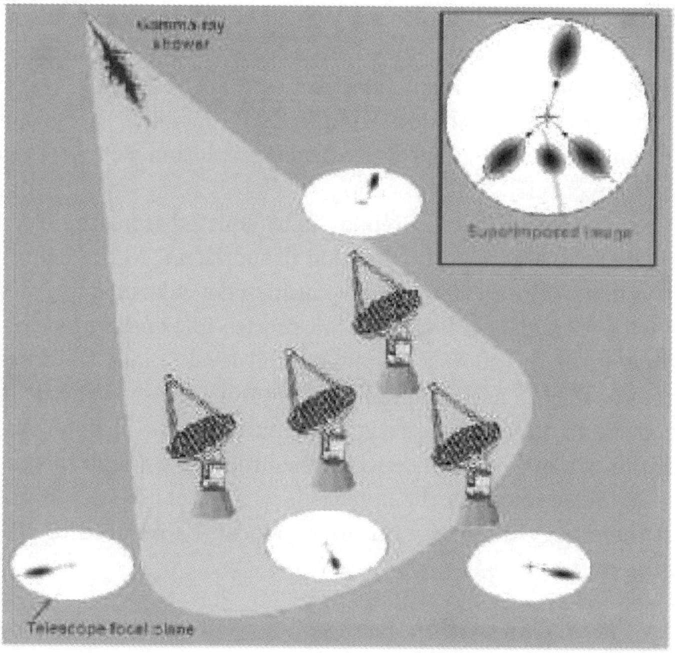

Figure 7. Cartoon showing response of array of four detectors to air shower whose axis is parallel to the optical axes of the telescopes and some 30 m displaced from the center of array. (Figure courtesy of P.Cogan)

telescopes within the light pool of the Cherenkov light pool, $\approx$ a circle of diameter 200 m. The greater the separation, the better the angular resolution but increasing the separation beyond 100 m begins to reduce the effective gamma-ray collection area. A simple array of imaging ACTs can provide a source location of $\approx 0.05°$ for a relatively strong source with angular resolution of $\approx 0.1°$ for individual events. This is a factor of two improvement over that for a single telescope. An angular resolution of an arc-min or better appears feasible ultimately.

Background Discrimination: Multiple views of the same air shower from different angles obviously improves the signal-to-noise ratio when the images are combined. However in reducing the background of hadronic events the gain is not as large as might appear at first glance. Hadronic showers which develop like typical showers are easily identified and rejected, even in a single telescope. More subtle are the hadronic events which develop like an electromagnetic cascade (an early interaction channels much of the energy into an electron or gamma ray). Such events cannot be identified no matter how many views are provided on the cascade development. Similarly the cascades initiated by cosmic electrons are an irreducible background. However the array approach does completely remove the background from single local muons and the improved angular resolution narrows the acceptable arrival directions.

Energy Resolution; The Cherenkov light emitted from the electromagnetic cascade is to a first approximation proportional to the energy of the initiating gamma ray and thus can be considered a calorimetric component. However with a single ACT there is no precise information as to the impact parameter of the shower axis at ground level. Since the intensity of the Cherenkov light is a function of distance from the shower axis, the lack of information on this parameter is the limiting factor in determining the energy of the gamma ray. The energy resolution of a single imaging ACT is $\approx$ 30-40%. With an array the impact parameter can be determined to $\approx$ 10 m and the energy resolution, in principle, can be reduced to 10%.

4.3. *The Third Generation Arrays*

This third generation of ACTs has seen the formation of four large collaborations formed to build arrays of large telescopes: a largely German-Spanish collaboration that is building two 17 m telescopes on La Palma in the Canary islands (MAGIC)[42]: an Irish-British-Canadian-USA collaboration that is building an array of four 12 m telescopes in Arizona (VERITAS)[43]; an Australian-Japanese collaboration that has built four 10 m telescopes in Australia (CANGAROO-III)[44]; a largely European collaboration that has built an array of four 12 m telescopes in Namibia (HESS)[45] and plans to add a fifth telescope of 28 m aperture at the center of the array. The fact that two of the arrays are in each hemisphere is somewhat fortuitous but ensures that there will be good coverage of the entire sky and that all observations can be independently verified. Three of arrays are discussed elsewhere at this workshop; here the VERITAS observatory will be briefly described.

The sensitivity of these new arrays is probably not dissimilar; HESS and MAGIC has demonstrated what can achieved in the actual detection of known and new sources. With the second generation of ACTs (Whipple, HEGRA), it was possible to detect a source that was 5% of the Crab Nebula in 100 hours of observation. With HESS this is reduced to one hour and in principle in 100 hours it should be possible to detect a source as weak as 0.5% of the Crab. HESS has also demonstrated an energy resolution of 10% and an angular resolution of an arc-min.

5. VERITAS

The configuration chosen for VERITAS was a filled hexagon of side 80 m; in the first phase funding was available for only four telescopes so the hexagon has three non-adjacent vertices missing[43]. The four telescopes and cameras of VERITAS are identical and are now at an advanced state of construction. The first two telescopes and cameras were installed at a temporary site (the Whipple Observatory Basecamp at an elevation of 1.3 km) and saw first "gamma-ray light" in February, 2005 (Figure 8). The properties of the first telescope have been described elsewhere[46,47].

Telescope: The VERITAS telescopes are of the Davies-Cotton optical design with 12 m aperture and 12 m focal length. The mechanical structure consists of an altitude-azimuth positioner and a tubular steel optical support structure (OSS). The design is closely modeled on the existing Whipple 10 m optical reflector but with the added feature of a mechanical bypass of the upper quadrapode arm which transfers the load of the camera to the counterweight support. Completion of the first two telescopes has allowed the properties and sensitivities of the individual telescopes to be measured[47].

The 350 individual mirror facets on each telescope are hexagonal, each with an area of 0.322 m^2, providing a total mirror area of $\sim$110 m^2. They are made from glass, slumped and polished; the glass facets are aluminized and anodized at the VERITAS optical coating laboratory on-site. The reflectivity of the anodized coating is typically $>$ 90% at 320 nm. Each facet has a 24 m radius of curvature. They are located on a three point mounting on the spherical front surface (radius 12 m) of the OSS. The point spread function (PSF) at the position of Polaris (elevation 31°) was measured to be 0.06° FWHM; with bias alignment it is anticipated that this PSF will be achieved over most of the operating range of VERITAS.

Camera: The VERITAS cameras are closely modeled on those used previously by the group at the Whipple telescope but incorporate much more advanced triggering, electronics readout and data acquisition systems [31]. The instrumentation in the focal plane is a 499 element photomultiplier tube (PMT) camera, with 0.15° angular spacing giving a field-of-view of 3.5°. The camera is shown in Figure 8. The PMTs are Photonis XP2970/02 with a quantum efficiency $>$ 20% at 300 nm, currently operated at a gain of

Figure 8. **Left:** The first two VERITAS Telescopes. **Right:** The 499 pixel PMT camera.

$\sim 2 \times 10^5$. The PMT signals are amplified by high bandwidth preamplifiers integrated into the PMT base mounts. The signals are sent via ~50 m of RG59 stranded cable to the telescope trigger and data acquisition electronics, at which point the observed pulse for an input delta function has a rise time (10% to 90%) of 3.3 ns and a width of 6.5 ns.

The PMT signals are digitized using custom-built VME boards housing Flash ADCs with 2 ns sampling and a memory depth of 32μs. The trigger system is multi-level. At the telescope each channel is equipped with a programmable constant fraction discriminator (CFD) for each PMT, the output of which is passed to a pattern recognition trigger system which is programmed to recognize triggers resembling true compact Cherenkov light flashes. Individual telescope triggers are delayed and combined to form an overall array trigger. The FADCs permit the telescopes to operate at a lower threshold than would otherwise have been possible[47].

Future Program: The scientific program will concentrate on the study of extragalactic objects including AGN, radio galaxies, starburst galaxies and clusters, compact galactic objects including pulsars, binaries and microquasars, extended objects such as supernovae remnants, unidentified sources discovered in future space missions, and signatures of dark matter in the center of galaxies. A sky survey will be undertaken and the study of gamma-ray bursts will have high priority.

Acknowledgments

Over the past 40 years ground-based gamma-ray astronomy at the Smithsonian's Whipple Observatory has been supported by times by the Smithsonian Astrophysical Observatory, the U.S. Department of Energy, the

National Science Foundation and NASA. D.J Fegan is thanked for helpful comments on the manuscript.

References

1. Porter, N.A., *Proc. "Very High Energy Gamma Ray Astronomy"* Ooty, India, Publ. Tata Institute, 68 (1982).
2. Weekes, T.C., *Physics Reports*, **160**, 1 (1988).
3. Aharonian, F.A., Akerlof, C.W., *Ann. Rev. Nucl. Part. Sci.* **47** 273 (1997).
4. Fegan, D.J., *J. Phys. G: Nucl. Part. Phys.*, **23**, 1013 (1997).
5. Ong, R.A., *Physics Reports*, **305**, 93 (1998).
6. Weekes, T.C., *"Very High Energy Gamma Ray Astronomy"*, Publ. I.O.P. (U.K.) (2003).
7. Blackett, P.M.S., *Phys. Abst.* **52**, 4347 (1949).
8. Galbraith, W., Jelley, J.V., *Nature*, **171**, 349 (1953).
9. Jelley, J.V., *Phil. Trans. Roy. Soc.*, **A301**, 611 (1981).
10. Galbraith, W., Jelley, J.V., *J. Atmos. Phys.* **6**, 304 (1955).
11. Jelley, J.V., Galbraith, W., *J. Atmos. Phys.* **6**, 250 (1955).
12. Cocconi, G. *Proc. Int. Cosmic Ray Conf. (Moscow)*, **2**, 309 (1959).
13. Chudakov, A.E., Dadykin, V.I., Zatsepin, and Nestrova, N.M., *Transl. Consultants Bureau, P.N. Lebedev Phys. Inst.* **26**, 99 (1965).
14. Fruin, J.H. et al., *Phys. Lettr.* **2**, 176 (1964).
15. Gould, R.J., *Phys. Rev. Lett.*, **15**, 577 (1965).
16. Weekes, T.C., et al., *ApJ* **342**, 379, (1989).
17. Bhat, P.N. et al., *26th ICRC, (Salt Lake City)*, **5**, 191 (1999).
18. Resvanis, L., et al., *Proc. Workshop "Very High Energy Gamma Ray Astronomy"*, Publ.: D.Redeil, NATO ASI Series **199**, 225 (1986).
19. Grindlay, J.E. et al., *ApJL* **197**, L9 (1975).
20. Vladimirsky, B.M. et al., *Proc. Workshop on VHE Gamma Ray Astronomy*, Crimea (April, 1989), 21 (1989).
21. Jelley, J.V., Porter, N.A., *M.N.R.A.S.* **4**, 275 (1963).
22. Jelley, J.V., *Prog. Elem. Part. Phys. & Cos. Ray Phys.* Publ.: North Holland **IX**, 40 (1967).
23. Boley, F.I., *Rev. Mod. Phys.*, **36**, 792 (1964).
24. Hill., D.A., Porter, N.A., *Nature*, **191**, 690 (1960).
25. Grindlay, J. et al. *ApJL*, **559**, 100 (1996).
26. Chadwick, P.M., McComb, T.J.L. & Turver, K.E. *J. Phys. G.; Nucl. Part. Phys.* **16**, 1773 (1990).
27. Weekes, T.C., *Space Sci. Rev.* **59**, 315 (1993).
28. Weekes, T.C., Turver, K.E., *Proc. 12th ESLAB Symp. (Frascati)*, 279 (1977).
29. Cawley, M.F. et al., *Exp. Astron.* **1**, 173 (1990).
30. Cawley, M.F, et al. *19th ICRC (La Jolla, California)* **1**, 131 (1985).
31. Gibbs, K., *"The Application of Imaging to the Atmospheric Cherenkov Technique: Observations of the Crab Nebula"*, Ph.D. Dissertation, University of Arizona, (unpublished) (1987).
32. MacKeown, P.K. et al., *Proc. 18th ICRC (Bangalore)* **9**, 175 (1983).

33. Hillas, A.M., *Proc. Proc. 19th ICRC (La Jolla)*, **3**, 445 (1985).
34. Punch, M., *"New Techniques in TeV Gamma-ray Astronomy"*, Ph.D. Dissertation, National University of Ireland, (unpublished) (1993).
35. Punch, M., et al., *Nature*, **358**, 477 (1992).
36. Quinn, J., et al., *ApJL* **456**, L83 (1996).
37. Holder, J. et al., *ApJ* **583**, L9 (2002).
38. Catanese, M. et al., *ApJ* **501**, 616 (1998).
39. Horan, D. et al., *ApJ* **571**, 753 (2002).
40. Kildea, J. et al., (in preparation) (2005).
41. Konopelko, A. et al., *Astropart.Phys.* **10**, 275 (1999).
42. Lorenz, E., *"GeV-TeV Astrophysics"* (Snowbird, Utah), AIP Conf. Proc. **515**, 510 (1999).
43. Weekes, T.C., et al., *Astropart. Phys.* **17**, 221 (2002).
44. Matsubara, Y., *"Towards a Major Atmospheric Cherenkov Detector"* (Kruger Park), ed. O.C. de Jager, 447 (1997).
45. Hofmann, W., *"GeV-TeV Astrophysics: Towards a Major Atmospheric Cherenkov Detector IV" (Snowbird)* 500 (1999).
46. T. Weekes et al., *Proc. "Cherenkov2005",Palaiseau, (April, 2005)*, (2006), 3.
47. J. Holder et al., Astroparticle Phys. (in press), (2006).

JEM-EUSO MISSION TO ATTACH JEM/EF OF ISS

TOSHIKAZU EBISUZAKI (RIKEN), FUMIYOSHI KAJINO (KONA UNIV.) , MOTOHIKO NAGANO (FUKUI INSTITUTE OF TECH), YOSHIYUKI TAKIZAWA, YOSHIYA KAWASAKI, MITSUTERU SATO, M. E. BERTAINA (RIKEN) , TOSHIYUKJ SAWABE (KONAN UNIV.) , TORU SHIBATA, NAOTO SAKAKI (AOYAMA GAKUIN UNIV.), NAOYA INOUE (UNIV. SAITAMA), YUKIO UCHIBORI (NIRS), AND EUSO COLLABORATION

JEM-EUSO mission is the science mission to detect extreme energy particles with the energy above 10^{20} eV from the orbit. It is attached to Japanese experiment module of International Space Station. The outline of the mission is presented in the present paper

1. Introduction

EUSO (Extreme Universe Space Observatory) is a super wide-field telescope to observe extreme energy particles with the energy above 10^{20} eV from the International Space Station in the orbit of 430 km altitude from the ground. An extreme energy particle from space collides with a nucleus in the Earth's atmosphere and produces an Extensive Air Shower (hereafter, EAS) that is consist of numerous number of electrons, positrons, and photons. EUSO takes images of fluorescence UV photons emitted by nitrogen molecules excited in EAS with every several micro-seconds to observe three-dimensional development of EAS.

Fresnel lenses and records the track of and EAS in the atmosphere with a time resolution of 2.5μ s and a spatial resolution of about 0.75×0.75 m (Corresponds to 0.1×0.1 degree). These time-sliced images allow us to reproduce energies and directions of primary particles. The focal surface of the EUSO telescope is made by about 6,000 multi-anode photomultipliers. The number of pixel is about two hundred thousand in total.

EUSO can determine the direction of EECR with a resolution less than several degrees. It observes the atmosphere in the area of a circle with a radius of 250km. The sensitivity of EUSO is higher than Pierre Auger Observatory by a factor of 20 and than Telescope Array by a factor of one hundred.

EUSO was originally selected by European Space Agency (ESA) as a mission attached to European Columbus module: The phase-A study is successfully completed in June 2004. However, because of financial problems in ESA and European countries, the start of phase-B is postponed for a long time. Japanese and U.S. teams re-define EUSO as a mission attached to Japanese Experiment

Module/Exposure Facility (JEM/EF) of ISS. They renamed it as JEM-EUSO (figure 1) and started the preparation targeting the launch of 2012 in the frame work of second phase of JEM/EF utilization. It will be launched in 2012 by H2B and conveyed by HTV (H-II transfer Vehicle) to ISS.

Figure 1. JEM-EUSO telescope attached to Japanese Experiment Module of International Space Station: left Nadir mode: right Tilted mode

2. Improvements in JEM-EUSO

JEM-EUSO reduces the threshold energy down to around 10^{19} eV and increase the effective area by the advances in technology and to superior features of JEM/EF. The reduction in threshold energy is realized by

1) Increased diameter of the telescope,
2) Improved lens material and new optical design,
3) Improved detectors in higher quantum efficiency, and
4) Improved algorism for event trigger.

The increase in effective area is realized by inclining the telescope from nadir (tilted mode; figure 1). In this tilted mode, the threshold energy gets higher since the mean distance to EAS and atmospheric absorption both increase. First half of the mission lifetime is devoted to lower energy in nadir mode and second half of the mission to high energy by tilted mode.

3. Scientific Targets

JEM-EUSO will open up new field of astronomy of extreme energy particles with, by far, improved statics (more than 1,000 events). It will detect almost one thousands events above 10^{20} eV in its five years operation. GZK (Greisen-Zatsepin-Kuz'min) process [1,2] makes trans-GZK complex in the energy spectrum [3,4]: It consists of

(1) Steeper slope at $(5 - 10)\times10^{19}$ eV, so called GZK-cutoff

(2) GZK bump at (4 - 8)×10^{19}eV. It is formed by particles that came from the distance of several or several dozen D_{GZK} , loose energy and deposit around it.

(3) Ankle region feature around (0.3 - 3)×10^{19}eV. Highly red-shifted GZK-bump of made by the particles came from cosmological distances.

(4) GZK recovery, that starts from 3×10^{20}eV. Those are made from with in distance smaller then D_{GZK}.

Trans-GZK complex is, as a whole, reflects the history of the Universe. JEM-EUSO will determine (1), (2), and (4) of trans-GZK complex, will by far a higher statics than other ground based experiments. These three features give us the absolute energy calibration for the fluorescence method. The detailed comparison with the results of ground-based experiments allows us to unite the energy spectra in the ankle region: the energy spectra reported by different experiments are significantly different each other. JEM-EUSO will figure-out this trans-GZK complex.

JEM-EUSO also determines the origin of EECR particles by arrival direction analysis. Arrival directions of EECR are determined with the accurately better than several degrees. A particle with energy of 10^{20}eV is reflected less than one degree (in the case of proton) by the galactic magnetic field and directly arrives to the Earth from its origin. Therefore, if we trace back to the arrival direction, we can reach the point to production of that particle. The arrival direction analysis is divided into point source analysis and global analysis.

First, in the point source analysis, events are clustered into within the circle of the instrumental resolution; In fact, AGASA experiment reported that significant tendency to make clusters in the events over 4×10^{19}eV [5]. If they came from to isotropically distributed point sources in three-dimensional space, 100 events into several dozen cluster.

Next, in global anisotropy analysis, arrival directions are integrated for spherical harmonics and study the significance in each component. This analysis reveals the belonging of sources of EECR to our galaxy or the local super cluster. In such analysis, the exposure must be uniform over all the sky. JEM-EUSO attached to ISS with an inclination of 51.6degree, observes both north and south sky and gives us an uniform exposure for all sky.

Furthermore, JEM-EUSO can detect extreme energy neutrinos through a horizontal EAS that takes place deep in the atmosphere (Horizontal Air Shower: HAS) and upward going EAS (Upword-going Airshowers: UAS). Later one is produced by tau-neutrino interacting with nucleus in the interior of the earth. Since an extreme energy particle certainly exists, extreme energy neutrinos must exit. Neutrino with much smaller interaction cross section with matter can escape from emitting region not blocked by matter nor magnetic field and can propagate for cosmic distance: We can distil the information from them.

Finally, JEM-EUSO also can observe atmospheric luminous phenomena such as lightning, nightglow, and meteors. The night glow in the wavelength between 330nm-400nm is dominated by the emission by oxygen molecules in Herzberg I band around the boundary region at an altitude of 95 km between mesosphere and thermosphere. This emission is reported to have a strong correlation with the green line (557.7 nm) of oxygen atom [6].The stripes (width of 40 km) of the emission of this green line are observed in the sky from the ground [7] . These stripes are believed to be produced by the gravity wave formed in troposphere and propagated to the upper atmosphere [8]. This propagation of gravity wave may affect to the energy and angular momentum transfer to the mesosphere and thermosphere. In order to study these phenomena, rockets and satellite observations are planed actively [9].

4. Activities in Japan

Development activities in Japan are summaries as follows.

- Weakly Focused Electron Multi-Anode Photomultiplier with Hamamatsu Photonics Inc.
- Elementary Cell (EC) Module with four PMTs and its vibration test against 20 G (rms).
- Photo-detector Module (PDM) with nine ECs and its vibration test against 20 G (rms).
- High Voltage Circuits that supply to 36 PMTs in a PDM
- A circuit to protect MAPMT from excessively strong light sources such as lightning or city light.
- Aging test of PMTs: no significant change in PMT performance in for the total amount photons in entire mission period
- End to End simulation code including shower generation, ray-trace, shower detection, and reconstruction.
- A Balloon experiment with the same PMT that use in EUSO (2005.8) .

References

1. Greisen, K., *Phys. Lett.* 16, 148 (1966).
2. Zatsepin, G. T. and Kuz'min, V. A., *JETP Phys. Lett.* **4**, 78 (1966).
3. Stecker, F.W. *Nature*, **342**, 401, (1989)
4. Berezinsky,V.S. and Grigorieva, S.I, *Astron.Astrophys.* **199**, 1 (1988)
5. Takeda, M. et al., *Astrophys. J.*, **522**, 255 (1999).
6. Thomas, R. J. *Journal of Geophysical Research*, **86**, 206 (1981).
7. Onoma, F. et al. *Annales Geophysicae*, 2005, **23**, (2385).
8. Horinouchi, T. Nakamura, T., and Kosaka, J. *Geophysical research Letters*, 2007, **29**, (2007).
9. Iwagami, N. et al. *Advance in Space Research*, 2005, **35** (1964).

PROPAGATION OF ULTRA-HIGH ENERGY COSMIC RAYS ABOVE 10^{19} EV IN A STRUCTURED EXTRAGALACTIC MAGNETIC FIELD AND GALACTIC MAGNETIC FIELD

HAJIME TAKAMI AND HIROYUKI YOSHIGUCHI

Department of Physics, School of Science, the University of Tokyo,
7-3-1, Hongo, Bunkyo-ku, Tokyo, 113-0033, Japan
E-mail: takami@utap.phys.s.u-tokyo.ac.jp

KATSUHIKO SATO

Department of Physics, School of Science, the University of Tokyo,
7-3-1, Hongo, Bunkyo-ku, Tokyo, 113-0033, Japan
and Research Center for the Early Universe, School of Science,
the University of Tokyo, 7-3-1, Hongo, Bunkyo-ku, Tokyo, 113-0033, Japan
E-mail: sato@phys.s.u-tokyo.ac.jp

We develop a new numerical method for simulations of the arrival distribution of Ultra-high Energy Cosmic Rays (UHECRs) above 10^{19} eV, taking their propagation in a structured magnetic field into account. This method enables us to save the CPU time greatly. Using this method, we calculate the propagation of UHE protons considering both a structured extragalactic magnetic field (EGMF) and the Galactic magnetic field, and simulate the arrival distribution. Our models of EGMF and UHECR source distribution reflect structures observed around the Milky Way. As an application, we compared our simulated arrival distribution with that observed by Akeno Grand Air Shower Array (AGASA) statistically. From this comparison, we find that the most appropriate number density of UHECR sources that best reproduces the AGASA observation is $10^{-4} - 10^{-5}$ Mpc^{-3}, dependent on source types.

1. Introduction

The nature of Ultra-High Energy Cosmic Rays (UHECRs) is poorly known. One of the unsolved problems on UHECRs is their origin. The small-scale anisotropy of the UHECR arrival distribution above 4×10^{19} eV observed by Akeno Giant Air Shower Array (AGASA) [1] has thought to have information on their sources. On the other hand, High Resolution Fly's Eye (HiRes) has not observed the anisotropy [2]. This disagreement is expected to be solved by next generation detectors such as Pierre Auger Observatory (Auger) and

Telescope Array (TA).

In many simulations on the arrival distribution of UHECRs, small-scale anisotropy is expected if their sources have discrete spatial distribution. However, there is no simulation considering both extragalactic magnetic field (EGMF) and Galactic magnetic field (GMF), and spatial distributions of EGMF and sources of UHECRs.

Our developed numerical method enables us to calculate the arrival distribution of UHECRs under those conditions [3]. We assume UHECRs are protons above 10^{19} eV and simulate the arrival distribution at the Earth.

2. Models of Magnetic Field

We use the *IRAS* PSCz catalog of galaxies [4] in order to construct our model of a structured EGMF. The *IRAS* catalog consists of 14,677 galaxies with redshift and infrared fluxes > 0.6 Jy and covers about 84 % of the sky. Galaxies in uncovered sky are assumed to have uniform distribution and the selection effect is corrected with luminosity function as the structure of galaxy distribution is not lost. The details are written in Ref.3.

In order to construct our EGMF model, we first cover the universe with cubes of side $l_c = 1$Mpc, which is the correlation length of the EGMF. Magnetic fields in each cube are assumed to be turbulent with the Kolmogorov spectrum. Magnetic strengths, $|B|$, are assumed to be $\propto \rho^{2/3}$ from a simulation of the evolution of a galaxy cluster, where ρ is the density of matter [5]. In addition, we assume that matter density is proportional to luminosity density, ρ_L, constructed from our galaxy sample, $|B| \propto \rho^{2/3} \propto {\rho_L}^{2/3}$. $|B|$ is normalized to $0.4\mu G$ in a cube with the center of the Virgo cluster. This structured EGMF model is applied within 100 Mpc from the Earth. Outside 100 Mpc, we assume uniform turbulent EGMF, whose strength is 1 nG.

We adopt the model of Galactic magnetic field (GMF) used in Ref.6. This model consists of a spiral field and a dipole field.

3. Method for Calculation of the Propagation

We introduce our method for simulations of the arrival distribution of UHECRs. considering the propagation of UHECRs in the Galactic space and the intergalactic space. We assume that source distribution is a part of our galaxy sample. Only parameter is number density of UHECR sources.

In the intergalactic space, we consider not only the deflections by EGMF but also energy loss processes. We consider pair creation, photopion production with the cosmic microwave background (CMB) and adiabatic energy loss due to the cosmic expansion. In Galactic space, all energy loss processes are neglected.

It costs much CPU time to construct the arrival distribution from calculation of the propagation of UHECRs in magnetic fields because magnetic deflections of UHECRs decrease number of cosmic rays that can arrive at the Earth. In order to solve this problem, we have developed a method for construction of the arrival distribution from trajectories of UHE protons calculated by following the inverse processes of the propagation.

First, we inject 2,000,000 UHECRs with charges of -1 and a spectral index of -1 (flat in $\log_{10}$ E space) from the Earth and record their trajectories. Second, for each trajectories we calculate factors,

$$P_{\text{selec}} \propto \sum_i \frac{L_{i,j}}{(1+z_{i,j}){d_{i,j}}^2} \frac{dN/dE_g(d_{i,j}, E_i)}{E^{-1}} \frac{dE_g}{dE}. \tag{1}$$

Here i labels sources on each trajectory, while $z_{i,j}$ and $d_{i,j}$ are the redshift and distance from the Earth that is passed by the jth proton, respectively. $L_{i,j}$ is luminosity of each source galaxy for the case of luminosity weighted source scenario or unity for the case of number weighted source scenario. $dN/dE(d_{i,j}, E_i) = dN/dE_g dE_g/dE \propto E^{-2.6}$ is the energy spectrum of protons at the ith source. The quantity $E_g = E_g(E, d)$ is the energy of a cosmic ray at a source, which has the energy E at the Earth. The quantity dE_g/dE represents the variation of the shape of the energy spectrum through propagation. A method for calculation of dE_g/dE is written in Ref.3 for detail.

Finally, we select a hoped number of events from these trajectories corresponding to the probabilities. The selected events are UHE protons arriving at the Earth. If we have to select the same trajectory more than once, we generate a new event whose arrival angle is calculated by adding a normally distributed deviation with zero mean and variance equal to the experimental resolution of AGASA to the original arrival angle.

4. Results

The arrival distribution of UHECRs obtained by our method is compared with the distribution observed by AGASA statistically. We use the two-point correlation function as a statistical quantity since the small-scale anisotropy has been thought to have information on UHECR sources. In order to compare the function calculated from simulated event distribution with that from AGASA observation, we use χ^2 method, χ_{10}.

Figure 1 shows this quantity as a function of source number density with two different source scenarios. From this figure, the most appropriate source number density that reproduces the AGASA observation is $10^{-4}\mathrm{Mpc}^{-3}$ for luminosity weighted source scenario and $10^{-5}\mathrm{Mpc}^{-3}$ for number weighted source scenario with large uncertainty.

The disagreement about the small-scale anisotropy between AGASA and HiRes is expected to be solved by Auger and TA near the future. In that time, we will be able to obtain more strong constraints on UHECR source. We also want to investigate the propagation in cosmic magnetic field more and more using our developed numerical method.

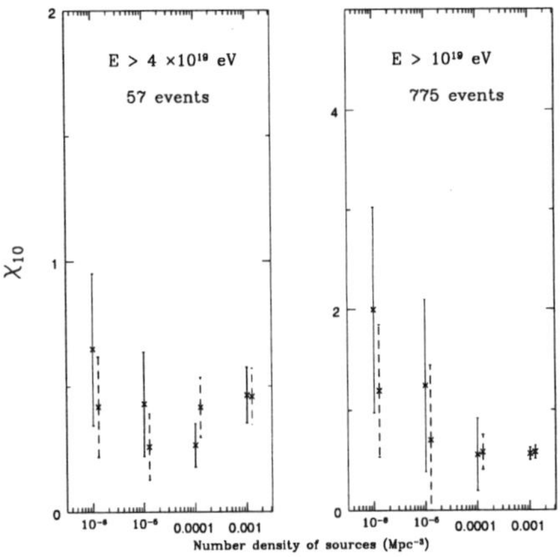

Figure 1. Values of χ_{10} as a function of number density of source. The error bars represent the statistical fluctuations due to the source selection from our galaxy sample. The left and right panels are calculated in the energy ranges of $E > 4 \times 10^{19}$ eV and $E > 10^{19}$ eV, respectively. The solid lines and the dashed lines represent χ_{10} in the case of luminosity weighted source model and number weighted source model, respectively.

References

1. M. Takeda et al., *ApJ* **522**, 225 (1999)
2. R.U. Abbasi et al., *ApJ* **610**, L73 (2004)
3. H. Takami, H. Yoshiguchi and K. Sato, *ApJ* **639**, 803 (2006).
4. W. Saunders, et al., *MNRAS* **317**, 55 (2000).
5. K. Dolag, M. Bartelmann and H. Lesch, *A&A* **387**, 383 (2002).
6. U. Alvarez-Muniz, R. Engel and T. Stanev, *ApJ* **572**, 185 (2002).

HIGH ENERGY NEUTRINO EMISSION FROM GAMMA-RAY BURSTS

KOHTA MURASE
E-mail: kmurase@yukawa.kyoto-u.ac.jp

SHIGEHIRO NAGATAKI
E-mail: nagataki@yukawa.kyoto-u.ac.j

Yukawa Institute for Theoretical Physics, Kyoto University
Oiwake-cho, Kitashirakawa, Sakyo-ku,
Kyoto, 606-8502, Japan

In the standard model of GRBs, protons can be accelerated as well as electrons. Through the photomeson reaction, such protons can produce high energy neutrinos, which may be detected by large Čherenkov detectors such as IceCube. We have carried a systematic and much more thorough calculation than past work analyses of this neutrino emission to determine the high energy neutrino background from GRBs. By executing the Monte Carlo simulation kit Geant4, we can take account of pion-multiplicity and proton-inelasticity, and investigate those effects. We also find that the obtained neutrino background can be comparable with the prediction of Waxman & Bahcall without supposing the enough large nonthermal baryon loading factor necessary for the assumption that GRBs are the main sources of UHECRs. We will discuss the constraints on the nonthermal baryon-loading factor and the effects of magnetic fields.

1. Introduction

Gamma-ray bursts (GRBs) are one of the most energtic phenomena in the universe. In the standard model [2], the prompt emission is explained by internal shocks and needs the strong magnetic field. To explain such radiation, there should be relativistic electrons. In such environments, not only electrons but also protons may be accelerated. Under the assumption that ultra-high-energy cosmic rays (UHECRs) come from GRBs, Waxman & Bahcall predict neutrino bursts from internal shocks, which may be detected by future large neutrino detectors [3]. Ice Čherenkov detectors such as AMANDA at the South Pole have already been constructed and taking data. Now, the future detectors such as IceCube is being constructed [4]. If the prediction is correct, these detectors may detect these neutrinos correlated with GRBs in the near future.

2. Neutrino Burst from GRBs

We are considering the epoch which internal shock occurs. We focus on long GRBs, whose duration is typically ~ 30 s. We consider collision radii in the range of $(10^{13} - 10^{16})$ cm. The shell width is typically given by $l \approx r/\Gamma$, although the uncertainty remains. Hence, the dynamical time scale is $t_{\rm dyn} \approx l/c$. The number of collisons N will be related with observed pulses. We take $N \sim (10 - 100)$. The typical isotropic energy is $(10^{52} - 10^{54})$ ergs, so the energy per subshell is estimated by $(10^{50} - 10^{53})$ ergs. We fix the total, geometrically corrected radiated energy 1.24×10^{51} ergs. Hence the beaming factor is estimated by $E_{\gamma,\rm tot} = f_{\rm b}^{-1} E_{\gamma,\rm tot}^{\rm iso}$. Since the acceleration mechanism is poorly known, we cannot avoid some assumption. To get the maximum energy of proton, we compare the acceleration time scale estimated optimistically, $t_{\rm acc} \sim \varepsilon_p/eBc$ with the cooling time scale which includes the synchrotron cooling, the adiabatic cooling, the inverse-Compton cooling and the photomeson cooling. The photomeson cooling time scale can be evaluated through GEANT4, by which we can take into account multiplicity and inelasticity. The energy density of accelerated protons can be written by $U_p = \epsilon_{\rm acc} U_\gamma$ using the nonthermal baryon loading factor. We set $\varepsilon_p = 10$ GeV, because the relative bulk Lorentz factor will be $\sim$ a few. Using the obtained results of proton's cooling time scale, we can calculate pion spectra and resulting neutrino spectra, taking into account cooling processes of pion and muon. Finally, assuming that the GRB rate traces the star formation rate (SFR), we can evaluate a diffuse neutrino background [5,6].

3. Concludings and Discussions

We calculate proton's photomeson cooling efficiency and resulting neutrino spectra from GRBs more quantitatively than previous works. We can include multiplicity and inelasticity by executing GEANT4 and study these effects, which enhance the proton cooling efficiency in the high energy region, so they help prevent protons from accelerating up to the ultra-high energy region in some cases. But in many cases, the synchrotron loss time scale and the dynamical time scale determine proton's maximum energy. In our cases, GRBs can be optically thick to photomeson production at

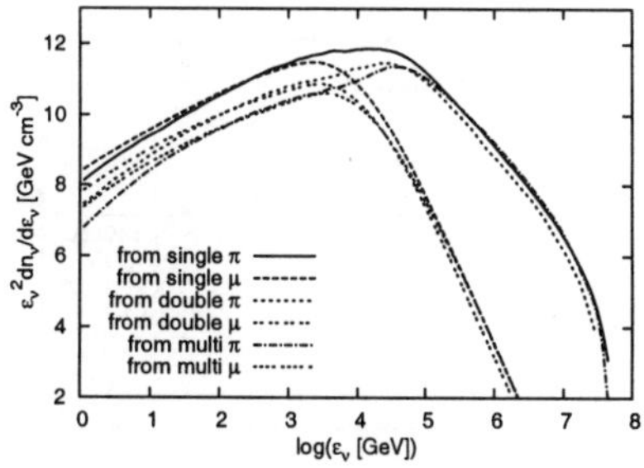

Figure 1. Muon-neutrino ($\nu_\mu + \bar{\nu}_\mu$) spectra in the comoving frame. A result for $r = 2 \times 10^{13}$ cm and ϵ_B is set to 1.

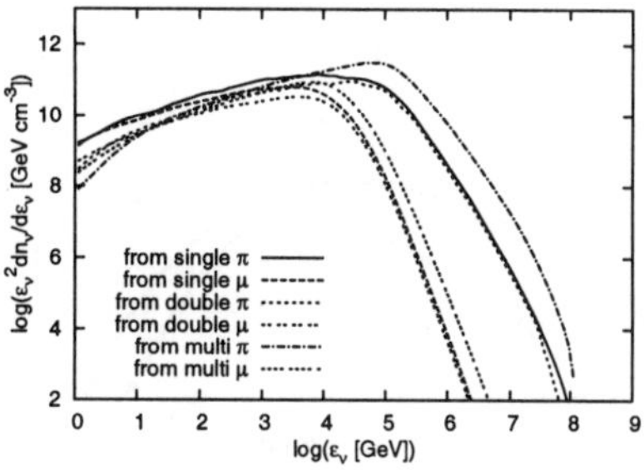

Figure 2. The same as Fig. 1. But a result for $r = 1.8 \times 10^{14}$ cm and ϵ_B is set to 0.1.

$r \lesssim 10^{14}$ cm, while at $r \gtrsim 10^{14}$ cm GRBs can be optically thin to it, so the production of UHECRs is possible at larger radii [7]. The effects from the multi-pion production on resulting spectra are also calculated. We show that the contribution of multi-pion is usually negligible at inner radii but can be larger than that of single-pion at outer radii by a factor. This case is shown in Fig. 1. But such a contribution can be significant for the flatter photon spectrum, even though mesons lose their energy through these cooling processes. This case is shown in Fig. 2. Radiative cooling of pions and muons plays a crucial role in resulting spectra. Neutrino spectra are suppressed above the high-energy break energy. If the magnetic field is strong, such a suppression becomes large and vice versa, if the magnetic field is weak. We also calculate the neutrino background from GRBs for some parameter sets, that are shown in Fig. 3 and Fig. 4. Here $\epsilon_{\rm acc} = 10$ corresponds to the case where the fraction of accelerated protons is similar with that of electrons. If $\epsilon_{\rm acc} = 100$ is possible, higher flux than previous works can be possible. Neutrino observations by IceCube can expect a few or a few tens order of neutrinos per year, although it is important which parameter set is fiducial. Even when the extrapolation of SFR to high redshifts may not be valid, our results would not be so much affected.

To raise neutrino flux, GRBs require the larger nonthermal baryon-loading factor. This value is unknown, but there may be some clues. First, UHECRs observations can give the upper limit to $\epsilon_{\rm acc}$. Furthermore, the large baryon-loading factor suggests a significant contribution of the accelerated protons in the observed hard radiation through secondaries produced in photomeson production. Such emission may be observed in the multi-GeV energy range by electromagnetic cascades by GLAST. Third, $\epsilon_{\rm acc}$ will be

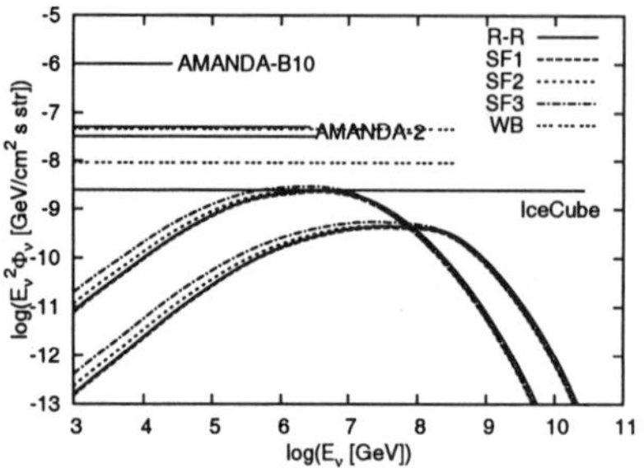

Figure 3. The diffuse neutrino background from GRBs for $z_{\rm max} = 20$ on several SFR models. R-R means the case using Rowan-Robinson SFR. The upper WB bound is for z-evolution of QSOs. The lower is for no z-evolution. $\epsilon_{\rm acc} = 10$ and $\epsilon_B = 1$.

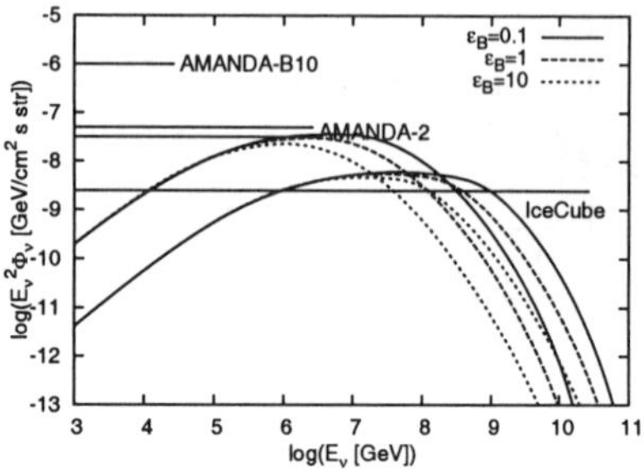

Figure 4. The diffuse neutrino background from GRBs for $z_{\rm max} = 20$ with ϵ_B changing. But the large baryon-loading factor, $\epsilon_{\rm acc} = 100$, is assumed. The upper lines are for set A, while the lower lines are for set B. All lines use the SF3 model.

constrained by the GRB total explosion energy which is still unknown. If IceCube can detect neutrinos from GRBs and confirm that they have the expected level of neutrino flux by our calculation, this will be one of the strong evidences that a significant fraction of protons can be accelerated and our employed internal shock model is valid. We should take into account the respective distributions of parameters to execute the most refined calculaion. Unfortunately, many parameters have large uncertainty at present. For this reason, we calculate for a wide range of these parameters. More and more observations in the near future and more refined theoretical models will allow our results to be improved.

Acknowledgments

We thank the organizing committee of the international workshop on Energy Budget in the High Energy Universe.

References

1. K. Murase and S. Nagataki, Phys. Rev. D, 73, 063002 (2006).
2. B. Zhang and P. Mészáros, IJMP.A, **19**, 14 (2004).
3. E. Waxman and J. Bahcall, Phys. Rev. Lett., **78**, 2292 (1997).
4. F. Halzen, astro-ph/0602132.
5. C. Porciani and P. Madau, ApJ, **548**, 522 (2001).
6. D. Guetta, T. Piran, and E. Waxman, ApJ, **619**, 412 (2005).
7. K. Asano, ApJ, **623**, 967 (2005).

SIMULATION OF SALT NEUTRINO DETECTOR PERFORMANCE FOR ULTRA HIGH-ENERGY NEUTRINO DETECTION*

YUSUKE WATANABE, MASAMI CHIBA, YASUHIRO TAKAYAMA, MASATOSHI FUJII, OSAMU YASUDA, FUMIAKI YABUKI, YUJI SHIBASAKI

Department of physics, Tokyo Metropolitan University, 1-1 Minami-Ohsawa Hachioji-shi, Tokyo, 192-0397, Japan

TOSHIO KAMIJO

Department of Electrical Engineering, Tokyo Metropolitan University, 1-1 Minami-Ohsawa Hachioji-shi, Tokyo, 192-0397, Japan

YUICHI CHIKASHIGE, TADASHI KON, AKIO AMANO, YOSHITO TAKEOKA, YUTAKA SHIMIZU, SATOSHI MORI, SOSUKE NINOMIYA

Faculty of Science and Technology, Seikei University, 3-3-1 Kichijyoji Kitamachi Musasino-shi, Tokyo, 180-8633, Japan

MICHIAKI UTSUMI

Department of Energy Science and engineering, School of Engineering, Tokai University, 1117 Kitakaname Hiratsuka-shi, Kanagawa, 259-1292, Japan

Detection possibility of ultra high-energy (UHE) neutrino (E $>10^{15}$ eV) in natural huge rock salt formation has been studied. Collision between the UHE neutrino and the rock salt produces electromagnetic (EM) shower. Charge difference (excess electrons) between electrons and positrons in EM shower radiates radio wave coherently (Askar'yan effect). In this paper, detection possibility of Salt Neutrino Detector (SND) was studied using SND simulator including attenuation length, background noise and bandwidth of antennae.

1. Introduction

Ultra high-energy (UHE) neutrinos (E$>10^{15}$ eV) can travel without energy loss over astronomical distance. UHE neutrinos give us information about early stage of the universe. Salt Neutrino Detector (SND) aims to detect UHE neutrinos, especially produced by decay of charged pions originated from

* Work partially supported by a Grant in Aid for Scientific Research for Ministry of Education, Science, Technology and Sports and Culture of Japan, and Funds of Tokubetsu Kenkyuhi, at Seikei University

interactions between UHE cosmic rays and the cosmic microwave background (Greisen, Zatsepin, Kuz'min) [1].

The UHE neutrino can be detected using a natural huge rock salt (>10 Gton). Interaction between an UHE neutrino and rock salt produces a hadron shower in which many π^0s are included. The π^0 decays into 2γ. Thus, a gigantic electromagnetic (EM) shower is generated. Charge difference (excess electrons) between electrons and positrons in an EM shower radiates Cherenkov-radio wave coherently (Askar'yan effect) [2]. The track length of the excess electrons is 26% against that of total electrons.

F.Halzen, E.Zas, and T.Stanev calculated the electric field (E-field) strength of radio wave radiated from high-energy ($\leq 10^{15}$eV) EM shower in ice [3]. We made SND simulator, which is able to calculate E-field strength of radio wave radiated from EM shower with energy more than 10^{15} eV in rock salt using structure function. That is space distribution of excess electrons for longitudinal direction (1 dimensional structure function model). This paper presents neutrino detection sensitivity of SND using SND simulator.

2. Calculation method

The simulation of UHE-EM shower takes more than five days to finish using program code Geant4 by a computer with 3.4 GHz clock frequency. [4] We calculated E-field strength using 1 dimensional structure function model. This model treats EM shower as 1 dimensional charge distribution. Thus, Geant4 simulation of EM shower could be replaced by using the structure function. E-field strength was calculated by the following equation. [5, 6]

$$E(R,\omega,\theta) = i\frac{e\omega}{\sqrt{2\pi}R}\left(\frac{n}{c}\right)^2 e^{ikR}\sin\theta \int Q(r)e^{ipr}dr. \qquad (1)$$

$$p = (1 - n\cdot\cos\theta)\omega / c. \qquad (2)$$

Where e is elementary electric charge, ω is angular frequency, R is distance from the origin of EM shower to the observation point, n is refractive index for rock salt (=2.4), k is wave number, and θ is observation angle against EM shower axis. The structure function (Q(r)) is introduced by us.

$$Q(r) = \frac{0.31\times 0.26}{\sqrt{\alpha}}\exp\left[\beta t(1-\gamma\times\ln\left\{\frac{3\times\beta t}{\beta t+2\alpha}\right\})\right]. \qquad (3)$$

Where t is distance from origin of EM shower and α, β, and γ having energy dependent were determined using Geant4.

3. SND Simulation

We simulated neutrino detection possibility of SND including attenuation length, background noise, and bandwidth of antennae. Attenuation length has dependence on the frequency. We used complex permittivity by SND group. [7-11]

$$\tan\delta = \frac{\varepsilon''}{\varepsilon'},\ L = \frac{\lambda}{\pi\sqrt{\varepsilon'}\tan\delta}. \tag{4}$$

Where $\varepsilon', \varepsilon''$, $\tan\delta$, λ and L are real permittivity, imaginary permittivity, loss tangent, wavelength of the radio wave, and the traveling distance, respectively. We used the measured value of Asse (Germany, ε'=5.97, $\tan\delta$ =5.28×10^{-4}). Background noise in the rock salt dome was consistent with blackbody emission (300K) only. [12] The bandwidth is from 0.1 to 0.5GHz optimum. Fiducial volume of SND is a Cube of $3\times3\times3$ km^3. Number of antenna installed in the rock salt is 5000 ($10\times10\times50$) as figure 1. Interval distance of antennae is 250 m in x, y direction, and 50m for depth. We simulated E-field strength of antenna from EM shower for energy 10^{16}-10^{19}eV.

4. Result

Display of antenna array with hit antennae is shown figure 1. Counting condition of hit antenna is that the ratio between Askar'yan radio wave and 300K-blackbody emission (E>6.9 [V/m]) is 1. The hit antennae are shown cone at two layers. Cherenkov angle is about 65.7° for rock salt. The number of hit antennae is 6, 34, and 176 for energy of 10^{17}, 10^{18}, and 10^{19} eV, respectively.

5. Summary

We calculated the neutrino detection sensitivity of SND. Neutrino detection sensitivity is shown figure 2. SND can detect radio wave radiated from EM shower more than energy 10^{17}eV. The number of interaction between a neutrino with 10^{18} eV and rock salt is 6 event against 10^4 of incident event. GZK neutrino of 8-62 per year would be detected by SND. Thus, SND is suitable to detect GZK neutrinos. We should continue the research by the simulation to take into flavor mixing.

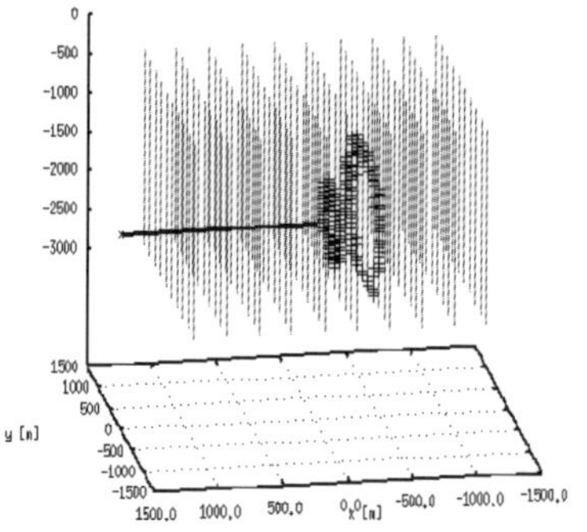

Figure 1. Display of antenna array with hit antennae for 10^{19} eV. Cross point (*) is hit antenna. Line is trajectory of incident neutrino.

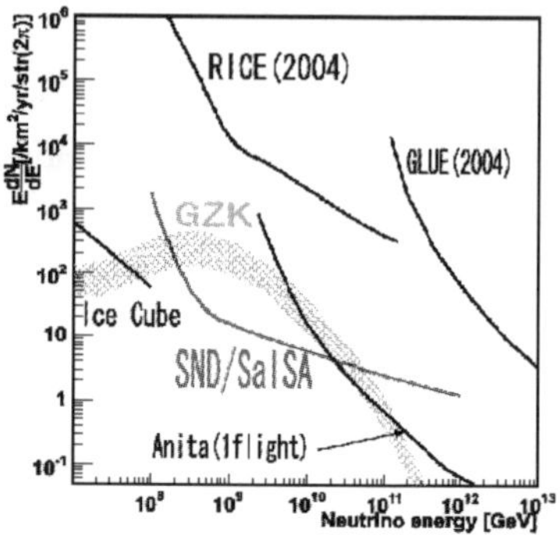

Figure 2. Neutrino detection sensitivity per 1 year. The line of SND is drown, when number of hit antenna is larger than or equals five.

References

1. G K. Greisen, *Phys. Rev. Lett.* **16**, 748 (1966); G.T. Zatsepin, V.A. Kuz'min, Zh. Eksp. Teor. Fiz., Pis' ma Red. **4**, 114 (1966) [*Sov. Phys.-JETP Lett.* **4**, 78 (1966)].
2. A. Askar'yan, Zh. Eksp. & Teor. Fiz. **41**, 616 (1961) [*Sov. Phys. JETP* **14**, 441 (1962)]; G.A. Askar'yan, *Sov. Phys. JETP* **48**, 988 (1965) [**21**, 658 (1965)].
3. E.Zas, F.Halzen, T.Stanev, *Phys. Rev.* **D45**,362(1992)
4. S.Agostinelli *et al.,* Nucl. Instrum. & Methods, *NIM* **A 506**(2003), 250-303
5. J.D. Jacson, Classical Electrodynamics (Wiley, NewYork, 1975)
6. J. Alvaretz-Muniz *et al., Phys. Rev* **D62**, 2000, 063001
7. M. Chiba, T. Kamijo, O. Yasuda, Y. Chikashige, T. Kon, Y. Takeoka and R. Yoshida, Physics of Atomic Nuclei **67**, 2050-2053(2004); P.W.Gorham *et al.*, arXiv:astro-ph/0412128 v2 17 Dec 2004; D.Saltzberg, D.Besson, P.Gorham, A.Odian, R.Milincic, and D.Williams, in Proc. of SPIE **4858** Particle Astrophysics Instrumentation, edited by Peter W. Gorham, (SPIE, Bellingham, WA, p.191 (2003).
8. M. Chiba, T. Kamijo, M. Kawaki, H. Athar, M. Inuzuka, M. Ikeda, O. Yasuda, Proc. 1st Int. Workshop for Radio detection if High Energy Particles [RADHEP-2000], UCLA, AIP Conf. Proc. **579**, p.204 (2000)
9. T. Kamijo, M. Chiba, Memoirs of Faculty of Tech., Tokyo Metropolitan University, No.**51 2001**, 139(2002)
10. M. Chiba *et al.*, Proc .of the First NCTS Workshop Astroparticle Physics, Taiwan, World Scientific Publishing Co. Ltd. P.99 (2002)
11. T. Kamijo, M. Chiba, in Proc. Of SPIE **4858** Particle Astroparticle Physics Instrumentation, edited by Peter W. Gorham, (SPIE, Bellingham, WA) p.151 (2003)
12. P.W.Gorham et al., Phys. Rev. D72, 023002 (2005)

MEASUREMENT OF ATTENUATION LENGTH FOR UHF RADIO WAVE IN NATURAL ROCK SALT SAMPLES CONCERNING ULTRA HIGH ENERGY NEUTRINO DETECTION *

MASAMI CHIBA, YUSUKE WATANABE, YASUHIRO TAKAYAMA, MASATOSHI FUJII, OSAMU YASUDA, FUMIAKI YABUKI, YUJI SHIBASAKI

Department of Physics, Tokyo Metropolitan University, 1-1 Minami-ohsawa, Hachioji-shi, Tokyo, 192-0397, Japan

TOSHIO KAMIJO

Department of Electrical and Electric Engineering, Tokyo Metropolitan University, 1-1 Minami-ohsawa, Hachioji-shi, Tokyo, 192-0397, Japan

YUICHI CHIKASHIGE, TADASHI KON, AKIO AMANO, YOSHITO TAKEOKA, YUTAKA SHIMIZU, SATOSHI MORI, SOSUKE NINOMIYA

Faculty of Science and Technology, Seikei University, 3-3-1 Kichijyoji Kitamachi, Musashino-shi, Tokyo, 180-8633, Japan

MICHIAKI UTSUMI

Department of Energy Science and Engineering, School of Engineering, Tokai University, 1117 Kitakaname Hiratsuka-shi, Kanagawa, 259-1292, Japan

Ultra high energy (UHE) neutrinos ($E>10^{15}$ eV) exist at any rate due to presence of the cosmic microwave background and UHE cosmic rays implied by Greisen, Zatsepin and Kuz'min (GZK). The low rate of GZK neutrinos requires us to utilize a large mass (>50 Gton) of detection medium. The UHE neutrino generates a huge number of unpaired electrons in rock salt. They would emit sensible radio wave by coherent Cherenkov effect (Askar'yan effect). Attenuation lengths of natural rock salt samples including synthesized one at 0.3 and 1.0 GHz were measured to find a suitable site constructing a salt neutrino detector. The result indicates a possibility for constructing the salt neutrino detector with economical antenna spacing.

* Work partially supported by a Grant in Aid for Scientific Research for Ministry of Education, Science, Technology and Sports and Culture of Japan, and Funds of Tokubetsu Kenkyuhi, at Seikei University.

1. Introduction

Several theoretical models predict emission of ultra-high energy (UHE) cosmic neutrinos ($E>10^{15}$ eV) from active galactic nuclei etc. Meanwhile UHE protons lose the energy while traveling 163 Mly (50Mpc) due to collision with the 2.7 K cosmic microwave background. The process is called Greisen, Zatsepin and Kuz'min (GZK) cut-off [1]. The energy of GZK neutrinos generated by the effect ranges 10^{15} eV-10^{20} eV, and the flux is as low as 1 ($km^{-2}day^{-1}$). An enormous detector made of rock salt with the mass of 50 Gton or the volume of $(3km)^3$ is needed and desirable having sensitivities of the energy, the direction, the time and the flavor of the GZK neutrinos [2].

The huge detection medium requires long-range transmission wave with a large attenuation length, which carries the information of the neutrino interaction. Radio wave would have a long range through natural rock salt. G. A. Askar'yan [3] had proposed detection of radio wave emission with coherent amplification produced by excess electrons in an electro-magnetic shower in dense materials. Askar'yan effect was confirmed using a bunched electron beam at SLAC [4]. While for low-density medium, radio emission was calculated in an atmospheric shower by M. Fujii and J. Nishimura and recently confirmed experimentally [5]. Rock salt domes are distributed widely and there would be a suitable site [6]. We have been studied rock salt samples having long radio wave transparency [7].

2. Attenuation length for radio wave in rock salt

We have measured complex permittivity in rock salt samples by a perturbation method [8] using cylindrical cavity resonators of 0.3 GHz (749 mm$\phi \times$ 100 mm) and 1.0 GHz (225 mm$\phi \times$ 30 mm). Both have the Q value about 10,000.

We obtained real part (square of refractive index) and imaginary part (absorption in a medium) of complex permittivity, by measuring decease of the resonance frequency and widening of the resonance width after insertion of the samples, respectively.

The attenuation length L is calculated by Eq. (1):

$$\tan\delta = \frac{\varepsilon''}{\varepsilon'}, \qquad L = \frac{\lambda}{\pi\sqrt{\varepsilon'}\tan\delta}. \tag{1}$$

Where $\varepsilon', \varepsilon''$, $\tan\delta$ and λ are real permittivity, imaginary permittivity, loss tangent and wavelength of the radio wave, respectively. At the traveling distance of L, the electric field strength is diminished to 1/e.

A. R. Hippel [9] gave the attenuation length or $\tan\delta$ at 10 MHz and 25 GHz, which were only lower limits of the attenuation length for rock salt. The attenuation lengths larger than 250 m were given in 150, 300 and 750 MHz by in situ measurements at United Salt's Hockley mine located near Houston, Texas [10]. We show recent measurements at 0.3 and 1.0 GHz in figure 1 using rock salt samples of Hockley (USA), Zuidwending (Netherlands), Asse (Germany), Heilbronn (Germany) and Lugansk (Ukraine).

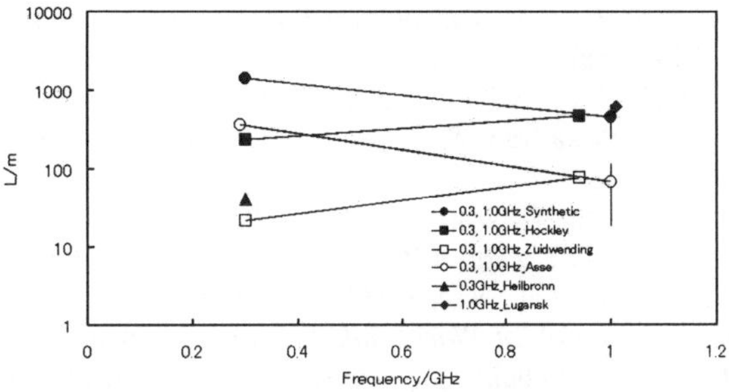

Figure 1. Two types of frequency dependence (proportional and inverse) are shown.

The attenuation lengths at 0.3 GHz of synthetic (1430 ± 216 m) and Asse (367 ± 51 m) rock salts are longer than those at 1.0 GHz (451 ± 34 m) and (67 ± 7 m), respectively. The tendency is consistent with a hypothesis as $\tan\delta$ being constant with the frequency. On the contrary, the attenuation length at 1GHz of Hockley (471 ± 56 m) and Zuidwending (77 ± 7 m) are longer than those at 0.3GHz (237 ± 33 m) and (22 ± 1 m), respectively. The frequency dependence is the same as the in situ measurement at Hockley [10].

The errors are estimated by deviation of the measured values. The cause is estimated as setting errors in position and inclination of the samples at the center of the cavity. Minute gap between the sample and caps to close the insertion holes would also add the error. They are mainly comes from imperfect carving of the samples. Lugansk sample was carved out from an almost single crystal block showing the long attenuation length.

3. Summary

We have measured complex permittivity in natural rock salt samples by a perturbation method using cylindrical cavity resonators of 0.3 GHz and 1.0 GHz The attenuation lengths of Asse (367 ± 51 m) at 0.3 GHz and Hockley (471 ± 34 m) at 1GHz indicate the realization of the salt neutrino detector. Using the attenuation length, a simulation resulted that an economical antenna spacing could detect 8-62 GZK neutrinos/year with a rock salt volume of $(3\ \mathrm{km})^3$ [11].

References

1. K. Greisen, *Phys. Rev. Lett.* **16**, 748 (1966); G.T. Zatsepin, V.A. Kuz'min, *Zh. Eksp. Teor. Fiz., Pis' ma* Red. **4**, 114 (1966) [*Sov. Phys.-JETP Lett.* **4**, 78 (1966)].
2. M. Chiba et al., *Physics of Atomic Nuclei* **67**, 2050-2053(2004); P.W.Gorham et al., Phys. Rev. **D72**, 023002 (2005); D.Saltzberg et al., in *Proc. of SPIE* **4858** *Particle Astrophysics Instrumentation*, edited by Peter W. Gorham, (SPIE, Bellingham, WA, p.191 (2003).
3. G.A. Askar'yan, *Zh. Eksp. & Teor. Fiz.* **41**, 616 (1961) [*Sov. Phys. JETP* **14**, 441 (1962)]; G.A. Askar'yan, *Sov. Phys. JETP* **48**, 988 (1965) [**21**, 658 (1965)].
4. D. Saltzberg, P. Gorham, D. Walz et al., *Phys. Rev. Lett.* **86**, 2802 (2001).
5. M. Fujii and J. Nishimura, *Proc. 11th Int. Conf. On Cosmic Rays*, Butapest, p.709 (1969); H. Falke et al., *Nature* **435**, 313 (2005).
6. J. L. Stanley, "*Handbook of World Salt Resources*", Plenum Press, New York (1969); T. H. Michel, "*Salt Domes*", Gulf Publishing Company, Houston (1979).
7. M. Chiba et al., *Proc. 1st Int. Workshop for Radio Detection of High Energy Particles [RADHEP-2000], UCLA, AIP Conf. Proc.* **579**, p.204 (2000); T. Kamijo and M. Chiba, *Memoirs of Faculty of Tech., Tokyo Metropolitan University*, No.**51 2001**, 139 (2002) ;M. Chiba et. al., *Proc. of the First NCTS Workshop Astroparticle Physics, Taiwan*, World Scientific Publishing Co. Ltd. p.99 (2002); Toshio Kamijo and Masami Chiba, *in Proc. of SPIE* **4858** *Particle Astrophysics Instrumentation*, edited by Peter W. Gorham, (SPIE, Bellingham, WA) p.151 (2003); M. Chiba et. al., *Proc. of the International Workshop (ARENA2005), DESY, Zeuthen*, World Scientific Publishing Co. Ltd. p.25 (2006); *ibid.*,Y.Watanabe et. al., p.50.
8. H. A. Bethe and J. Schwinger, *NDRC Report* D1-117 (1943); R.L.Sproull and E.G.Linder, *Proc. of I.R.E.*, **34**, 305 (1946); 1.J.C. Slater, *Rev. Mod. Phys.*, **18**, 441 (1946); G.Birnbaum and J. Franeau, *J.Appl.Phys.*, **20**, 817 (1949); N.Ogasawara, *J. Inst. Elect Eng., Japan*, **74**, 1486 (1954); R. Ueno and T. Kamijo, *IEICE Trans. Commun.* **E83B**, 1554 (2000).
9. A.R.von Hippel ed., *Dielectric Materials and Applications*, P.302, 361, John Wiley & Sons, INC, (1954); *Landolt-Boernstein, Zahlenwerte und Function aus Physik, Chemie, Astronomie, Giophysik und Technik, Eigenshaften der Materie in Ihre Aggeregatzustaenden, 6.Teil, Elektrische Eigenshaften I*, Herausgegeben von K.H.Hellwege und A.M. Hellwege, P.456, 505, Springer-Verlarg (1959); R.G.Breckenbridge, *J. Chem.Phys.* **16** (10) p.959 (1948).
10. P. Gorham, et al., *Nucl. Instrum. & Methods.* **A490**, 476 (2002).
11. Y. Watanabe et al., *Simulation of salt neutrino detector performance for ultra high-energy neutrino detection* in these proceedings.

PARTICLE ACCELERATION BY MAGNETOHYDRODYNAMIC TURBULENCE

J. CHO*

Dept. of Astronomy and Space Science, Chungnam National Univ.,
Daejeon, Korea
E-mail: jcho@cnu.ac.kr

A. LAZARIAN

Dept. of Astronomy, Univ. of Wisconsin
Madison, WI53706, USA
E-mail: lazarian@astro.wisc.edu

Recent advances in understanding of magnetohydrodynamic (MHD) turbulence call for revisions in the picture of particle acceleration. We make use of the recently established scaling of slow and fast MHD modes in strong and weak MHD turbulence to provide a systematic study of particle acceleration in magnetic pressure (low-β) and gaseous pressure (high-β) dominated plasmas. We consider the acceleration by large scale compressions in both slow and fast particle diffusion limits. We establish that fast modes accelerate particles more efficiently than slow modes. We find that particle acceleration by pitch-angle scattering and TTD dominates acceleration by slow or fast modes when the spatial diffusion rate is small.

1. Introduction

MHD turbulence is an important agent for particle acceleration as was pointed first by Fermi [5] and later was discussed by many other authors (see Chandran & Maron [2] and references therein). Second order Fermi acceleration by MHD turbulence was appealed for acceleration of particles in many astrophysical environments, e.g. Solar wind, Solar flares, the intracluster medium, gamma-ray bursts (see [1,6,8]). Naturally, properties of MHD turbulence (see Cho & Lazarian [3] and references therein) are essential for understanding the acceleration mechanisms. Here, we summarize recent studies of particle accleration by MHD turbulence (see Cho & Lazarian [4] for details).

*Work partially supported by research fund of chungnam national university in 2005.

2. Methods

When a particle moves inside an eddy of size l, the change of particle's momentum over a time Δt is

$$\Delta p \sim (dp/dt)\Delta t, \tag{1}$$

where p is the momentum of the particle. The momentum diffusion coefficient is

$$D_p \sim (\Delta p)^2/\Delta t \sim (dp/dt)^2 \Delta t. \tag{2}$$

The change of momentum depends on compression[a][7]:

$$\frac{dp}{dt} \approx -p\nabla \cdot \mathbf{v}_l. \tag{3}$$

Substituting eq. (3) into eq. (2), we obtain

$$D_p \sim p^2 (\nabla \cdot \mathbf{v}_l)^2 \Delta t. \tag{4}$$

Therefore we can determine D_p when we know $\nabla \cdot \mathbf{v}_l$ and Δt.

3. Scaling of Δt and $\nabla \cdot \mathbf{v}_l$

We assume that the fluid is magnetized and particles gyrate around magnetic field lines. From now on, the term "parallel" means "parallel" to the local mean magnetic field.

Scaling of Δt depends on the spatial diffusion rate. When the spatial diffusion rate is high (fast diffusion limit),

$$\Delta t \sim l_\parallel^2 / D_\parallel, \tag{5}$$

where $l_\parallel$ is the parallel size of an eddy and $D_\parallel$ the parallel spatial diffusion coefficient. On the other hand, when the spatial diffusion rate is small (slow diffusion limit),

$$\Delta t \sim l_\parallel / V_A, \tag{6}$$

where V_A is the Alfven velocity.

[a]We can understand it as follows. First, consider compression only in the directions perpendicular to the mean magnetic field. This will result in compression of magnetic field lines, which increases the perpendicular momentum of the particle. Second, consider compression only in the directions parallel to the mean magnetic field. This will cause adiabatic heating and increase the parallel momentum of the particle. Overall, the rate of momentum change depends on the rate of compression.

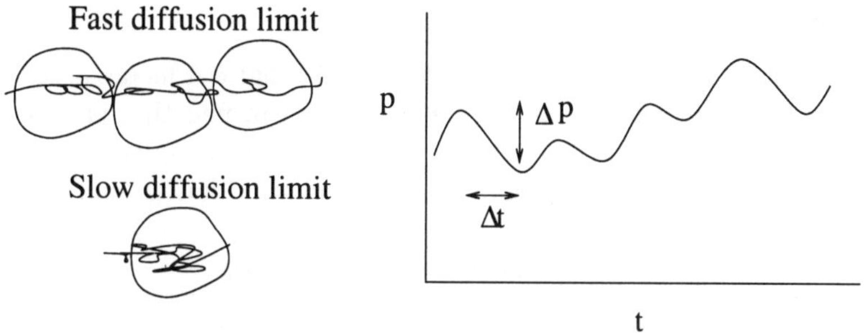

Figure 1. (*Left*) We consider both fast and slow spatial diffusion limits. In the fast spatial diffusion limit, a particle can move to different eddies. In slow fiffusion limit, the eddy changes its properties before the particle escape the eddy. (*Right*) Particle momentum shows a random walk-like behavior and, therefore, a diffusion process in momentum space. The diffusion coefficient in momentum space is $\sim (\Delta p)^2/(\Delta t)$.

Different MHD modes have different values of $\nabla \cdot \mathbf{v}_l$. Slow modes are elongated and compressional motions are neither parallel nor perpendicular to the mean magnetic field. The resulting $\nabla \cdot \mathbf{v}_l$ is $\sim v_{l,slow}/l_{\parallel}$ for low-β plasmas and $\sim v_{l,slow}/(\beta l_{\parallel})$ for high-β plasmas plasma, where β is the ratio of gas to magnetic pressure. On the other hand, fast modes are isotropic (see Cho & Lazarian [3]), and compressional motions are radial. The resulting $\nabla \cdot \mathbf{v}_l$ is $\sim v_{l,fast}/l$ for all values of β.

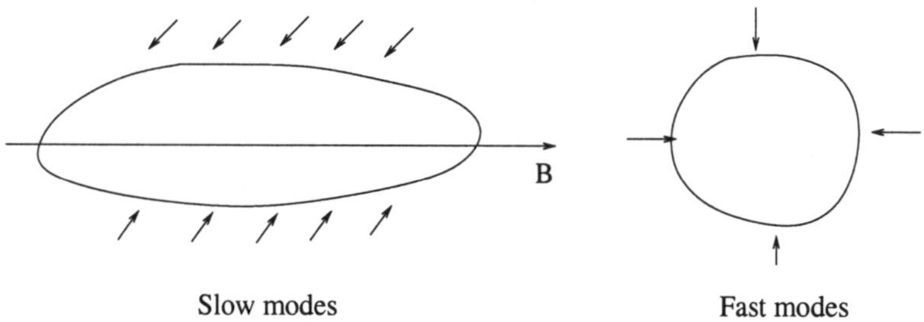

Figure 2. (*Left*) Slow mode eddies are elongated along the mean magnetic field and the direction of compressional motions depend on plasma β, the ratio of gas to magnetic pressure. (*Right*) Fast mode eddies are isotropic and compressional motions are also isotropic.

4. Results

We summarize the results in Table 1. We can see that fast modes accelerate particles more efficiently than slow modes. We can find that particle acceleration by pitch-angle scattering (and transit time damping) dominates acceleration by slow or fast modes when the spatial diffusion rate is small.

Table 1. Momentum diffusion coefficients.

D_p	Estimates	Properties of fluid
D_p^{slow}	$\sim \frac{p^2 v_A}{L}\left(\frac{v_{L,slow}}{v_A}\right)^2\left(\frac{Lv_A}{l_{mfp}v_{ptl}}\right)$	low-β, fast diffusion limit
	$\sim \frac{p^2 v_A}{L}\left(\frac{v_{L,slow}}{v_A}\right)^2 Q_{TD}$	low-β, slow diffusion limit
	$\sim \frac{p^2 v_A}{L}\left(\frac{v_{L,slow}}{v_A}\right)^2\left(\frac{Lv_A}{l_{mfp}v_{ptl}}\right)\beta^{-2}$	high-β, fast diffusion limit
	$\sim \frac{p^2 v_A}{L}\left(\frac{v_{L,slow}}{v_A}\right)^2 \beta^{-2} Q_{TD}$	high-β, slow diffusion limit
D_p^{fast}	$\sim \frac{p^2 v_A}{L}\left(\frac{v_{L,fast}}{v_A}\right)^2\left(\frac{Lv_A}{l_{mfp}v_{ptl}}\right)$	all-β, fast diffusion limit
	$\sim \frac{p^2 v_A}{L}\left(\frac{v_{L,fast}}{v_A}\right)^2\left(\frac{Lv_A}{l_{mfp}v_{ptl}}\right)^{1-2m}$	low-β, slow diffusion limit
	$\sim \frac{p^2 v_A}{L}\left(\frac{v_{L,fast}}{v_A}\right)^2\left(\frac{Lv_A}{l_{mfp}v_{ptl}}\right)^{1-2m}\beta^{-m}$	high-β, slow diffusion limit
D_p^{pitch}	$\sim \frac{p^2 v_A}{L}\left(\frac{Lv_A}{l_{mfp}v_{ptl}}\right)$	pitch-angle scattering
$D_p^{supersonic}$	$\sim \frac{p^2 v_A}{L}\left(\frac{v_L}{v_A}\right) M_s^{2\gamma}$	super-sonic fluids

Note: Notations used are as follows: L = the energy injection scale; l_{mfp} = particle's mean free path; v_{ptl} = particle's velocity; $Q_{TD} \sim O(1)$; m = the slope of the spectrum of fast modes; D_p^{pitch} = momentum diffusion coefficient by pitch angle scattering; M_s = sonic Mach number; $\gamma \sim O(1)$.

References

1. B. Chandran, *ApJ*, **599**, 1426 (2003)
2. B. Chandran and J. Maron, *ApJ*, **603**, 23 (2004)
3. J. Cho and A. Lazarian, *Theo. Comp. Fluid Mech.* **19**, 127 (2005)
4. J. Cho and A. Lazarian, *ApJ*, **638**, 811 (2006)
5. E. Fermi, *Phys. Rev.* **75**, 1169 (1949)
6. V. Petrosian and S. Liu, *ApJ*, **610**, 550 (2004)
7. V. Ptuskin, *Soviet Astron. Lett.* **14**, 255 (1988)
8. R. Schlickeiser and J. Miller, *ApJ*, **492**, 352 (1998)

SU(2)$_L$-TRIPLET DARK MATTER AND HEAT ANOMALY IN COSMIC POSITRON EXPERIMENT

SHIGEKI MATSUMOTO

Theory Group, KEK, Oho 1-1 Tsukuba, 305-0801, Japan
E-mail: smatsu@post.kek.jp

JUNJI HISANO, OSAMU SAITO AND MASATO SENAMI

ICRR, University of Tokyo, Kashiwa, 277-8582, Japan

Recently the HEAT collaboration has been reported the anomaly about the positron excess in the comic ray. The anomaly attracts attention because it may originate in the dark matter annihilation in the galactic halo. In this letter, I would like to address about the interesting fact that the SU(2)$_L$-triplet dark matter can explain the anomaly with satisfying the present dark matter abundance observed by WMAP. When the mass of the dark matter is around 2 TeV, which is favored from the thermal relic abundance, the non-perturbation effect significantly enhances the annihilation cross section into positrons in the non-relativistic limit. We show that the effect enables us to account for the HEAT anomaly.

1. Introduction

The existence of the non-baryonic cold dark matter has been established by the WMAP measurement of the cosmic microwave background[1] such as $\Omega_{\rm CDM}h^2 = 0.113^{+0.016}_{-0.018}$. However, the nature of the dark matter still remains a mystery. Many people believe that the dark matter is a Weakly Interacting Massive Particle (WIMP), because its thermal relic abundance is naturally within the observed range[2].

Various experiments for detecting a WIMP dark matter directly or indirectly have been performed or are planed in order to investigate its nature. Among those experiments, the indirect detection of dark matter using positrons attracts attention because of the recent result reported by the HEAT collaboration[3]. In this measurements, the anomaly about the positron excess from the expected background has been observed, and it may originate in the dark matter annihilation in the galactic halo. Though the excess can not be regarded as a dark matter signal due to poor statistics of the observation, upcoming experiments such as PAMELA[4] and AMS-02[5] may confirm this signal accurately with high statistics.

On the theoretical side, however, it is difficult to explain the positron excess from the dark matter annihilation with satisfying its thermal relic abundance consistent with the WMAP observation. To be more precise, if the abundance of dark matter is determined in the thermal relic scenario, the WMAP result requires the total annihilation cross section of dark matter to be $\langle \sigma v \rangle_{\rm tot} \sim 10^{-26}$ [cm^{3}sec^{-1}]. On the other hand, if the positron excess in the HEAT experiment originates in the dark matter annihilation in our galactic halo, the dark matter annihilation cross section into positrons is expected to be $\langle \sigma v \rangle_{e^+} \sim 10^{-24}$ [cm^{3}sec^{-1}], which is much larger than that expected from the relic abundance.

We have found that if the dark matter is the neutral component in the triplet representation of the Standard Model SU$(2)_L$ group, it can explain both the positron excess in the HEAT experiment and the relic abundance of dark matter observed by WMAP[6]. In the following sections, we show how this dark matter overcome the difficulty discussed above.

2. Annihilation cross section of SU$(2)_L$-triplet dark matter

When the dark matter has a SU$(2)_L$ charge and its mass is much larger than the weak gauge boson mass, an usual perturbative method can not be applied to calculate its annihilation cross section[6]. This is due to the threshold singularity caused by the mass degeneracy between the dark matter and its SU$(2)_L$ partner. In fact, we have to resum ladder diagrams at all orders, in which the weak gauge bosons are exchanged, to calculate the reliable annihilation cross section. This resummation can be interpreted as follows. Since the mass of the dark matter is much heavier than the weak gauge boson mass, the dark matter feels the long-range force induced from the weak gauge boson exchange. Due to the force, the wave function of the dark matter pair is significantly modified from the plane wave before the annihilation. Furthermore, the bound states, which are composed of the dark matter and its SU$(2)_L$ partner pairs, appear due to the long-range force if the dark matter mass (m) is large enough. The bound state has a almost zero binding energy when $m \sim 2, 8, \cdots$ TeV. In these cases, the annihilation cross section in the non-relativistic limit is enhanced by several orders of magnitude compared to the tree-level cross section due to the resonance, and has a strong dependence of the relative velocity (v) between incident dark matters.

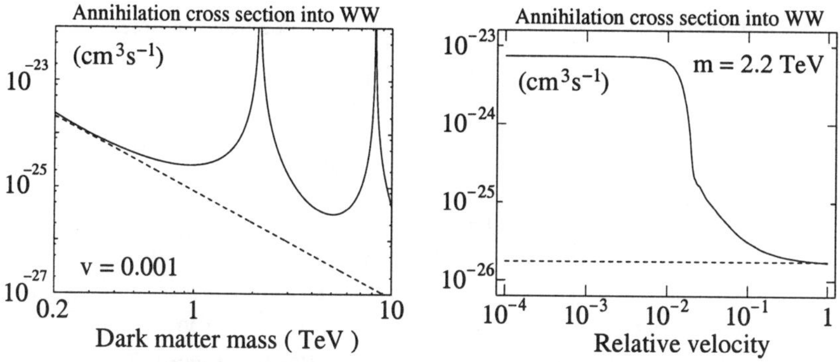

Figure 1. Annihilation cross section of $SU(2)_L$ triplet dark matter into W^+W^- (solid line) and that in the tree level calculation (dashed line).

The $SU(2)_L$ triplet dark matter annihilates mainly into weak gauge bosons due to its $SU(2)_L$ charge. In Fig.1, the annihilation cross section of $SU(2)_L$ triplet dark matter into W^+W^- is shown as a function of m with fixed $v = 10^{-3}$ (left figure) and a function of v with fixed $m = 2.2$ TeV (right figure). The mass difference between the dark matter and its $SU(2)_L$ partner is generated by the $SU(2)_L$ symmetry breaking, which is roughly 0.1 GeV. As expected from above discussion, resonances appear at $m = 2, 8$ TeV in the left figure. Also the strong v dependence is shown in the right figure when m is around the resonance. Thanks to the dependence, the dark matter can explain both the HEAT anomaly and the WMAP result as shown in the next section.

3. Relic abundance and HEAT anomaly

If the relic abundance of the dark matter is explained by the thermal scenario, the mass consistent with the WMAP observation is around 2 TeV as shown in Fig.2 (left figure). The effect of the resonance is not efficient in the calculation of the abundance, because the typical velocity of the dark matter at the freeze-out temperature is about 1/3 and the tree level calculation is good enough as shown in Fig.1[a]. On the other hand, the positron flux from the dark matter annihilation in the halo is significantly affected by the resonance, because the typical velocity of the dark matter at the present universe is about 10^{-3}. In the triplet dark matter case, positrons

[a]The abundance may be slightly modified due to the resonance. The calculation of the abundance including the resonance is now in progress, and will appear soon.[7]

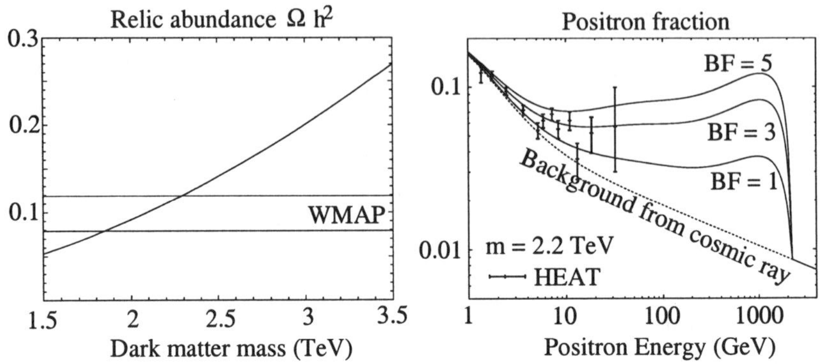

Figure 2. Relic abundance of $SU(2)_L$ triplet dark matter (left figure) and positron fraction from its annihilation (right figure).

are produced through leptonic and hadronic decays of weak gauge bosons. The positron fraction $\equiv$ positron flux/(electron + positron fluxes) is shown in Fig.2 (right figure). For this figure, the propagation of positrons in the galaxy was considered using a diffusion model. BF is an enhancement factor called the boost factor, which parametrize the effect of the inhomogeneity in the local dark matter distribution on the positron flux, whose existence is supported by the N-bodies simulations and expected to be 2-5.

As shown in Fig.2, the $SU(2)_L$ triplet dark matter with the mass $\sim$ 2 TeV naturally accounts for not only the dark matter abundance but also the HEAT anomaly. The possibility of this dark matter will be confirmed or rejected in upcoming experiments such as PAMELA and AMS-02.

References

1. D. N. Spergel *et al.* [WMAP Collaboration], Astrophys. J. Suppl. **148**, 175 (2003); C. L. Bennett *et al.*, Astrophys. J. Suppl. **148**, 1 (2003).
2. G. Jungman, M. Kamionkowski and K. Griest, Phys. Rept. **267**, 195 (1996).
3. S. W. Barwick *et al.* [HEAT Collaboration], Astrophys. J. **482**, L191 (1997); J. J. Beatty *et al.*, Phys. Rev. Lett. **93**, 241102 (2004).
4. M. Boezio *et al.*, Nucl. Phys. Proc. Suppl. **134**, 39 (2004).
5. F. Barao [AMS-02 Collaboration], Nucl. Instrum. Meth. A **535**,134 (2004).
6. J. Hisano, S. Matsumoto and M. M. Nojiri, Phys. Rev. Lett. **92**, 031303 (2004); J. Hisano, S. Matsumoto, M. M. Nojiri and O. Saito, Phys. Rev. D **71**, 063528 (2005); J. Hisano, S. Matsumoto, O. Saito and M. Senami, Phys. Rev. D **73**, 055004 (2006).
7. J. Hisano, S. Matsumoto, M. Nagai, O. Saito and M. Senami.

COSMIC GAMMA-RAY BACKGROUND ANISOTROPY FROM DARK MATTER ANNIHILATION

SHIN'ICHIRO ANDO

Department of Physics, School of Science, University of Tokyo, Tokyo 113-0033, Japan
E-mail: ando@utap.phys.s.u-tokyo.ac.jp

EIICHIRO KOMATSU

Department of Astronomy, University of Texas at Austin, Austin, TX 78712, USA
E-mail: komatsu@astro.as.utexas.edu

High-energy photons from pair annihilation of dark matter particles contribute to the cosmic gamma-ray background (CGB) observed in a wide energy range. The precise shape of the energy spectrum of CGB depends on the nature of dark matter particles. In order to discriminate between the signals from dark matter annihilation and other astrophysical sources, however, the information from the energy spectrum of CGB may not be sufficient. We show that dark matter annihilation not only contributes to the mean CGB intensity, but also produces a characteristic *anisotropy*, which provides a powerful tool for testing the origins of the observed CGB. We show that the expected sensitivity of future gamma-ray detectors such as GLAST should allow us to measure the angular power spectrum of CGB anisotropy, if dark matter particles are supersymmetric neutralinos and they account for most of the observed mean intensity. As the intensity of photons from annihilation is proportional to the density squared, we show that the predicted shape of the angular power spectrum of gamma rays from dark matter annihilation is different from that due to other astrophysical sources such as blazars, whose intensity is linearly proportional to density. Therefore, the angular power spectrum of the CGB provides a "smoking-gun" signature of gamma rays from dark matter annihilation.

1. Introduction

High-energy photons from annihilation of dark matter particles provide indirect means to probe the properties of dark matter. Annihilation signatures, especially gamma rays, have been searched for in regions where the dark matter density is expected to be high, as annihilation rate is proportional to the density squared, ρ_χ^2. Among some possibilities is the extragalactic background light, the cosmic gamma-ray background (CGB), which has been measured in a wide energy range.[1] It has been speculated that some fraction of the CGB may originate from annihilation of dark matter particles in halos distributed over cosmological distances.[2,3]

Dark matter annihilation may be a viable explanation for the CGB, but do we know for sure that the CGB does come from annihilation? We argue that *anisotropy* of the CGB may provide a smoking-gun signature.[4] Although the CGB is isotropic at the leading order, anisotropy should also exist if the CGB originates from cosmological halos. The future gamma-ray detectors with an enhanced sensitivity and angular resolution, such as the GLAST, should be able to see such anisotropy. We calculate the angular power spectrum in GeV region for supersymmetric neutralinos. We then discuss the detectability of CGB anisotropy by GLAST, showing that the predicted anisotropy can be easily measured by 1-year operation of this experiment.

2. Angular power spectrum

The full details for deriving the formulation of getting the angular power spectrum C_l are given in the other paper,[4] on which this paper is based; we refer the interested reader to that paper. We here note that C_l roughly corresponds to the correlation between two points on the sky separated by an angle $\theta \approx \pi/l$.

We assume that the neutralino mass is 100 GeV. In Figs. 1(a) and 1(b) of the left panel, we show the predicted angular power spectrum evaluated at the observed gamma-ray energy of $E_\gamma = 10$ GeV for $M_{\rm min} = 10^6 M_\odot$ and $10^{-6} M_\odot$, respectively. Here $M_{\rm min}$ is the minimum halo mass, below which no halos are assumed to be formed with. The "1-halo term" represents correlations between particles within the same halo, whereas the "2-halo term" represents correlations between particles in two distinct halos.

We compare the predicted power spectrum with the expected sensitivity of the GLAST experiment. We take the following specifications for GLAST: the field of view is $\Omega_{\rm fov} = 4\pi f_{\rm fov} = 2.4$ sr, the angular resolution is $\sigma_b = 0.115°$, and the effective area is $A_{\rm eff} = 10^4$ cm^2 at 10 GeV. In addition, for the diffuse gamma-ray observation, the background contamination can be reduced to 6% of the CGB, which is a promising characteristic [?]. Therefore, error of C_l is essentially determined by the Poisson noise of the cosmic signal. In Fig. 1 (*Right*), we show the predicted angular power spectrum at the observed gamma-ray energies of $E_\gamma = 3$, 10, and 20 GeV, assuming

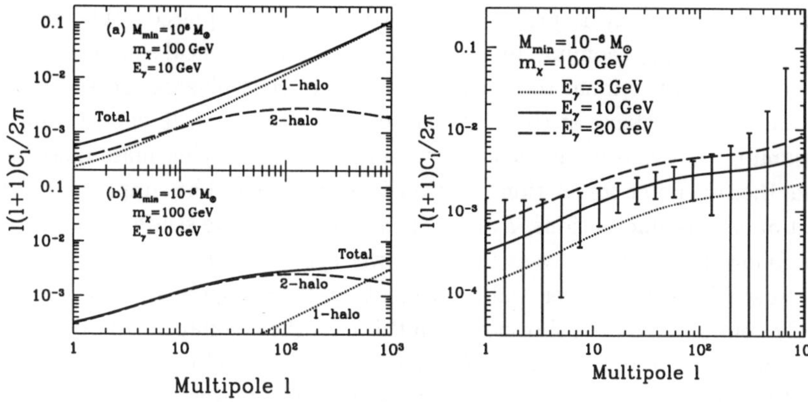

Figure 1. *Left.*—Angular power spectrum of the CGB, C_l, from annihilation of supersymmetric neutralinos, evaluated for (a) $M_{\rm min} = 10^6 M_\odot$ and (b) $M_{\rm min} = 10^{-6} M_\odot$. Note that C_l is dimensionless: the mean intensity squared should be multiplied in order to convert it to the units of intensity squared. The neutralino mass m_χ is assumed to be 100 GeV. The predicted angular spectrum is shown at the observed gamma-ray energy of $E_\gamma = 10$ GeV. Contributions to C_l from the 1-halo (dotted) and 2-halo (dashed) terms are shown as well as the total signal (solid). *Right.*—Angular power spectrum of the CGB, C_l, from annihilation of supersymmetric neutralinos, evaluated for $M_{\rm min} = 10^{-6} M_\odot$. The neutralino mass m_χ is assumed to be 100 GeV. The predicted angular spectrum is shown at the observed gamma-ray energy of $E_\gamma = 3$, 10, and 20 GeV. The 1σ error bars of C_l expected from GLAST for 1 year of operation are also shown at $E_\gamma = 10$ GeV.

$M_{\rm min} = 10^{-6} M_\odot$, with the expected 1σ errors of C_l at $E_\gamma = 10$ GeV for $t = 1$ yr of observations. We find that the GLAST should be able to measure the angular power spectrum of the CGB fairly easily for 1 year of observations, if the dark matter particle is the neutralino with mass around 100 GeV and its annihilation dominates the observed CGB in GeV region. We obtained the similar figure in the case of $M_{\rm min} = 10^6 M_\odot$, which is shown in the other paper.[4] Therefore, we conclude that, if dark matter particles are supersymmetric neutralinos and the observed CGB in GeV region is dominated by their annihilation, the GLAST should be able to measure the angular power spectrum of CGB anisotropy, regardless of the minimum mass.

3. Discussion

The angular power spectrum shown in Figs. 1 should be very characteristic of annihilating dark matter in extragalactic dark matter halos, as the gamma-ray intensity is proportional to the density squared. The intensity of gamma rays coming directly from other astrophysical sources should be linearly proportional to density. It is likely that blazars are the most dominant constituent of the GeV gamma rays over a wide energy range. Assuming that blazars are biased tracers of the underlying mass distribution, the two-point correlation function of blazars should be simply given by that of density fluctuations, $P(k)$. For more quantitative study, we perform the following simple analyses for the power spectrum of blazars.

First, if blazars are very rare objects, then their angular spectrum is entirely dominated by the shot noise. In this case the angular power spectrum does not depend on l, and thus $l(l+1)C_l$ is proportional to l^2 at $l \gg 1$. We see that the shot noise spectrum totally lacks the power on large angular scales (i.e., the spectrum is too steep), and can be easily distinguished from dark matter annihilation.[4]

Second, we take the other extreme limit where blazars are quite common and trace the underlying matter density field, δ, fairly well. In this simplified prescription, the average number of blazars in a halo linearly increases with the host halo mass. In reality, however, this may not be true and we may also need to take into account the difference between a central galaxy and satellite galaxies within a dark matter halo. Anyway, also in this case, the blazar anisotropy might be quite different from that due to dark matter annihilation.[4] A different approach to calculating the angular correlation function is adopted in another paper in the case of Type Ia supernovae.[5]

References

1. A. W. Strong, I. V. Moskalenko and O. Reimer, *Astrophys. J.* **613**, 956 (2004).
2. S. Ando, *Phys. Rev. Lett.* **94**, 171303 (2005).
3. K. Ahn and E. Komatsu, *Phys. Rev.* **D72**, 061301(R) (2005).
4. S. Ando and E. Komatsu, *Phys. Rev.* **D73**, 023521 (2006).
5. P. J. Zhang and J. F. Beacom, *Astrophys. J.* **614**, 37 (2004).

HIGH ENERGY COSMIC RAYS, NEUTRINOS, AND PHOTONS FROM GAMMA-RAY BURSTS

K. ASANO

Division of Theoretical Astronomy,
National Astronomical Observatory of Japan,
2-21-1 Osawa, Mitaka, Tokyo 181-8588, Japan
E-mail: asano@th.nao.ac.jp

We numerically simulate neutrino and photon emission originated from accelerated protons in gamma-ray bursts (GRBs). Pion and kaon production via photomeson processes results in characteristic spectra of neutrinos and photons, which inform us on the physical situations and proton-acceleration efficiency in GRBs.

1. Introduction

The rapid time variabilities and the compactness problem suggest that gamma-ray bursts (GRBs) should arise from internal shocks within relativistic flows. In the standard model, a strong magnetic field is generated, and electrons are Fermi-accelerated in shocked regions. The physical conditions in the shocked region imply that protons may be also Fermi-accelerated to energies $\sim 10^{20}$ eV. High-energy protons in the GRB photon field can create high-energy neutrinos via photopion production. Future observations of neutrinos will be important to prove the standard model of GRBs and the particle acceleration theory. The highest energy of neutrinos brings us information on physical conditions of GRBs. In addition, photons originating from accelerated protons may be observed with future detectors such as GLAST [1]. Since there are many ambiguous points in the GRB standard model, it is very important to observe such emissions.

We would like to emphasize that observations of high-energy photons and neutrinos tell us the physical condition of GRBs. Especially, the magnetic field, the bulk Lorentz factor Γ of the shocked shells, and the amount of accelerated protons are unknown parameters. Such parameters can be estimated from observations of high-energy particles in the near future. In this manuscript, using the Monte Carlo method, we show some interesting examples of neutrino and photon emissions originating from accelerated protons.

2. Method

Our method of simulation is essentially the same as in Asano (2005) [2] and Asano and Nagataki (2006) [3], but photon emission processes from electrons and positrons are included. The physical processes we inlude are photo-pion production, photo-kaon production, pion-decay, muon-decay, electron-positron pair creation, synchrotron emission, inverse Compton emission, and synchrotron self-absorption. We follow cooling processes of each particle produced in the shell via Monte Carlo method. The cascade process depends on photon spectrum given in advace. Therefore, we iterate our simulation until the photon spectrum converges. Pe'er and Waxman (2005) [4] also calculated GRB photon spectra including the effect of photo-pion production using a time-dependent numerical model. However, they did not consider photon emissions from pions, muons or protons. We discuss importance of such emission processes in the high-energy band.

3. Neutrino

Some examples of neutrino spectra from GRBs are shown in Asano (2005) [2]. Since high-energy charged pions will cool down before they decay into neutrinos, the neutrinos from pions do not bring us the information on the highest energy of protons. The highest energy neutrinos may originate from kaons in GRB internal shocks, because kaons are hard to cool in comparison with pions. In Fig.1 we plot neutrino spectra from our simulation [3], assuming a burst with the total energy of 10^{54} ergs, from 1000 shells of $\Gamma = 100$ at 10^{13} cm from the central engine. Above 10^{17} eV neutrinos from kaons dominate neutrinos from pions.

Although there is a little chance to detect such neutrinos, attempts of detection are very important to prove physical conditions in GRBs.

4. GeV-Photon

Our comprehensive results of photon spectra, taking into account the effects of accelerated protons, will be published in the near future [5]. We calculate photon spectra for a very wide range of parameters; energy in a shell, Γ, radius R from the center, and the magnetic field. The minimum energy of power-law injected electrons ($n(\varepsilon) \propto \varepsilon^{-3}$) is chosen to make the typical energy (break energy) of photon spectra 300 keV. The energy of accelerated

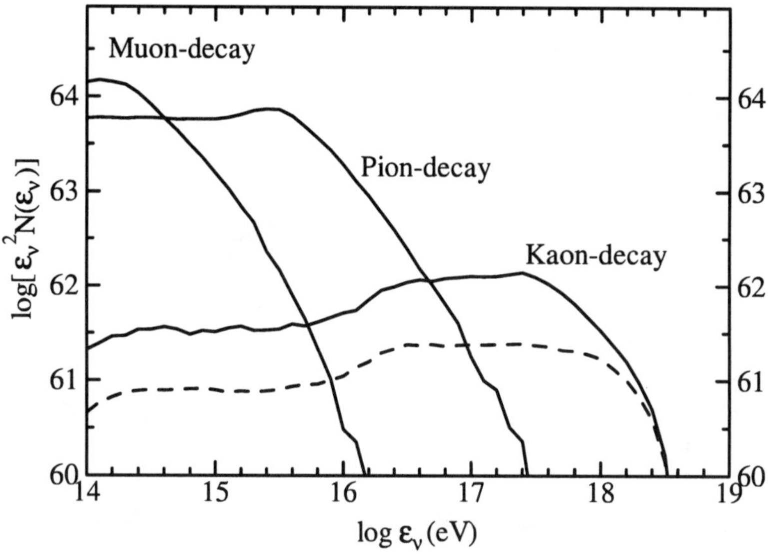

Figure 1. Spectra of neutrinos $\epsilon_\nu^2 N(\epsilon_\nu)$ (in units of eV). Dashed line is for neutrinos from K_L^0-decay.

protons is assumed to be the same as the energy of accelerated electrons. Behaviours of the photon spectra in the high-energy band obtained from our simulations are useful to compare with future observations. The most interesting and prominent cases are photon spectra from shells with very strong magnetic field. Although the energy density of the magnetic field, U_B, is assumed to be comparable to the electron energy density, U_e, in the standard model, there is possibility of $U_B \gg U_e$.

Fig.2 is the photon spectra from 100 shells of 10^{51} ergs (totally 10^{53} ergs), $\Gamma = 300$, and $R = 10^{13.5}$ cm at $z = 0.1$. The ratio of energy densities of the magnetic field to that of accelerated electrons is 30 in this example. Above $\sim$ 100 MeV, absorption due to electron-positron pair creation is crucial, and the synchrotron self-absorption depletes photons below $\sim$ 1 keV.

From Fig.2 one can see that muon synchrotron significantly contributes to the spectrum around $\sim$ 100 MeV. It is difficult to determine whether such a spectrum is due to muon synchrotron, inverse compton emission by electrons, or intrinsic deviation of the electron spectrum from the power-law. However, a detection of such a bump in the spectrum is a very important progress in GRB physics.

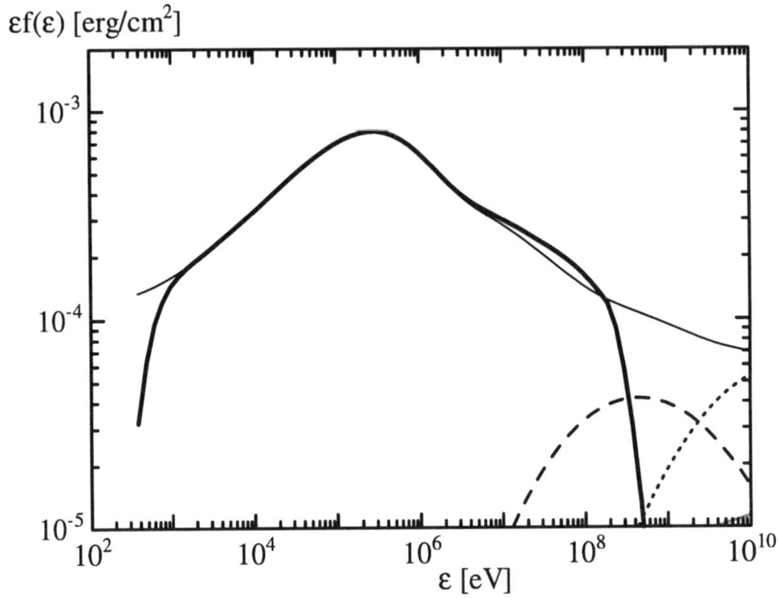

Figure 2. Photon spectra for a burst of 10^{53} ergs with strong magnetic field. Thick solid: Total spectrum. Thin solid: Synchrotron emission from electrons and positrons. Dashed: Synchrotron emission from muons. Dotted: Synchrotron emission from pions.

In addition, our simulations show other interesting feature of the spectra, such as proton synchrotron, change of spectral index in EUV band and so on. From future observations in MeV-GeV or EUV band, we can determine the efficiency of proton acceleration in GRBs.

References

1. K. Asano and F. Takahara, *PASJ* **55**, 433 (2003).
2. K. Asano, *ApJ* **623**, 967 (2005).
3. K. Asano and S. Nagataki, *ApJ* **640**, L9 (2006).
4. A. Pe'er and E. Waxman, *ApJ* **628**, 857 (2005).
5. K. Asano and S. Inoue, in preparation.

DAMPING OF FAST MODES OF MHD TURBULENCE AND ELECTRON ACCELERATION IN SOLAR FLARES

HUIRONG YAN

Canadian Institute for Theoretical Astrophysics, Toronto, ON, Canada
Email: yanhr@cita.utoronto.ca

We address the problem of turbulence damping and particle acceleration in Solar flares. We consider turbulent energy cascade of fast modes and their collisional and collisionless damping processes, which also include damping through the transfer of turbulent energy to non-thermal particles. We identify collisionless damping by thermal particles as the dominant process for damping of fast modes in Solar corona conditions. This makes it possible to decouple particle acceleration and turbulence damping, which simplifies the problem substantially. We estimate the effect of particle acceleration on the cascade of fast modes, and show that our approach provides sufficiently accurate results.

1. Introduction

The mechanism of energy release and the process of its transfer to heating and acceleration of nonthermal particles in many magnetized astrophysical plasmas in general, and solar flares in particular, are still matter of considerable debate. Recent research show that turbulence may play an essential role in these processes. In the case of solar flares, it is believed that the energy comes from release of stored magnetic energy via reconnection (see Priest & Forbes 2000, Lazarian, Vishniac & Cho 2004 and discussions therein). Both the ordinary and magnetic Reynolds numbers so that the plasma is susceptible to production of turbulence. More importantly, recent high resolution observations of solar flares by *Yohkoh* and *RHESSI* satellites have provided ample evidence that, at least from the point of view of particle acceleration, plasma turbulence and plasma waves appear to be the most promising agent not only for the acceleration mechanism but also the general energizing of flare plasma (see *e.g.* Petrosian & Liu 2004, hereafter PL04, and references cited there). This may also be true in other situations (see Lazarian et al. 2002). A substantial progress in understanding of incompressible (Goldreich & Sridhar 1995, hereafter GS95) and compressible

MHD turbulence (see Cho & Lazarian 2005 and references therein), as well as MHD-turbulence-particle interactions (Chandran 2000, Yan & Lazarian 2002, 2004, henceforth YL02, YL04, respectively) calls for revisiting the problems.

In general, one requires a self-consistent treatment of the generation of turbulence and its subsequent interactions with the background plasma. Particle acceleration rate depends on the wave spectrum and the wave damping rate are partially determined by the particle spectrum. The purpose of this paper is to briefly review recent progress in the understanding of the cascade process (§2) and evaluate the damping rates due to thermal particles in the background plasma and a nonthermal population representing the accelerated spectrum $N(E)$ (§3). In §4 we summarize our results and discuss their use in our future work on solving the coupled wave-particle kinetic equations.

2. Turbulence and its damping

Turbulence generated at large scales can cascade to small scales by non-linear interactions. One important characteristic of turbulence is its self-similarity. Power law spectra were obtained numerically for Alfvénic, fast and slow mode turbulence in CL02 and CL03 for the case when turbulent energy is injected at large scales. For instance, it has been found Alfvén (and slow) modes exhibit scale-dependent anisotropy and follow GS95 relations, that were obtained for incompressible turbulence. The mixing motions associated with Alfvénic turbulence induce the scale-dependent anisotropy on slow modes, which on their own would evolve on substantially longer time scale.

Fast modes in low β plasma, on the other hand, develop on their own, as their phase velocity is only marginally affected by mixing motions induced by Alfv́en modes. According to CL02 fast modes follow an isotropic "acoustic" cascade with $W(k) \sim k^{-3/2}$. For such a cascade in each wave-wave collision a small fraction of energy equal to v_{ph}/v_k is transfered to smaller scales so that the cascade time scale is characterized by (CL02):

$$\tau_{cas} = (v_{ph}/v_k)(l/v_k) = (L/\delta V)(v_{ph}/\delta V)(kL)^{-1/2} \quad (1)$$

where $v_{ph} = \omega/k$ is the phase velocity of fast mode and we have used the scaling $v_k = \delta V(kL)^{-1/4}$.

Alfv́en and slow modes modes are inefficient in scattering cosmic ray particles (Chandran 2000, YL02). YL02 identified isotropic fast modes as the dominant scattering agent. It is also possible to show (see Yan & Lazarian 2003) that fast modes are the dominant mechanism for acceleration of particles via *resonant* interaction.

The second important process determining the spectrum of turbulence is its damping rate. Damping becomes important when the damping time Γ_k^{-1} becomes comparable or shorter than the cascading time τ_{cas}. Various processes can damp the MHD motions. In fully ionized plasma, there are basically two kinds of damping: collisional and collisionless. Viscous damping is unimportant in solar flares.

All these facts above have been studied for linear waves. How can this be applied to strong turbulence? For instance, it is well known that the linear damping of waves is anisotropic, i.e. it depends on the angle θ between the wave vector $\mathbf{k}$ and local direction of magnetic field $\mathbf{B}$. As turbulence undergoes turbulent cascade or/and waves propagate in a turbulent medium the angle θ is changing. This problem of θ variations due to the randomization of wave vector $\mathbf{k}$ and the wandering of the magnetic field lines was discussed in YL04 in relation to damping of interstellar turbulence.

For small θ the randomization is $\delta\theta \sim (kL)^{-1/4}$ for quasi-parallel modes (YL04). For fast modes propagating in other directions, $\delta\theta$ scales as $\sim (kL)^{-1/6}$. This shows that randomization is important on large scales $\lesssim L$. In the presence of anisotropic damping these changes in θ results in redistribution of fast mode energy.

For a low β_p plasma, the cutoff scale of turbulence owing to the collisionless damping is,

$$k_c L = \frac{4\zeta^2 \cos^2\theta}{\pi\alpha\beta_p \sin^4\theta} \exp\left(\frac{2\alpha}{\beta_p \cos^2\theta}\right). \tag{2}$$

Due to the averaging within $\delta\theta$, the truncation scale of fast modes around 90° will be approximately equal to

$$\bar{k}_c = k_c|_{90^o - \delta\theta} = \frac{4\zeta^2 \sin^2\delta\theta}{\pi\alpha\beta_p \cos^4\delta\theta} \exp\left(\frac{2\alpha}{\beta_p \sin^2\delta\theta}\right). \tag{3}$$

The result is shown in Fig. 1, where both collisionless and viscous damping lengthscales are plotted. We see that indeed collisionless damping is dominant and truncates fast modes at scales much larger than the thermal proton gyroscale. We also see that the narrow cone of quasi-perpendicular modes disappears due to the randomization of θ.

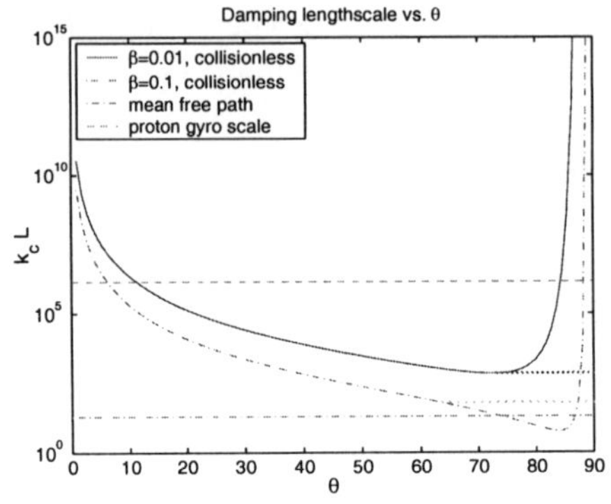

Figure 1. Collisionless damping scales of turbulence vs. the angle θ between $\mathbf{k}$ and $\mathbf{B}$ in wandering magnetic field. The angle variation $\delta\theta$ increases with scale, therefore the averaging around 90° is substantial (dotted line). The averaged damping scale is larger than original in $\beta = 0.01$ case. In $\beta = 0.1$ case, the averaged line is larger than original for $\theta > 87^o$ and smaller than original otherwise. The dashed line corresponds to thermal proton gyroscale, which defines the small scale limit for MHD regime.

3. Particle acceleration by fast modes

3.1. *Acceleration by undistorted spectrum*

For fast modes, there are basically two types of interactions: gyroresonance and transit time acceleration (TTD). Gyroresonance requires electromagnetic perturbations at a particular scale, e.g. $k_{res} \sim \Omega/v_{\|} \sim r_g^{-1}$. Given the parameters we adopt here, $k_{res}L > 10^9$ even for relativistic electrons. This is certainly beyond MHD regime (see Fig. 1). The largest contribution then comes from TTD, which happens at all scales.

Due to field line wandering, the acceleration is decreased in $\beta = 0.01$ case. In the case of $\beta = 0.1$, it is decreased for $E_k \gtrsim 0.6$KeV and increased for lower energies. There results can be explained from the curve of damping scale in Fig. 1. It is demonstrated there that the averaged cutoff wave number $\bar{k}_c$ is smaller than the original value $k_c(\theta)$ $(90^o - \delta\theta < \theta < 90^o)$ in the first case. On the contrary in the second case, the averaged k_c is larger than $k_c(\theta)$ for $\theta < 87^o$. These modes accelerate low energy particles.

3.2. *Back reaction of accelerated particles*

The wave damping rate calculated above assumes that the energy lost by waves with spectrum $W(\mathbf{k})$ goes into heating the plasma and that the plasma maintains its Maxwellian distribution. In reality, we need to include the damping of the waves by the nonthermal tail of the distribution accelerated by the waves as well. Here we derive the damping rate due to a power law distribution of electrons and protons with isotropic particle pitch angle distribution: $N(E) = N_0(E/E_0)^{-a}$ for $E > E_0$. We assume that the number of electrons in the nonthermal tail is equal to that of thermal ones at E_0. Observations indicate that $E_0 \gtrsim 3$keV.

For electrons, we obtain the truncation scale due to non-thermal-damping by electrons

$$\Gamma_{nonth}(\mathbf{k}) = \frac{\sqrt{2}\pi m_e^2 c^3 N_0 k}{4\rho_i}\left(1 - \frac{\beta_A^2}{\eta^2}\right)\frac{1-\eta^2}{\eta}(\gamma_0 - 1)^a \times$$
$$(\gamma_m - 1)^{-a+\frac{1}{2}}\left[\frac{\gamma_m - 1}{a - 3/2} - \frac{\beta_A^2}{(2a-1)\eta^2}\right] \quad (4)$$

The results are demonstrated in Fig. 2. In the case where nonthermal tail starts at energy $E_0 = 3KeV$, the damping wave number is larger than that of thermal damping (see Fig. 1) except for very large pitch angles. However, if taking into account the averaging due to field line wandering (dotted lines in Fig. 1), thermal damping is always dominant. In the case of $E_0 = 10KeV$, the damping scale is much smaller than thermal proton gyroscale, i.e., it is beyond the MHD regime so that nonthermal damping in this case can be ignored.

Comparing with thermal collisionless damping (Fig. 1), we see that the damping by high energy particles is less important except for the case when fast modes are propagating with $\theta \gtrsim 80^o$. This can be easily understood as $\cos\theta \simeq v_A/v_{\|} \sim 0$ for relativistic particles according to the Cherenkov resonance condition. Nevertheless, because of the averaging due to field line wandering, it is always the thermal damping that is dominant.

Comparing with electrons, the damping due to TTD by **ions** is not important. Gyroresonance with protons provides another way for damping the turbulence. The analytical results can be found in Petrosian, Yan & Lazarian (2006).

4. Summary

Energy injected on large scales can be transfered to smaller scales by turbulence cascade. This transfer is angle dependent because of damping. The most important mechanism for damping appears to be collisionless damping by thermal particles. For waves propagating at large pitch angles, the damping due to TTD with nonthermal particles appears to be dominant. However, if taking into account field line wandering in turbulence, it is always the thermal damping that determines the spectrum of turbulence on

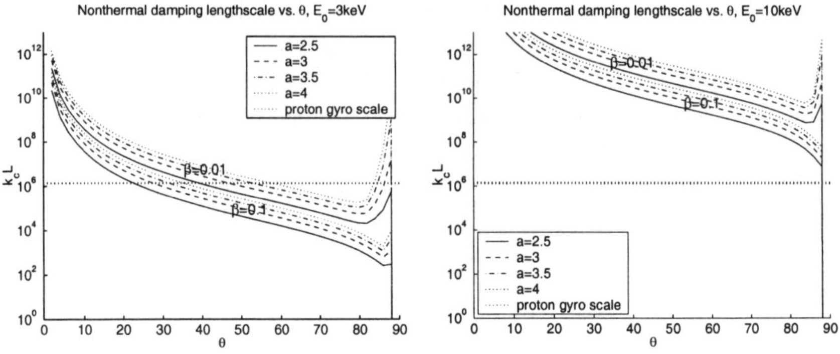

Figure 2. TTD damping by nonthermal electrons for two cases with the lower energy limit of the power law distribution at: a) $E_0 = 3KeV$ and b) $E_0 = 10KeV$ (see text). In a), the damping wave number is larger than that of thermal damping (see Fig. 1) except for very large pitch angles. However, if taking into account the averaging due to field line wandering (dotted lines in Fig. 1), thermal damping is always dominant; in b) we see that the damping scale is much smaller than thermal proton gyroscale, i.e., it is beyond the MHD regime so that nonthermal damping in this case can be ignored.

small scales. This indicates us that that particle acceleration and turbulence evolution can be decoupled. In other words, turbulence spectrum doesn't change due to the acceleration. This simplifies the problem substantially so that we don't have to rely on simulation to resolve the problem. Our result shows that up to 10% of total energy of turbulence is transfered to accelerate particles. Most energy is drained to thermal particles and heats the plasma.

References

1. Chandran, B. 2000, Phys. Rev. Lett., **85**(22), 4656
2. Cho, J. & Lazarian, A. 2002, Phys. Rev. Lett., **88**, 245001 (CL02)
3. Goldreich, P. & Sridhar, H. 1995, ApJ, **438**, 763
4. Lazarian, A., Vishniac, E., & Cho, J. 2004, ApJ, **608**, 180L
5. Lazarian, A., Petrosian, V., Yan, H., & Cho, J. 2002, Review at the NBSI workshop 'Beaming and Jets in Gamma Ray Bursts', ed. R. Ouyed, astro-ph/0301181
6. Petrosian, V. & Liu, S. 2004 ApJ, **610**, 550
7. Petrosian, V., Yan, H., & Lazarian, A. 2006 accepted to ApJ, astro-ph/0508567
8. Priest, E. & Forbes, T. 2000, *Magnetic Reconnection: MHD Theory and Applications*, CUP
9. Yan, H. & Lazarian, A. 2002, Phys. Rev. Lett., **89**, 281102 (YL02)
10. Yan, H. & Lazarian, A. 2003, ApJ, **592**, 33L
11. Yan, H. & Lazarian, A. 2004, ApJ, **614**, 757 (YL04)

OPTICAL MEASUREMENTS FOR CANGAROO-III

R. KIUCHI, M. YUASA, M. OHISHI, A. KAWACHI, M. MORI AND THE CANGAROO COLLABORATION

High Energy Cosmic Ray Division, ICRR, University of Tokyo, 5-1-5 Kashiwanoha, Chiba, 277-8582, Japan
E-mail: kiuchi@icrr.u-tokyo.ac.jp

CANGAROO-III consists of four telescopes installed near Woomera, South Australia to observe celestial gamma-ray sources by detecting Cherenkov light from air showers. Stereo observations have been performed since March 2004 with an improved angular resolution and a lower energy threshold. In this paper, we present some preliminary results from optical measurements with a cooled CCD camera on the reflectivity of the telescope reflector and the atmospheric transmittance.

1. Introduction

In October 2005, extensive maintenance work was carried out on the telescopes, including washing the mirror segments. In this period, we measured the reflectivity of the reflectors and the typical atmospheric transmittance at our site. For all measurements, we used a cooled CCD camera (ST-7XMEi, SBIG[1]) together with a camera lens (F1.4, focal length 50 mm) and Johnson-type optical filters to study the dependence on wavelength. The CCD chip (KAF-0402ME, KODAK[2]) has high quantum efficiency especially in the shorter waveband which makes it possible to do optical measurements with a U-band filter. We used three optical filters, V-band, B-band, and U-band. The central wavelength of transmittance is 533 nm (V-band), 434 nm (B-band), and 362 nm (U-band).

2. Reflectivity of the telescope reflector

The CANGAROO-III reflector is a tessellated parabola with a diameter of 10m, consisting of 114 spherical mirror segments each with a diameter of 78cm [3]. As described above, we washed all segments with water. The reflectivity of the telescope reflectors was measured before and after washing to determine how much the reflectivity had improved. There are several

methods to evaluate the reflectivity, but we used a cooled CCD camera since this method can estimate the reflectivity of the whole reflector very simply. The procedure was as follows: First, the telescope tracked a bright star and we took a reflected star image on the white screen[4] using the CCD camera (Fig. 1). Second, the tracking coordinates of the telescope were slightly changed so that the star came into the field of view of the CCD camera, and we took the direct star image (Fig. 2). The direct and the reflected star images were taken with as short an interval as possible so as to minimize the effect of the difference of the sky condition. This procedure was repeated for several stars with each optical filter, and all three optical filters (V,B,U) were used with each telescope. These measurements were made for the three telescopes of the CANGAROO-III stereo system (i.e., currently excluding telescope T1) to compare the relative reflectivity among the telescopes. The quantity of the flux of the reflected star image (in CCD counts) divided by that of the direct star image is proportional to the reflectivity of the whole reflector, so we compare the reflectivities using above quantities calculated from the data which were taken before and after the washing of the segments.
Fig. 3 shows the results for telescope 4, and only results of V- and B-band are shown because we do not have U-band data taken before washing. From the results for all three telescopes, we estimate that the reflectivity of the reflector after washing became about 1.3 times greater than that before washing.

3. Atmospheric transmittance

The atmospheric transmittance at our site is an important factor to estimate the γ-ray energy and flux from the observed Cherenkov light. The observation site is located in the desert, and we can assume that the atmo-

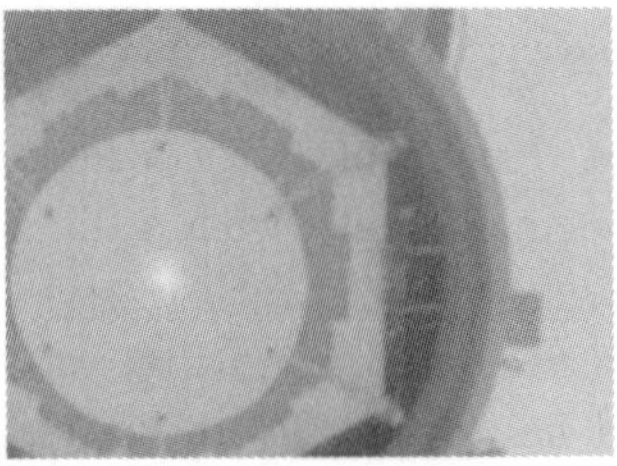

Figure 1. A reflected star image on the white screen taken by a cooled CCD camera.

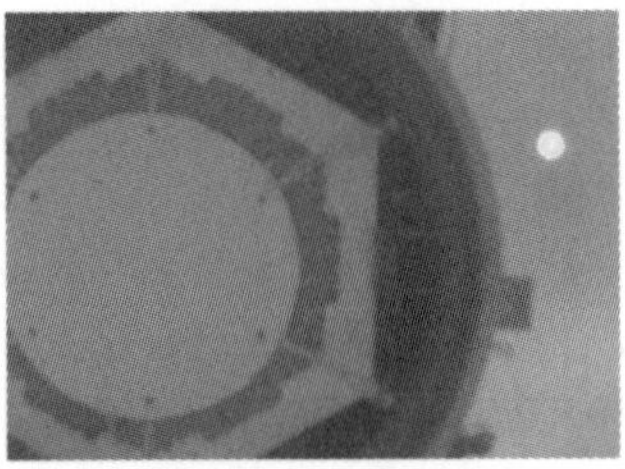

Figure 2. A direct star image. The star is the same in Fig1.

spheric transmittance is close to that including only Rayleigh scattering. In October 2005, we measured the atmospheric transmittance using bright stars. The concept of these measurements were as follows: By using the fact that the measured flux of a star changes according to its zenith angle, which is the effect of the atmospheric extinction scattering by air molecules and dust, etc., we can estimate the flux of the star above the atmosphere where the scattering is zero by extrapolation from the flux of stars measured at various zenith angles. We can thus calculate the atmospheric transmittance from observed star fluxes. The set-up of the measurement was the same as that used in reflectivity measurements except for the white screen, because this measurement used only direct star images (Fig. 2), and the images of a handful of bright stars at various zenith angles were taken. Fig. 4 is a plot of the atmospheric extinction effect on the flux of stars. Here the airmass represents the thickness of the atmosphere and the definition is $\frac{1}{\cos(\theta)}$, where θ is the zenith angle.

As shown in Fig. 4, we fit the results with a linear function. From the value of the flux of a star at the point that airmass equals to zero, we estimated the atmospheric transmittance for each measured points (Fig. 5). We used the atmospheric transmittance simulation code, Modtran[5], to compare our measurements and the "desert aerosol model" in Modtran, which is a model including little aerosol content in the atmosphere, and which is plotted to-

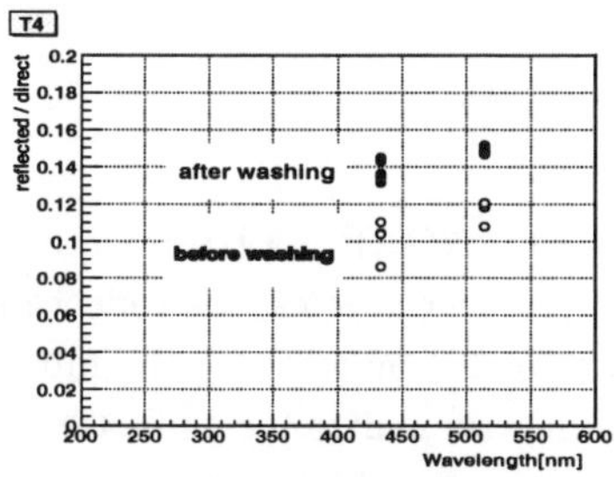

Figure 3. The relative reflectivity of the telescope 4 reflector. Before (open circles) and after (filled circles) the washing works . The vertical axis represents the quantity of the flux of the reflected star divided by that of the direct star. Since there are no data with U-band filter for before the washing works, only data with V,B band filter are plotted.

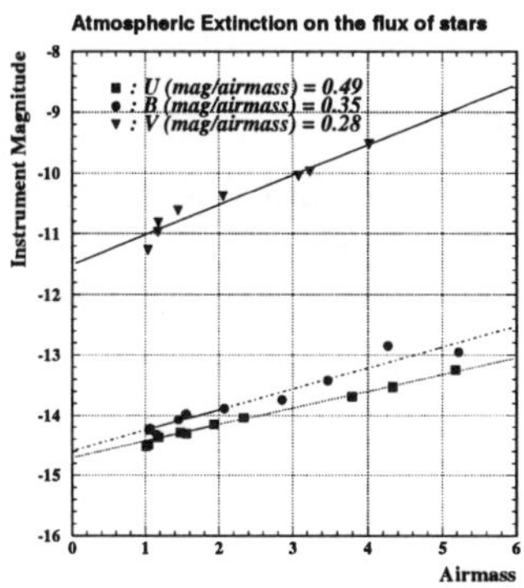

Figure 4. Flux of stars (in magnitude) versus the thickness of the atmosphere (in airmass). The origin of the vertical axis is arbitrary, and the fitted lines to the measured points are also plotted.

gether with measured points.
Considering the systematic errors such as the calculation of extrapolation and the changing the flux of stars due to instability in sky conditions, we can say that the "desert aerosol model" is similar to the atmosphere at our site.

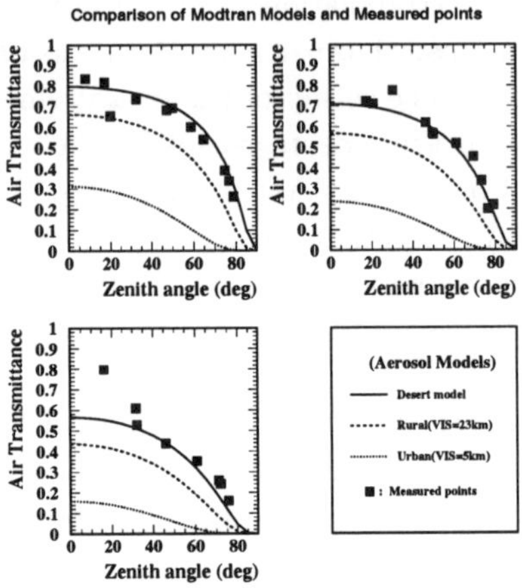

Figure 5. The calculated atmospheric transmittance. The square points are from measurement and the lines are from "desert aerosol model" in Modtran simulation

4. Summary

We have carried out optical measurements for CANGAROO-III in October 2005. From the measurements of the reflectivity of the telescope reflectors, the reflectivity has been increased by a factor of 1.3 by washing the mirror segments. Measurements of the atmospheric transmittance indicate the "desert aerosol model" in Modtran simulation is close to our environment.

References

1. http://www.sbig.com/
2. http://www.kodak.com/global/en/digital/ccd/products/fullframe/KAF-0402E/indexKAF-0402E.jhtml
3. Kawachi, A. et.al., Astropart. Phys. 13, 261-269 (2001)
4. Kiuchi, R. et al., Proc. 29th ICRC (Pune), 5, 315 (2005)
5. Berk, A., Bernstein, L.S., Robertson, D.C., 'MODTRAN: A Moderate Resolution Model for LOWTRAN 7', GL-TR-89-0122, 1989

CONSIDERATION OF CASSEGRAIN IMAGING ATMOSPHERIC CHERENKOV TELESCOPES

YOHEI YUKAWA

MASAKI MORI
TAKANORI YOSHIKOSHI

Institute for Cosmic Ray Research, University of Tokyo, 5-1-5 Kashiwa-no-Ha Kashiwa City, Chiba 277-8582, Japan

We studied Cassegrain-type telescope design for IACTs to reduce the weight and cost of the telescope, keeping currently achieved optical performance of Imaging Air Cherenkov Telescopes (IACTs) in spot size, field of view, light collection performance. In addition to currently used prime-focus type telescope design, Classical Cassegrain and Ritchey-Chrétien Cassegrain optical system have been evaluated by using of an aberration theory and a ray-tracing method. We found that, in some ideal cases, the Ritchey-Chrétien optics can reduce the length of telescope down to 20% of the prime-focus optics of the same spot size at 1 degree incident angle, whereas the loss of light by the secondary mirror shadow and the second reflection is about 20% assuming a reflectivity of 85%.

1. Introduction

There is an unobserved energy range in gamma-ray astronomy around 10 GeV to 100 GeV, which correspond to Energy thresholds of celestial gamma-ray observatories and ground-based gamma-ray detectors such as IACTs, respectively.

An empirical cost estimation function [1] that the cost of building a telescope is proportional to the diameter to the power of 2.7, predicts that for IACTs the challenge to explore the "unopened" 10 GeV to 100 GeV regions will be very expensive if we enlarge the aperture of telescopes to lower their energy threshold.

We considered some Cassegrain optics as a possible solution to keep the modest optical performance and to save the building cost. Relationships between F-values and diameters of current and planed large aperture optical telescopes including IACTs are plotted in figure 1. Our aim is not simply enlarge the aperture but also achieve smaller F-values of the primary mirror to reduce the cost.

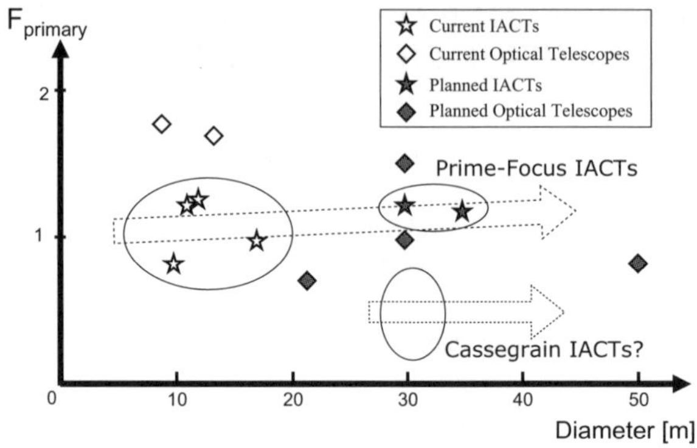

Figure 1. This figure shows future and current large aperture telescopes in the view of F-value (of primary mirror) and its aperture size. Note that all IACTs which plotted here are prime-focus type and all optical telescopes are Cassegrain-like design (they use plural reflections). Major IACTs have similar aperture sizes and F-values to current optical telescopes and this relationship seems to be applicable in future telescopes.

2. Geometrical Aberration

We estimated spot sizes of 4 optical systems by integrating aberration function [2] over entrance pupil and regarding RMS of aberration as the spot size. Their optical characteristics can be summarized as follows:

- Parabolic
  - No spherical aberration.
  - Good timing resolution.
- Davies-Cotton [3]
  - Less coma aberration than parabolic design.
  - Timing resolution may be critical for large aperture IACTs.
- Classical Cassegrain
  - No spherical aberration.
- Ritchey-Chrétien
  - No spherical and 3rd coma aberration.

Spot size is one of important parameters for IACTs to obtain Cherenkov image parameters in order to separate gamma-ray shower events from background events. Thus we can derive telescope design limitations from the restriction of spot size through the relationships between spot size and F-value. Figure 2 shows that the parabolic design needs $F \geq 1.2$ to achieve less than 0.05 degrees in spot size at the incident angles of 1.0 degree. This means that a 30 m aperture parabolic IACT should set the camera at least 36 m away from the center of mirror. A larger F-value means a rigid and expensive support structure for prime-focus telescopes.

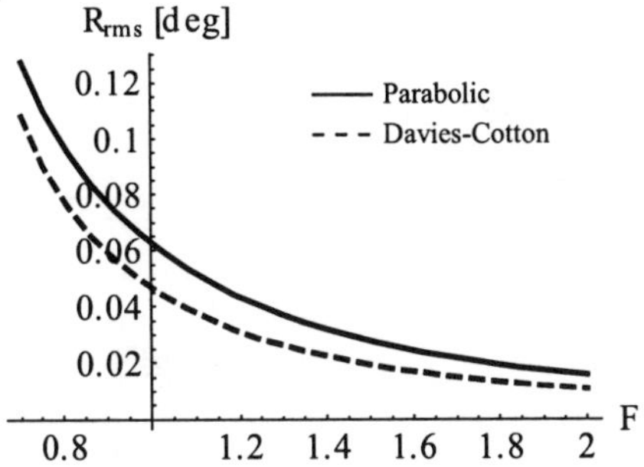

Figure 2. Relationships between spot sizes at incident angle of 1.0 degree and F-values calculated from geometrical aberration theory.

3. Ray-tracing simulation

We executed a ray-tracing simulation to search for suitable configurations for IACTs of classical and Ritchey-Chrétien Cassegrain. We assumed following conditions:

- Camera is set at the vertex of primary mirror to simplify the model and to reduce the moment of camera.
- All mirrors are made of single segments and have a constant blur (gaussian sigma of 0.017 degrees)
- The spot size should be less than 0.05 degrees within the incident angle of 1.0 degree.

From the simulation, we found the Ritchey-Chrétien design can have smaller F-value for the primary mirror to satisfy these conditions than the classical design. Two set of parameters were selected as candidates as Ritchey-Chrétien designs.

A) Ritchey-Chrétien, the F-value of primary mirror is 0.43; the aperture size of the secondary mirror is 17% of that of the primary.

B) Ritchey-Chrétien, the F-value of primary mirror is 0.25; the aperture size of the secondary mirror is 29% of that of the primary.

Figure 3 shows optical characteristics of these two designs compared to F/1.0 and F/1.2 parabolic ones. The total amount of light collection depends on mirror reflectivity; (A) 77.6% (reflectivity 80%) ~ 87.3% (reflectivity 90%) and (B) 73.4% (reflectivity 80%) ~ 82.5% (reflectivity 90%).

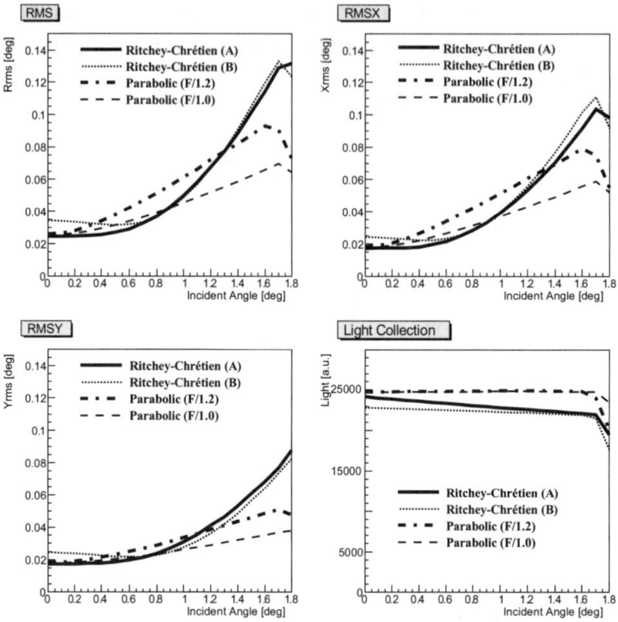

Figure 3. Spot sizes and light collection amount (assuming reflectivity is 100%) at various incident angles for two Ritchey Chrétien designs and two parabolic designs. We assumed all mirrors are made of single segments and each mirror have 0.017 degrees gaussian blur.

4. Conclusion

We obtained useful relations to predict spot sizes for several types of optics to be used for IACTs. We also obtained acceptable configurations of Cassegrain-type IACTs: F-values of Ritchey-Chrétien design can be less than those of prime-focus design, keeping modest spot size of less than 0.05 degrees. However, more study for the case of segmented mirrors is necessary for the realization.

References

1. GEMINI Project. Enabling a Giant Segmented Mirror Telescope for the Astronomical Community, the AURA New Initiatives Office. (2002).
2. Murk Bottema and R. A. Woodruff. Third order aberrations in Cassegrain type telescopes and coma correction in servo-stabilized images. Applied Optics, 10, 1971.
3. J. M. Davies and E.S. Cotton. Design of the quartermaster solar furnace. Solar Energy, 1, 1957.

INFLUENCE OF A NON-DIPOLE MAGNETIC FIELD ON THE PEAK ENERGIES OF CYCLOTRON ABSORPTION LINES

OSAMU NISHIMURA

Department of Electronics and Computer Science,
Nagano National College of Technology, 716 Tokuma, Nagano 381–8550,
E-mail: nishi@ei.nagano-nct.ac.jp

We calculate cyclotron lines in a neutron star slab assuming a non-dipole surface magnetic field. We study the influences of the non-dipole surface field on the properties of cyclotron resonant scattering lines. When the magnetic field strength decreases with height in the line-forming region, the ratios of the higher harmonics to the fundamental at the peak energies of the cyclotron lines become more than the integer values. On the other hand, when the magnetic field strength increases with height in the line-forming region, the ratios at the peak energies of the cyclotron lines become less than the integer values. The nonharmonicity, which has actually been observed in some accretion-powered X-ray pulsars, is more significant than that expected from relativistic effect in cyclotron resonant energy. This may suggest the line-forming region threaded by the strong nondipole magnetic field in some accreting X-ray pulsars. In observations, the ratios are more than or less than the integer values. This could imply that the magnetic field strength in the line-forming region decreases or increases with the altitude or the horizontal distance from an emission region.

1. Introduction

Cyclotron lines in the spectra of neutron stars provide a powerful tool to measure directly the magnetic field strength of neutron stars. They have been detected in the spectra of more than 10 accretion-powered X-ray pulsars, indicating commonly broad and shallow features [1]. More than one cyclotron absorption feature were detected in some accreting X-ray pulsars. The ratios of the line energies seems to be larger than (4U1907+09 [2] and Vela X-1[3]) or less than (4U 0115+63[4–5]) the integer values. In the present paper, the field distribution $B(z)$ is assumed to be given by

$$B(z) = B_0 \pm \frac{z}{z_{max}} \Delta B. \qquad (1)$$

Here, z is the height from the bottom of the slab, z_{max} is the height of the slab, B_0 is the magnetic field strength at the bottom of the slab and ΔB is the total variation in the magnetic field strength. In our works, the width of the cyclotron energies, $\hbar\Delta\omega_B$, is taken to be 10 keV, which varies linearly between 20 and 30 keV. We refer the magentic field that decreases with height as "decreasing magnetic field" and one that increases with height as "increasing magnetic field". In one dimension, we solve the radiative transfer through the slab near the surface of the neutron star with non-dipole magnetic field, which varies linearly with the height of the slab. We consider the 1-0 geometry, a slab illuminated from below, as the line-forming regions[6].

2. Results

We investigate therefore the properties of the cyclotron resonant scattering lines for both cases of "decreasing magnetic field" and "increasing magnetic field" in the 1-0 geometry. In this section, the height of the slab is taken to be 5m.

Figure 1 shows the cyclotron lines formed through the slab threaded by the magnetic field which decreases linearly with height. The ratios of the peak energies of the higher harmonic lines to the fundamental are more than the integer values primarily due to the difference in the phycial process of line formation between each harmonic. The absorption line in the first harmonic becomes deeper at the redward, since the peak of the first harmonic absorption line forms at the top of the slab due to the effect of scattering in cyclotron resonance. In the second harmonic, the process of cyclotron resonance can be mostly considered as pure absorption due to resonant Raman scattering. In the third harmonic, the features of the absorption line are similar to those in the second harmonic but the absorption line become deeper at the blueward since the optical depth is proportional to the strength of the magnetic field in the limit $n(B/B_c) \ll 1$.

When the magnetic field strength increases with height, the shapes of the cyclotron lines in the first harmonic become broad and shallow as shown in Fig. 2. The shapes of the lines at the first harmonic are similar to those observed in the spectra of accreting X-ray pulsars. Moreover, the ratios of the centroid energies of the second harmonic lines to the fundamental are less than 2, which have actually been observed in 4U 0115+63 [4–5]. When the magnetic field strength increases with height, the ratios between the peak energies of the lines with respect to the fundamental become smaller

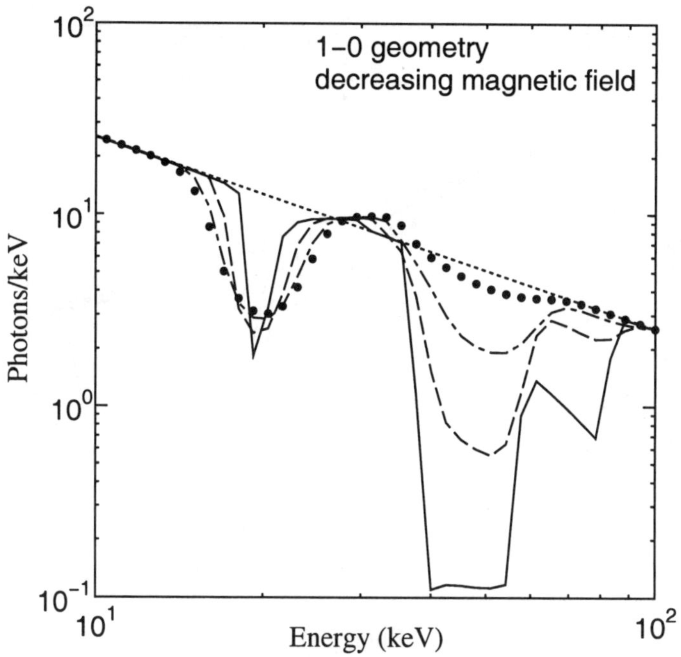

Figure 1. Spectra for slab thickness of 5×10^2 cm, an electron number density of $N_e = 3.14 \times 10^{21}$ electron cm^{-2}, the temperature kT=5 keV, the 1-0 geometry and "decreasing magnetic field". The solid, dashed, dot-dashed, thick-dotted curves represent spectra for cosine of the photon propagation anlges with respect to the magnetic field, $\mu = 0.1834, 0.5255, 0.7967$, and 0.9603, respectively. The dotted line denotes the power-law spectrum with $\alpha = 1$ emitted at the source plane.

than the classical integer values. It is interesting that this is agreement with Bulik et al.'s[7] results, in which stronger magnetic field components have a significant weight in fitting cyclotron lines in spectra of 4U 1538−52. Thus, the scattering events in cyclotron first harmonic make the peak energy only of the first harmonic line noticeably deviate from the integer ratio in the peak energies of all harmonic cyclotron lines. The higher harmonic lines would therefore almost have a harmonic relationship with a spacing of half of the peak energy of the second harmonic line. The way of the deviation in the ratios is agreement with the results observed in some acrreting X-ray pulsars, such as 4U 0115+63.

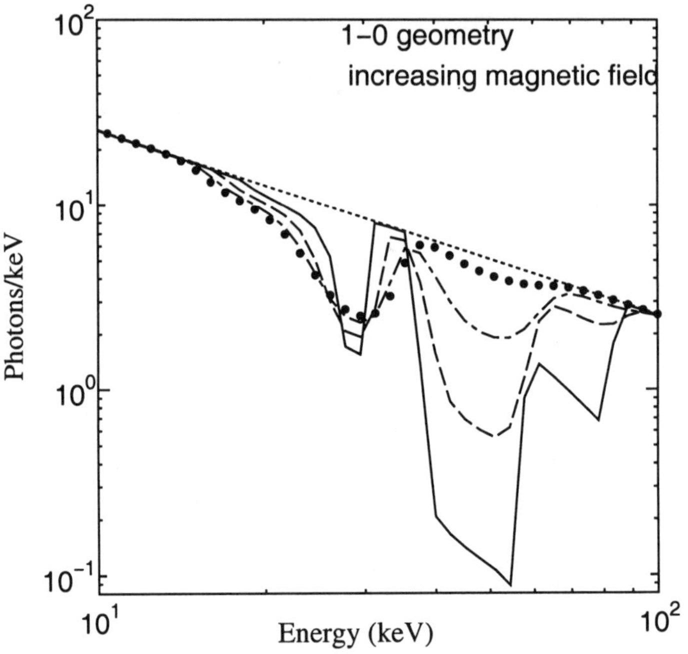

Figure 2. Same as figure 1, but for "increasing magentic field". The shapes of the lines at the first harmonic are similar to those observed in the spectra of accreting X-ray pulsars. The ratios of the peak energies of the higher harmonic lines to the fundamental are less than the integer values.

References

1. Coburn, W., Heindl, W. A., Rothschild, R. E., Gruber, D. E., Kreykenbohm, I., Wilms, J., Kretschmar, P., & Staubert, R. 2002, *Ap. J.*, **580**, 394
2. Cusumano, G., Di Salvo, T., Burderi, L., Orlandini, M., Piraino, S., Robba, N., & Santangelo, A. 1998, *Astr. Ap.*, **338**, L79
3. Kreykenbohm, I., Kretschmar, P., Wilms, J., Staubert, R., Kendziorra, E., Gruber, D. E., Heindl, W. A.,Rothschild, R. E. 1999, *Astr. Ap.*, **341**, 141
4. Heindl, W. A., Coburn, W., Gruber, D. E., Pelling, M. R., Rothschild, R. E., Wilms, J., Pottschmidt, K. & Staubert, R. 1999, *Ap. J.*, **521**, L49
5. Santangelo, A. et al. 1999, *Ap. J.*, **523**, L85
6. Isenberg, M., Lamb, D. Q., & Wang, J. C. L. 1998, *Ap. J.*, **493**, 154
7. Bulik, T., Meszaros, P., Woo, J. W., Hagase, F., & Makishima, K. 1992, *Ap. J.*, **395**, 564

MAGNETOROTATIONAL COLLAPSE OF VERY MASSIVE STARS: FORMATION OF JETS AND BLACK HOLES

YUDAI SUWA

Department of Physics, School of Science, the University of Tokyo, 7-3-1 Hongo, Bunkyo-ku, Tokyo 113-0033, Japan
E-mail: suwa@utap.phys.s.u-tokyo.ac.jp

TOMOYA TAKIWAKI

WDepartment of Physics, School of Science, the University of Tokyo, 7-3-1 Hongo, Bunkyo-ku, Tokyo 113-0033, Japan

KEI KOTAKE

Science & Engineering, Waseda University, 3-4-1 Okubo, Shinjyuku, Tokyo, 169-8555, Japan

KATSUHIKO SATO

Department of Physics, School of Science, the University of Tokyo, 7-3-1 Hongo, Bunkyo-ku, Tokyo 113-0033, Japan
Research Center for the Early Universe, School of Science, the University of Tokyo, 7-3-1 Hongo, Bunkyo-ku, Tokyo 113-0033, Japan

Population III stars are thought to be very massive stars. However, their properties are not clarified yet. In this research, we investigate the features of magnetorotational dynamics of such stars with computer simulations. We find jet-like explosion in a few models and reveal what type of initial model undergo explosion. In addition, we inquire features of newborn black holes as a remnant of core-collapse.

1. Introduction

Recently, great attention has been paid to the first stars, so-called Population III (hereafter Pop III) due to the discovery of hyper metal poor stars such as HE 0107-5240 and HE 1327-2326, which contain less than 1/100,000 of the iron observed in the Sun. These Pop III are predicted to have been predominantly very massive with $M \gtrsim 100M_{\odot}$. In this research we study the evolution of such very massive stars using a two-dimensional magnetohydrodynamics code.

2. Method

All simulations were performed with a modified version of the explicit magnetohydrodynamic (MHD) ZEUS-2D code[1], which is an Eulerian code based on the finite-difference method and employs an artificial viscosity of von Neumann and Richtmyer to capture shocks. In so doing, the code utilizes the so-called constrained transport (CT) method, which ensures the divergence free($\nabla \cdot \vec{B} = 0$) of the numerically evolved magnetic fields at all times. Furthermore, the method of characteristics (MOC) is implemented to propagate accurately all modes of MHD waves. The self-gravity is managed by solving the Poisson equation with the incomplete Cholesky decomposition conjugate gradient (ICCG) method. Axial symmetry and reflection symmetry across the equatorial plane are assumed. Spherical coordinates (r, θ) are employed with logarithmic zoning in the radial direction and regular zoning in θ. One quadrant of the meridian section is covered with 300 $(r)\times$ 30 (θ) mesh points. We made several major changes to the base code to include microphysics. First we added an equation for electron fraction to treat electron captures and neutrino transport by the so-called leakage scheme[2]. We extend the scheme to include all 6 species of neutrino $(\nu_e, \bar{\nu}_e, \nu_X)$. ν_X means $\nu_\mu, \bar{\nu}_\mu, \nu_\tau$ and $\bar{\nu}_\tau$. Neutrino losses are included with thermal losses taken from Itoh[3]. The cooling rate, L_ν, is also estimated by the scheme. Second we have incorporated the tabulated equation of state (EOS) based on relativistic mean field theory[4] instead of the ideal gas EOS assumed in the original code.

We start simulations with unstable $180M_\odot$ He core of a $300M_\odot$ star. We prepare polytropic star for density, whose polytrope index is $n = 3$, and calculate the energy on the assumption that core is isentropic and electron fraction $Ye = 0.5$ with Shen EOS. We assumed the core's entropy is $\sim 10k_B$ per nucleon[5]. With these procedures, we calculate numerically the hydrostatic density and energy distribution of a $300M_\odot$ star. From this star we produce a He core, $180M_\odot$. In this paper we treat this core as the initial model. We assume in this study the differential rotation law. In addition, we assume that the initial magnetic field is nearly uniform (B_0 Gauss) in the core and dipole on the outside. We compute 9 models changing the total rotational energy and strength of magnetic field by varying the value of angular velocity and central magnetic fields.

3. Numerical Results

All models of very strong initial magnetic field ($B_0 = 10^{12}$G) form jets because the magnetic pressure exceeds the gas pressure by field wrapping and compression in the collapsing core. Such jets propagate to the outside of the core. Fig. 1 shows the evolution of jet. The jet is obviously magneto-driven jet and leaves a very high entropy region behind the surface of shock. The more rapidly rotating models broaden the jet due to centrifugal force. Other models, which initially have weak magnetic fields, don't explode because the collapsed materials form a black hole before exploding.

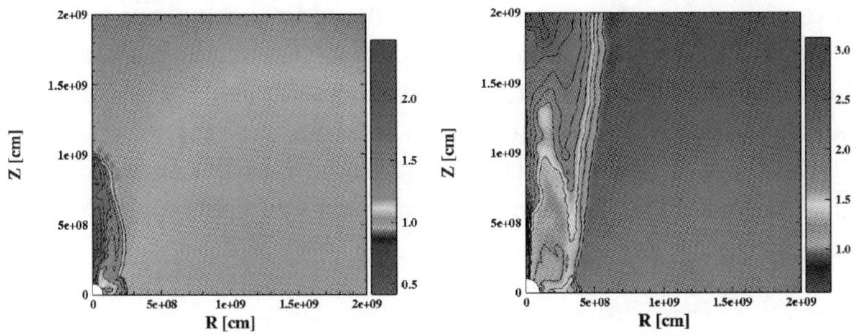

Figure 1. Time evolution of shock wave. They show the color coded contour plots of logarithm of entropy (k_B) per nucleon.

We also investigate the initial black hole mass of each model. We determine black hole formation with marginally stable orbit of Schwarzschild black hole. The initial mass of black holes for each model is summarized in Table 1. In this table, the mass of black holes are normalized in solar mass. Table 1 shows that the more rapid initial rotation, the bigger the black hole. As for the rapidly rotating star, the matter stop falling because of centrifugal force, thus more mass is necessary to form black hole. In addition to the rotation, the initial magnetic field affects the initial mass of black holes. The stronger magnetic field makes the initial black hole's mass smaller. This feature is due to the generation of the jet. When the jet arises, back-reaction overwhelm the matter behind the jet. Such a phenomenon quickens black hole formation time. Consequently, the initial black hole's mass gets smaller when the initial magnetic field is strong and jet-like explosions occur.

Table 1. Initial mass of black holes for differential rotating models. The masses are normalized by $M_{\odot}$.

B_0(Gauss) \ $T/\|W\|(\%)$	1%	2%	4%
10^{10}G	70.4	87.3	106.6
10^{11}G	70.4	87.3	106.6
10^{12}G	57.9	75.8	96.6

4. Summary

We have performed time-dependent two-dimensional MHD simulations of the rotational core collapse of magnetized very massive stars. In this study, we systematically investigated how strong magnetic field and rapid rotation affect dynamics from the onset of core collapse to shock propagation in the core and black hole formation. We find that very massive stars can eject materials by the effects of rotation and magnetic field. The formation of black holes are also investigated. Both rotation and magnetic fields affect the initial mass of newborn black holes. Rapid rotation makes black holes big due to centrifugal forces and strong magnetic fields make black holes smaller due to back-reaction of jet.

References

1. Stone, J. M. & Norman, M. L., *ApJS*, **80**, 753 (1992)
2. Epstein, R. I. & Pethick, C. J., *ApJ*, **243**, 1003 (1981)
3. Itoh, N., Adachi, T., Nakagawa, M., Kohyama, Y., & Munakata, H., *ApJ*, **339**, 354 (1989)
4. Shen, H., Toki, H., Oyamatsu, K., & Sumiyoshi, K., *Nucl. Phys.*, **A637**, 435 (1998)
5. Fryer, C. L., Woosley, S. E., & Heger, A., *ApJ*, **550**, 372 (2001)

RELATIVISTIC JETS IN POPULATION III SUPERNOVA AND EXTREMELY METAL-POOR STARS

N. TOMINAGA, H. UMEDA AND K. NOMOTO

Department of Astronomy, School of Science, University of Tokyo, Bunkyo-ku, Tokyo 113-0033, Japan; E-mail: tominaga@astron.s.u-tokyo.ac.jp

K. MAEDA

Department of Earth Science and Astronomy, College of Arts and Sciences, University of Tokyo, Tokyo 153-8902, Japan

N. IWAMOTO

Nuclear Energy Basic Engineering Research Sector Japan Atomic Energy Agency, Tokai, Ibaraki 319-1195, Japan

We calculate explosive nucleosynthsis induced by relativistic jets in Population III stars with 2-dimensional special relativistic hydrodynamics code. The resulting yields are compared with the abundance patterns of extremely metal-poor stars. Extremely metal-poor stars in Galactic halo, whose metallicities are much smaller than the Sun ([Fe/H]<-3), have been extensively observed. These stars are classified by their metallicities, i.e., extremely metal-poor (EMP: $-4 <$[Fe/H]< -3 and [C/Fe]~ 0), carbon-rich EMP (CEMP: $-4 <$[Fe/H]< -3 and $+1 <$[C/Fe]), and hyper metal-poor (HMP: [Fe/H]< -5 and [C/Fe]$\sim +4$) stars. The abundance patterns of these stars have been reproduced by 1-dimensional mixing-fallback model in the previous studies. Though the differences among the EMP, CEMP, HMP stars have been well-explained by the different mixing region and ejection factor, it has not been clear what causes these differences. In this study, we show that the yields of 2-dimensional jet-like explosions give good agreements with the abundance patterns of EMP, CEMP, HMP stars and that an energy injection rate from the central engine plays an important role in producing the differences among EMP, CEMP, HMP stars.

1. Introduciton

The universe contained very small amount of metals at the birth. First chemical enrichment was due to supernovae (SNe) explosions of population (Pop) III stars that are made from the metal-free gases. In early universe, the metal pollusion by a single SN can dominate preexisting metal contents.

Second generation stars are formed from the enriched gases. Their abundance patterns reflect nucleosynthesis in the Pop III SN[1,2,3,4]. Low-mass stars among them will survive until present days and be observed as metal-poor stars. Therefore the abundance patterns of metal-poor stars can give constraints on the nucleosynthesis in the Pop III SN. In particular the abundance patterns of metal-poor stars with low metallicity ([Fe/H] < -3) are important because chemical evolution studies suggested that the universe is unmixed and inhomogeneous at those metallicities[5,6].

The metal-poor stars are classified into three groups by their metallicities and abundance patterns. First is extremely metal-poor (EMP) stars[7,8] with $-4 <$ [Fe/H] < -3 and [C/Fe] ~ 0. Second is C-enhanced EMP (CEMP) stars[9,10] that has same metallicity as EMP stars ($-4 <$ [Fe/H] < -3) but large C/Fe ([C/Fe] $\sim +1$). Third is hyper metal-poor (HMP) stars[11,12,13] with low Fe/H ([Fe/H] < -5) and quiet large C/Fe ([C/Fe] $\sim +4$).

In this study, we reproduce the abundance patterns of EMP, CEMP, HMP stars by aspherical explosions with relativistic jets (recently the SN explosion with relativistic jets are observed as SNe associated with gamma-ray bursts). Especially we investigate dependences of the nucleosynthesis in aspherical HNe on energy deposition rates by relativistic jets ($\dot{E}_{\rm jet} = E_{\rm jet}/t_{\rm jet}$, where $E_{\rm jet}$ is total deposited energy and $t_{\rm jet}$ is duration of the energy deposition). In particular, each models are assumed to be exploded with different $\dot{E}_{\rm jet}$ and $t_{\rm jet}$ but with same $E_{\rm jet} = 1.5 \times 10^{52}$erg (more details will be described in [14]). We point out that the difference of $\dot{E}_{\rm jet}$ causes the differences of EMP, CEMP, and HMP stars.

2. Dependences on Energy Deposition Rate

2.1. *The Fe-peak elements*

The smaller amount of Fe is ejected for a smaller energy deposition rate as suggested in previous studies[15,16,17]. Fe can not be ejected from above the initial mass-cut ($M_{\rm cut}$(ini) $= 1.6 M_\odot$) for $\dot{E}_{\rm jet,51} \equiv \dot{E}_{\rm dep}/(10^{51}{\rm ergs/sec}) \lesssim 1$ (the left panel of Figs. 1). Fe is dominated from the materials above the initial mass-cut for $\dot{E}_{\rm jet,51} \gtrsim 1$, but by the deposited materials for $\dot{E}_{\rm jet,51} \lesssim 1$. In such case, the detailed nucleosynthesis calculations around the central remnant below the initial mass cut is necessary, but in this paper we assumed the fallen-back materials are deposited as jets with same abundances as in the progenitor.

The dependence of the ejected Fe mass stems from two reasons as follow:

one is that smaller energy deposition rate causes lower peak temperature[16] and decreases the amount of materials whose peak temperature is high enough to synthesize Fe. Another is that the fallback is enhanced by the smaller energy deposition rate. Since Fe is mainly synthesized in the inner region just above the initial mass cut, the smaller ejection from such region leads the smaller ejection of Fe.

2.2. *The α-elements*

The dependences of the ratios between C and other elements (O, Mg, and Fe) are shown in the right panel of Figures 1. The smaller $\dot{E}_{\rm jet}$ causes larger C/O, C/Mg, and C/Fe. This is because smaller $\dot{E}_{\rm jet}$ causes larger fallback. The inner materials should be fallen back more efficiently than the outer materials. Therefore, the amount of the inner materials (Fe, Mg and O) is smaller than the outer materials (C) with the smaller $\dot{E}_{\rm jet}$.

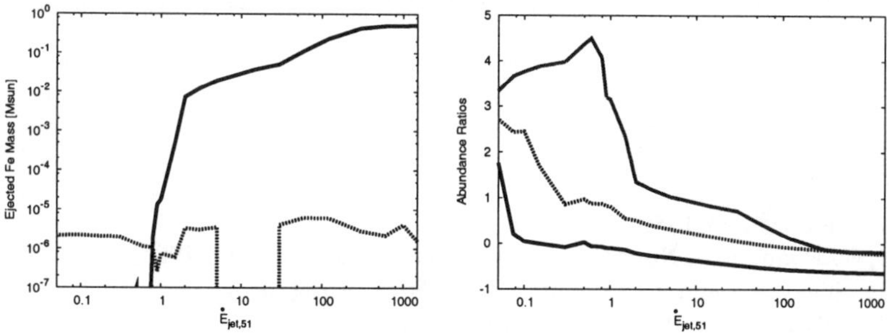

Figure 1. The dependences on the energy injection rates. (Left) the ejected Fe masses ejected from above the initial mass cut (*solid line*) and deposited as jets (*dashed line*). (Right) the abundance ratios, [C/O] (*solid line*), [C/Mg] (*dashed line*), and [C/Fe] (*dotted line*).

3. Comparison with observations

Previous studies suggested that the abundance patterns of EMP[18,19,20] and CEMP[19] stars can be reproduced well by a model with $M(\rm Fe) \sim 0.1 M_\odot$ and $\sim 10^{-3} M_\odot$, respectively. According to $M(\rm Fe)$, $\dot{E}_{\rm jet,51} = 120$ and 1.5 models are suitable to reproduce the abundance patterns of EMP and CEMP stars, respectively (the left panel of Figs. 2).

According to ratios [C/O] and [C/Mg] being independent on the assumption on the jet materials, HMP stars could be reproduced by the

$\dot{E}_{\rm jet,51} = 0.15$ and 0.05 models (the right panel of Figs. 2). N in the progenitor of HE 1327-2326 is assumed to be enhanced by a rotational mixing and/or an overshooting as in [21]. We should note that these models cannot eject ^{56}Ni from above the initial mass cut at all and that the abundances of Fe-peak elements are uncertain because these elements are dominated by the jet materials.

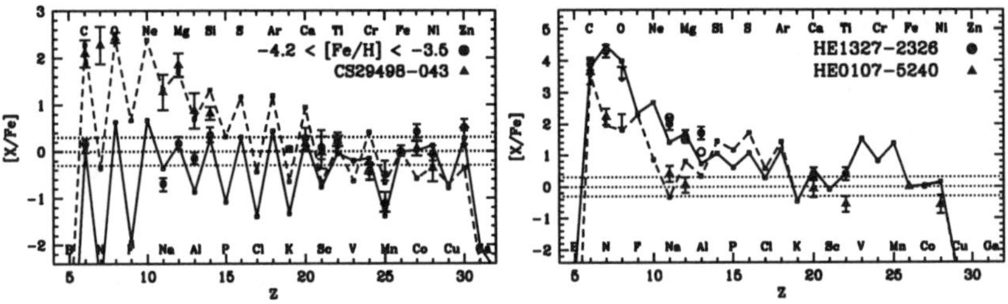

Figure 2. Comparisons with the observations, (left) EMP stars[7] (*circle*) and the CEMP star[10] (CS 29498-043, *triangle*), and (right) HMP stars [HE 1327-2326[13] (*circle*) and HE 0107-5240[11,12] (*triangle*)].

References

1. J. Audouze and J. Silk, *ApJ*, **451**, L49 (1995).
2. S. G. Ryan, J. E. Norris, and T. C. Beers, *ApJ*, **471**, 254 (1996).
3. T. Shigeyama and T. Tsujimoto, *ApJ*, **507**, L135 (1998).
4. T. Nakamura, et al., *ApJ*, **517**, 193 (1999).
5. D. Argast, et al., *A&A*, **356**, 873 (2000).
6. J. Tumlinson, *ApJ*, **641**, 1 (2006).
7. R. Cayrel, et al., *A&A*, **416**, 1117 (2004).
8. S. Honda, et al., *ApJ*, **607**, 474 (2004).
9. E. Depagne, et al., *A&A*, **390**, 187 (2002).
10. W. Aoki, et al., *ApJ*, **608**, 971 (2004).
11. N. Christlieb, et al., *Nature*, **419**, 904 (2002).
12. M. S. Bessell, N. Christlieb, and B. Gustafsson, *ApJ*, **612**, L61 (2004).
13. A. Frebel, et al., *Nature*, **434**, 871 (2005).
14. N. Tominaga, et al., in preparation (2006).
15. S. Nagataki, et al., *ApJ*, **596**, 401 (2003).
16. K. Maeda and K. Nomoto, *ApJ*, **598**, 1163 (2003).
17. S. Nagataki, et al., *ApJ*, submitted (2006).
18. H. Umeda and K. Nomoto, *ApJ*, **565**, 385 (2002).
19. H. Umeda and K. Nomoto, *ApJ*, **619**, 427 (2005).
20. N. Tominaga, H. Umeda, and K. Nomoto, *ApJ*, submitted (2006).
21. N. Iwamoto, H. Umeda, N. Tominaga, et al., Science, **309**, 451 (2005).

CORE-COLLAPSE VERY MASSIVE STARS: EVOLUTION, EXPLOSION, AND NUCLEOSYNTHESIS OF POPULATION III 500 – 1000 $M_\odot$ STARS

T. OHKUBO, H. UMEDA, K. NOMOTO, AND T. SUZUKI
Department of Astronomy, School of Science, The University of Tokyo,
7-3-1 Hongo, Bunkyo-ku,
Tokyo 113-0033, Japan
E-mail: ohkubo, umeda, nomoto, suzuki@astron.s.u-tokyo.ac.jp

K. MAEDA
Department of Earth Science and Astronomy, College of Arts and Sciences, The Univresity of Tokyo,
3-8-1 Komaba, Megro-ku,
Tokyo 153-8902, Japan
E-mail: maeda@esa.c.u-tokyo.ac.jp

S. TSURUTA
Department of Physics, Montana State University,
MT 59717-3840,
Bozeman, USA,
E-mail: uphst@gemini.msu.montana.edu

M. J. REES
Institute of Astronomy, Cambridge University,
Madingley Road,
Cambridge CB3 0HA, England,
E-mail: mjr@ast.cam.ac.uk

We calculate evolution, explosion, and nucleosynthesis of $500 M_\odot$ and $1000 M_\odot$ stars. Even such massive stars may explode at the end of their lives if they rotate. We use a 2 dimensional hydrodynamical code to take aspherisity into account. Our results show that (1) abundance pattern of ejected matter by explosion is consistent with observational data of intracluster medium gas, and M82 hot gas, (2) such massive stars can supply sufficient UV photons when one considers their contribution to the re-ionization of the universe with chemical evolution, and (3) final black hole mass is 500 solar-mass for $1000 M_\odot$ star model, which is consistent of the mass scale of intermediate-mass black hole (IHBH) identified in M82.

1. Introduction

It has been noticed that population III stars may have been very-massive. There may have existed over 300 solar-mass stars. Research of such massive stars is important in relation to (1) chemical enrichment in early phase galaxy, (2) supply of UV photons contributing to re-ionization of the universe, and (3) the origin of intermediate-mass black hole (IMBH), which has been identified as 700 solar-mass or more in M82 recently[1]. Motivated by these points, We calculate evolution, explosion, and nucleosynthesis of 1000 solar-mass stars. In section 2, we describe methods and models. Results for these three themes are shown in section 3.

2. Methods

We carry out the calculation of evolution in spherical models[2]). In hydrodynamical simulation, we use 2-dimentional hydrodynamical code based on bipolar jet model[3].

We compare the abundances with the observational data of extremely metal-poor (EMP) stars in the Galactic halo, gas in M82, intracluster matter (ICM), and intergalactic medium (IGM).

3. Results

Our findings are summarized as follows: (1) Our results of nucleosynthesis have similar patterns of [α/Fe] to the abundance pattern of ICM, the gas in the central region in M82 (figure 1). In the eft panel of figure 1 our theoretical abundance pattern is compared with that of ICM (bars[4]), and hot gas in M82 (pentagons[5]). The right panel shows for comparison with EMP stars[6]. Resulting small [O/Fe], [Ne/Fe] and large [Mg/Fe], [Si/Fe], [S/Fe] are consistent with the observational data of M82 and ICM.

For iron-peak elements, the main feature of the yields of very-massive stars is that [Mn/Fe] is small while [Zn/Fe] is large. This is consistent with the observed ratios in the EMP stars. The oversolar ratios of [Mg/Fe] and [Si/Fe] are also consistent. We need more data of [O/Fe] in EMP stars to see whether very-massive stars contributed to the early Galactic chemical evolution.

[C/Fe] of our results are $-0.86 \sim -0.68$, consistent in order of magnitude with IGM abundance ratio, -0.77[7], rather than the yields by PISNe ($-2.0 \sim -1.7$[8], [9])).

(2) Re-ionization by UV photon and chemical enrichment of IGM has been discussed. [10] estimate the efficiency of supplying UV photons and

chemical enrichment of IGM simultaneously. These authors suggest that the IMF that essentially formes less than 100 $M_\odot$ is favorable. However, this conclusion is due to the assumption that all CVMSs collapse entirely to a black hole. [11] considered the relation between the reionizing radiation and metal enrichment of IGM, using stellar atmosphere models and model yields available at that time. For the model yields, they assumed no metal ejection by stars of $\sim 30 - 130 M_\odot$ and also $M > 300 M_\odot$. Following their argument, here we compute the reionization efficiency for our CVMSs using model yields in the present work.

Adopting the mass of heavy elements ejected by our 1000 $M_\odot$ star model, $M_Z \sim 50 M_\odot$, the conversion efficiency ($\eta_{\rm Lyc}$) of energy produced in the HI ionizing radiation divided by the energy produced in the rest mass of metals ($M_Z c^2$) is $\eta_{\rm Lyc} \sim 0.05$. (We used Eq. 1 of [11]). Here we use the timescale of $t_{\rm ms} = 2 \times 10^6$ years for the 1000 $M_\odot$ Pop III star. With these values, the number of ionizing photons per baryon in the universe generated in association with the IGM metallicity $Z_{\rm IGM} \sim 10^{-4}$, obtained for our model, is $N_{\rm Lyc}/N_{\rm b} \sim 150$. (We used Eq. 2 of [11]). Note that this value well exceeds the value required for reionization of inter galactic hydrogen, $1 < N_{\rm Lyc}/N_{\rm b} \sim 10$ (see [12]). Therefore, contrary to the earlier results, our conclusion is that CVMSs can contribute significantly to reionization of IGM in the early epochs.

(3) Final black hole mass is $500 M_\odot$, which is consistent with the mass scale of IHBH found in M82 ($700 M_\odot$). Core-collapse type very-massive stars ($> 300 M_\odot$) can be the origin of IMBH.

References

1. H. Matsumoto, T. G. Tsuru, K. Koyama, K. Awaki, C. R. Canizares, N. Kawai, S. Matsushita and R. Kawabe, *ApJ* **547**, L25 (2001)
2. H. Umeda, K. Nomoto and T. Nakamura, *The First Stars* 150 (1999)
3. K. Maeda and K. Nomoto *ApJ* **598**, 1163 (2003)
4. J. R. Peterson, S. M. Kahn, F. B. S. Paerels, J. S. Kaastra, T. Tamura, J. A. M. Bleeker, C. Ferrigno and J. G. ernigan, *apj*, **590**, 207 (2003)
5. P. Ranalli, A. Origlia, A. Comastri, R. Maiolino and K. Makishima, astro-ph/0511021 (2005)
6. Cayrel, R., et al., *A&A* **416**, 1117 (2004)
7. A. Aguirre, J. Schaye, T.-S. Kim, T. Theuns, M. Rauch and W. L. W. Sargent, *ApJ* **602**, 38 (2004)
8. H. Umeda and K. Nomoto, *ApJ* **565** 385, (2002)
9. A. Heger and S. E. Woosley, *ApJ* **567**, 532 (2002)
10. F. Daigne, K. A. Olive, E. Vangioni-Flam, J. Silk and J. Audouze, *ApJ* **617**, 693 (2004)

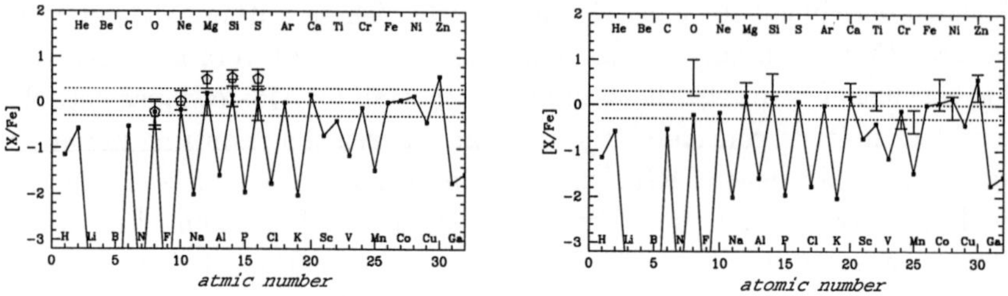

Figure 1. Abundance pattern of our $1000 M_{\odot}$ star model.

11. A. Venkatesan and J. W. Truran, *ApJ* **594**, L1 (2003)
12. R. S. Somerville, J. S. Bullock and M. Livio, *ApJ*, **593**, 616 (2003)

THE PROPERTIES OF THE UNIQUE TYPE Ib SUPERNOVA 2005bf AND IMPLICATIONS FOR THE DIFFERENCE BETWEEN TYPE Ib/c SUPERNOVAE

MASAOMI TANAKA, NOZOMU TOMINAGA, KEN'ICHI NOMOTO

Department of Astronomy, Graduate School of Science, The University of Tokyo
Hongo 7-3-1, Bunkyo-ku, Tokyo 113-0033, Japan
E-mail: mtanaka@astron.s.u-tokyo.ac.jp

KEIICHI MAEDA

Department of Earth Science and Astronomy, Graduate School of Arts and Science, University of Tokyo
Meguro-ku, Tokyo 153-8902, Japan

PAOLO A. MAZZALI

Max-Planck Institut für Astrophysik
Karl-Schwarzschild Str. 1, D-85748 Garching, Germany;
Istituto Nazionale di Astrofisica-OATs
Via Tiepolo 11, I-34131 Trieste, Italy

AND

JINGSONG DENG

National Astronomical Observatories, Chinese Academy od Science
20A Datun Road, Chaoyang District, Beijing 100012, China

The properties of the unique Type Ib supernova (SN) 2005bf are studied through the theoretical modeling of the early phase spectra and the light curve. The high velocity FeII lines in the earliest spectra are not the consequence of a jet-like explosion, but are naturally explained by a lower ionization state of Fe in the outermost layer due to the existence of the hydrogen. The velocities of the He lines measured from the position of the absorption minimum increase with time. Such a behavior is unprecedented, and can be understood if ^{56}Ni is not fully mixed with the He layer. An increasing deposition of γ-rays in the He layer is necessary to produce more and more non-thermal electrons. Therefore it is also suggested that the C+O layer between ^{56}Ni and the He layer is not very thick. The progenitors of Type Ib supernovae (SNe) and Type Ic SNe are briefly discussed by comparing SN 2005bf with other Type Ib/c SNe. We suggest that the mixig of ^{56}Ni plays an important role to make the difference between Type Ib SNe and Type Ic SNe.

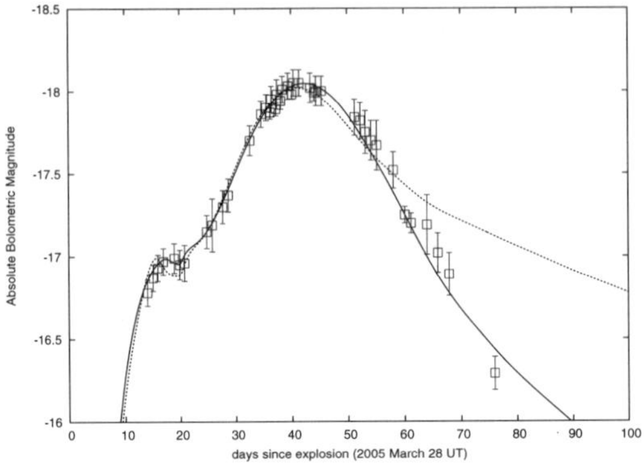

Figure 1. Bolometric light curves (open squares) and the synthetic LC (dotted line) and the synthetic LC with reduced γ-ray opacity (solid line).

1. Introduction

Type Ib supernovae (SNe) are defined as SNe that have He lines but no H lines, and Type Ic SNe are SNe that have neither He nor H lines. Supernova (SN) 2005bf is an unique SN that underwent an unprecedented transition from Type Ic to Type Ib [1]. The light curve (LC) of SN 2005bf has a peculiar double-peaked profile, and the luminosity is higher than usual for Type Ib/c SNe and the main peak is very late ($\sim$ 40 days after the explosion). These interesting properties made SN 2005bf the target of the observational and theoretical study [2,3,4]. Tominaga et al. (2005) [3] provided the theoretical modeling of the LC and the spectra and indicated the progenitor of the SN is a WN star, which was initially massive ($\sim 25 - 30 M_\odot$). Folatelli et al. (2006) [4] suggested a jet-like explosion because of the presence of the high velocity Fe lines, and denoted this SN as an event between Type Ib/c SNe and GRBs.

2. Light Curve Models

From the LC synthesis, the sets of the ejecta mass ($M_{\rm ej}$) and the explosion energy (E) are derived. Figure 1 shows the bolometric LC (open squares) and the synthetic LC (solid and dotted lines) computed with the model with ($M_{\rm ej}/M_\odot$, $E/10^{51}$erg) = (7,1.3). We also obtain a reasonable agreement with the models with ($M_{\rm ej}/M_\odot$, $E/10^{51}$erg) = (8,1.7), (6,1.0), (5,0.6) and it is found from the spectrum synthesis that the models except the most massive one are acceptable. A double-peaked shape of the LC requires a

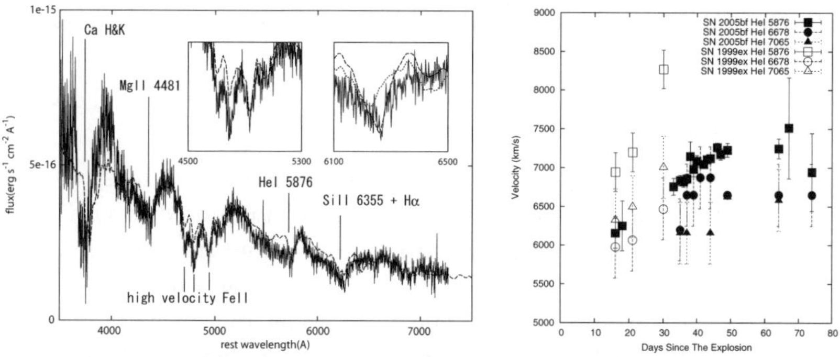

Figure 2. (left) Observed spectrum of SN 2005bf on 2005 April 9 compared with the synthetic spectrum (dashed line). The insets show the spectra around the high velocity FeII lines and Hα, respectively. (right) The temporal evolution of the HeI line velocities.

double-peaked ^{56}Ni distribution as long as it is assumed that the energy source is ^{56}Ni. The LC declines more rapidly than the synthetic LC [4,5], which indicates that the γ-rays from the decay of ^{56}Ni and ^{56}Co escape more efficiently than usual. Another possibility is the fallback on ^{56}Ni onto the central remnant [5], but the timescale of the fallback may not be long enough to reproduce the LC (K. Maeda et al. in preparation).

3. Spectroscopic Models

Here we show a result of the spectrum synthesis (the 2005 April 9th spectrum) and the rest is shown in Tominaga et al. (2005) [3] and M. Tanaka et al. (in preparation). Figure 2 (left) shows the comparison between the observed spectrum (solid line) and the synthetic spectrum (dashed line). Around 6200Å, the absorption feature is obtained as a blend of high velocity Hα and Si II λ6355. The fact that the center of the observed absorption is redder than the feature of photospheric Si II λ6355 (dotted line in the right inset) is a strong support of our identification of Hα.

The three dips around 4700 Å in the observed spectrum are identified as FeII λ4924, 5018 and 5169, respectively. They have a higher velocity ($v \sim 13,000$ km s^{-1}) than the photospheric velocity. We can reproduce these features with solar abundance Fe in a hydrogen rich layer at $v \sim 13,000$ km s^{-1} (see the left inset of Figure 2). Since the presence of hydrogen promotes the recombination of Fe, the line opacities of the FeII lines are increased in the high velocity layer. Although Folatelli et al. (2006) suggested these high velocity FeII lines result from the presence of a jet, our detailed analysis indicates such a jet is not necessary to explain these features.

4. Discussion

Type Ib/c SNe are thought to be core collapse explosions of massive stars, and the difference between Type Ib and Ic can be interpreted as a difference in the degree of mass loss from the progenitors before the explosion [6] or the degree of the mixing of ^{56}Ni [7]. SN 2005bf is an unique event that experiences the transition from Type Ic to Type Ib. In this sense, SN 2005bf can be classified as one of the transient events between Type Ib SNe and Type Ic SNe like SN 1999ex (see Hamuy et al. 2002 [8]). Recently it has been reported that SN 1999ex also has a thin H envelope [9]

Figure 2 (right) shows the He line velocities of SN 2005bf and SN 1999ex. Interestingly, the velocities in SN 1999ex also increase with time. It is known that the He lines in Type Ib SNe are produced mainly by non-thermal processes, i.e. the non-thermal electrons produced by γ-rays from the decay of ^{56}Ni and ^{56}Co. The fact that the He lines are initially quite weak indicates that γ-rays do not deposit in the He layer at the earliest epochs. In this context, the strange behavior of the He lines can be explained if the ^{56}Ni is not fully mixed with the He layers and the C+O layer is not very thick. The existence of such transient events may suggest that the mixing of ^{56}Ni is a primary difference between Type Ib and Ic SNe.

Acknowledgments

We are grateful to M. Modjaz, M. Hicken, P. Challis, R. P. Kirshner, G.C. Anupama, D.K. Sahu T.P. Prabhu, and M. Hamuy for providing the LC and the spectra.

References

1. Modjaz, M., Kirshner, R., & Challis, P. 2005, IAUC 8522
2. Anupama, G. C.,et al. 2005, *ApJ*, **631**, L125
3. Tominaga, N., et al. 2005, *ApJ*, **633**, L97
4. Folatelli, G., et al. 2006, *ApJ*, **641**, 1039
5. Woosley, S.E., & Weaver, T.A. 1995, *ApJS*, **101**, 181
6. Wheeler, J.C., et al. 1987, *ApJ*, **313**, L69
7. Woosley, S.E., & Eastman, R.G. 1997, *in Thermonuclear Supernovae*, ed. P. Ruiz-Lapuente, R. Canal, and J. Isern., ASI Series C, **486**, 821
8. Hamuy, M., et al. 2002, *ApJ*, **124**, 417
9. Elmhamdi, A., et al. 2006, *A&A*, **450**, 305

TIME DEVELOPMENT OF RELATIVISTIC TWO-TEMPERATURE PLASMA WITH ELECTRON-POSITRON PAIR PRODUCTION

MYONGGWAN KIM and FUMIO TAKAHARA

Department of Earth and Space Science, Graduate School of Science
Osaka University, Toyonaka OSAKA 560-0043

We investigate thermal and dynamical behavior of an optically thin two-temprature accretion plasma in a simplified one-zone model. We compute time development of a plasma which is located at a certain radius suffering from a constant frictional heating and bremsstrahlung cooling. When pair production is not taken into account, we find that equilibrium states are possible only for a certain range of surface density for a given heating rate. For a small surface density the proton temperature becomes so high that the plasma escapes from the gravitational potential of the central black hole to form an outflow. For a large surface density, the plasma cannot bear gravity force and collapses towards a low temperature and high density state. When the pair production is taken into account, pair concentration increases with heating rate, maintaining the balance between pair production and pair annihilation even for high heating rates. This result is in contrast to that of Kusunose and! Takahara(1988), where no pair equilibrium states were found for high accretion rates. We discuss possible reasons for this difference and suggest that pair concentration in two-temperature accretion plasma depends sensitively on detailed plasma profile.

1. Introduction

Accretion disks around compact objects are generally postulated as the energy source of active galactic nuclei(AGN) jet. After the problem of stability of the standard α-model near the central object was indicated, Shapiro et al.(1976) proposed that disks become optically thin and take two-temperature state. Then the ion temperature and the electron temperature become $\sim 10^{12}$K and $\sim 10^{9}$K, respectively. Lightman(1982) and Svensson(1982) investigated the equilibria of relativistic, thermal plasmas and showed the probability that copious electron-positron pairs might be produced in the plasma. Furthermore, Kusunose and Takahara(1988) investigated the effects of electron-positron pairs on a two-temperature accretion disk and showed that for an accretion rate higher than a few percent of the

Eddington accretion rate, no steady-state solutions exist. These studies indicate the probability that electron-positron pairs are copiously produced in accretion disk and ejected as an outflow, when accretion rate is so high that the temperature of the plasma in an accretion disk is relativistic.

2. Formulation

We calculate numerically the time development of pair concentration, disk thickness, electron and proton temperature. Here we deal with a part of accretion disk, at a distance r from central object, which rotate around Schwarzschild Black Hole of solar mass. And we simplify that as one-zone two temperature plasma. We follow the standard formulation of accretion disks (Shapiro et al. 1976) about radial motion and the transfer of angular momentum. But we assume that the plasma expand or contract dynamically toward vertical direction depending on gas pressure, radiation pressure, and gravity. Furthermore, we assume that plasmas consist of proton, electron and positron keeping electrical neutrality. In the plasmas, protons are suffering from a constant frictional heating, $3/8\pi \dot{M}\Omega_k^2 C$,and protons and electrons can exchange energy by Coulomb collision (Kusunose and Takahara 1988). Here $C = 1 - (6GM/rc^2)^{1/2}$, $\dot{M}$ is the rate of mass accretion, and Ω_k is Kepler angular velocity. We take into account bremsstrahlung through the collisions of $p-e^{\pm}$, $e^{\pm}-e^{\pm}$ and $e^{+}-e^{-}$ (Gould 1982) as radiative process, and neglect the effects of Compton scattering for simplicity. To determine radiation pressure, we calculate equation of radiative transfer simultaniously.

3. Results

We present the numerical results of the two cases. First, we don't take into account pair creation and annihilation, and just calculate thermal and dynamical behavior of plasma suffering from a constant frictional heating and bremsstrahlung cooling. In this case, we take $\dot{M}_*$ and τ_p as free parameter. Here $\dot{M}_* \equiv \dot{M}/(L_{Edd}/\eta c^2)$ is normalized mass accretion rate, $\tau_p \equiv \sigma_T n_p H$ is normalized surface density. $\sigma_T, n_p, H, L_{Edd}, \eta = 1/12$, and c are Thomson cross section, the number density of proton, the half disk thickness, the Eddintong luminosity, radiative efficiency, and the velocity of light.

In figure 1–3, the timedevelopment of the half disk thickness is shown for given $\dot{M}_*$ and τ_p. Here $H_* \equiv H/5r_g$, and r_g denotes Schwarzschild radius. In figure 1, for a small surface density, proton temperature becomes so high that the plasma escapes from the gravitational potential of the central

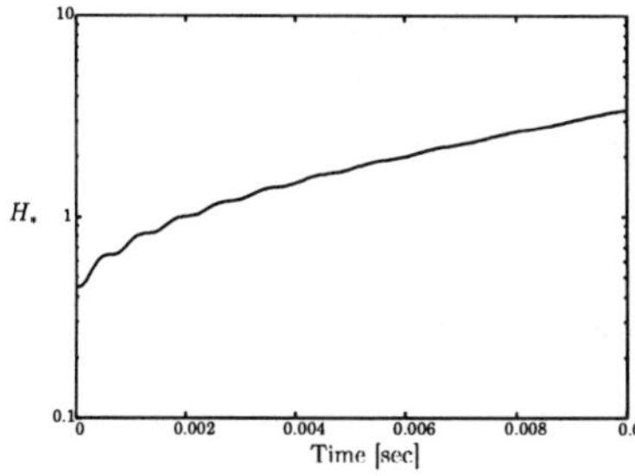

Figure 1. Time development of H_* for $\dot{M} = 0.01, \tau_p = 1.02$.

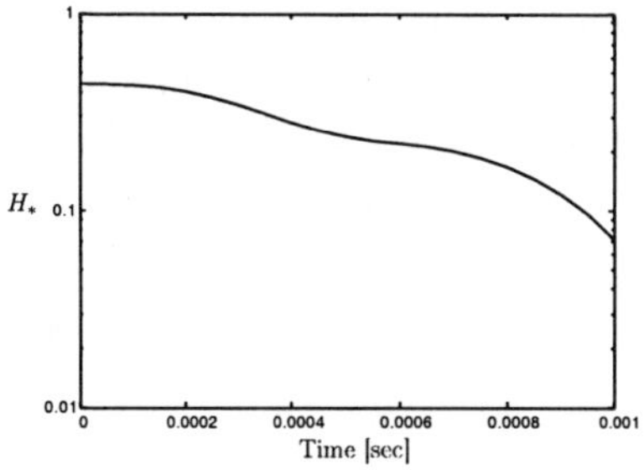

Figure 2. Time development of H_* for $\dot{M} = 0.01, \tau_p = 8.28$.

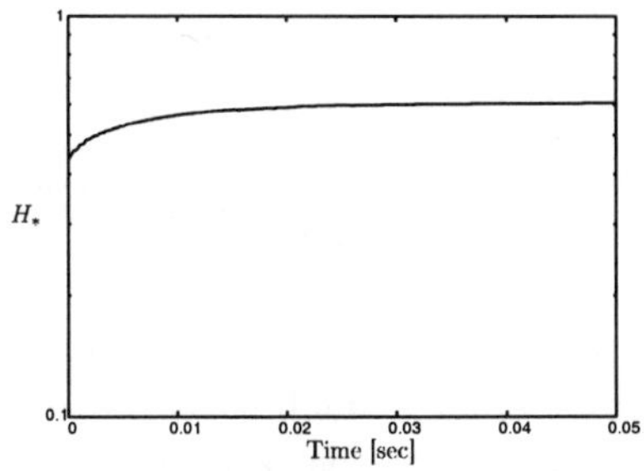

Figure 3. Time development of H_* for $\dot{M} = 0.01, \tau_p = 2.50$.

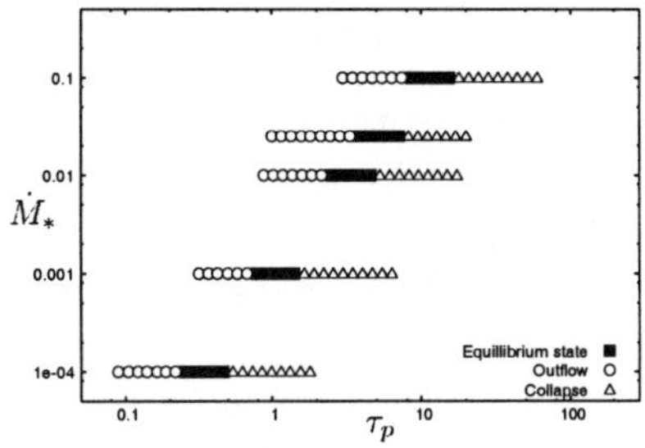

Figure 4. The states of plasmas for given $\dot{M}, \tau_p$.

black hole to form an outflow. In figure 3, for a large surface density, plasma cannot bear gravitational force and collapse towards a low temperature and high density state. We prove in figure 4 that the equilibrium states, figure 3, are possible only for a certain range of surface density for a given heating rate. The results for various $\dot{M}_*$ and τ_p are shown in figure 2.

Second, we take into account pair creation and annihilation (Svensson 1981), and investigate time development of pair concentration. In this paper, we only take into account photon-photon pair creation (Gould 1966). In table 1, for given $\dot{M}_*$ and τ_p, the plasma converges into equilibrium state and the pair concentration, the number density of proton, pair concentration, the half disk thickness, proton and electron temperatures have converged to a constant value. Here $z \equiv n_+/n_p$, n_+ denote pair concentration, the number density of positron, respectively. And $n_{p*} \equiv n_p/\sigma_T 5r_g$, $T_{p*} \equiv T_p/m_e c^2$, $T_* \equiv T_e/m_e c^2$. Futhermore, m_e, T_p, T_e denote the mass of the electron, the proton and electron temperature, respectively. In table 1, pair concentration increases with heating rate, maintaining the balance between pair production and pair annihilation even for high heating rate. This results are in contrast to that of Kusunose and Takahara(1988), where no pair equilibrium states were found for high accretion retes.

Table 1. Converged values for given $\dot{M}_*$ and τ_p.

$\dot{M}_*$	τ_p	n_{p*}	z	H_*	T_{p*}	T_*
0.01	3.9	19.3	1.08×10^{-7}	0.203	0.327	0.278
0.025	3.9	6.62	3.57×10^{-2}	0.582	3.17	1.15

4. Summary and Discussion

We investigated the time development of thermal and dynamical behavior of an optically thin two-temperature accretion plasma suffering from frictional heating and bremsstrahlung cooling. The main results are summarized as follows. When we don't take into account pair creation and annihilation, (1)the equilibrium states are possible only for a certain range of surface density for a given heating rate, (2)for a small surface density, the proton temperature becomes so high that the plasma forms an outflow, (3)for a large surface density, the plasma cannot bear gravitational force and collapse towards a low temperature and high density state. When we take into account pair creation and annihilation, pair concentration increases with heating rate, maintaining the balance between pair production and pair annihilation even for high heating rates. We discuss the possible reason for the difference between our results and that of Kusunose and Takahara(1988) and suggest that pair concentration in two-temperature accretion plasma depends sensitively on detailed plasma profile.

References

1. A. P. Lightman, *Astrophys. J.* ,**253**, 842 (1982).
2. M. Kusunose and F. Takahara, *Publ. Astron. Soc. Japan*, **40**, 435 (1988).
3. R. J. Gould and G. P. Schreder, *Phys. Rev.* **155**, 5 (1967).
4. R. J. Gould, *Astrophys. J.* **254**, 755 (1982).
5. R. Svensson, *Astrophys. J.* **258**, 321 (1982).
6. R. Svensson, *Astrophys. J.* **258**, 335 (1982).
7. S. L. Shapiro, A. P. Lightman, D. M. Eardlay, *Astrophys. J.*, **204**, 187 (1976).

CURRENT STATUS OF CLIO FOR DETECTION OF GRAVITATIONAL WAVES

T. AKUTSU[1], S. MIYOKI[1], T. UCHIYAMA[1], K. YAMAMOTO[1], M. OHASHI[1], K. KURODA[1], S. KAMAGASAKO[1], N. NAKAGAWA[1], M. TOKUNARI[1], K. KASAHARA[1], S. TELADA[2], T. TOMARU[3], T. SUZUKI[3], N. SATO[3], T. SHINTOMI[3], T. HARUYAMA[3], A. YAMAMOTO[3], D. TATSUMI[4], M. ANDO[5], A. ARAYA[6], A. TAKAMORI[6], S. TAKEMOTO[7], H. MOMOSE[7], H. HAYAKAWA[7], W. MORII[8] and J. AKAMATSU[8]

[1] *Institute for Cosmic Ray Research, The University of Tokyo, Japan*
[2] *National Institute for Advanced Industrial Science and Technology, Japan*
[3] *High Energy Accelerator Research Organization, Japan*
[4] *National Astronomical Observatory of Japan, Japan*
[5] *Department of Physics, University of Tokyo, Japan*
[6] *Earthquake Research Institute, The University of Tokyo, Japan*
[7] *Department of Geophysics, Kyoto University, Japan*
[8] *Disaster Prevention Research Institute, Kyoto University, Japan*

E-mail: tomo@icrr.u-tokyo.ac.jp

The CLIO is a 100 m baseline cryogenic laser interferometer for the detection of the gravitational waves, which is under construction in Kamioka mine, Japan. This is for the investigation the technical feasibility for the Large-scale Cryogenic Gravitational wave Telescope (LCGT), which is planned to be constructed in the same Kamioka mine with 30 times longer baseline than the CLIO. We successfully operated CLIO, whose three mirrors were cooled around 20K, as a gravitational wave detector using a locked Fabry-Perot control scheme.

1. Introduction

There are several gravitaional wave detectors (GWs) in the world [1,2,3,4]. To enhance their detectors strain sensitivities at the typical level of $\sim 3 \times 10^{-24}$ $\mathrm{Hz}^{-1/2}$ around 100 Hz, the advanced GWDs, such as Advanced LIGO [5], EGO [6] and LCGT [7] were proposed. Although such ultimate sensitivity is designed to be dominated only by an optical quantum noise, shot noise and thermal noise of an optical coating film that is evaporated on the mirror should be reduced well below the targeted sensitivity. One of approaches to reduce the coating thermal noise is to utilize cryogenic sapphire mirrors as the LCGT selects. Although there were many technical issues to realize a cryogenic laser interferometer, R&Ds that had been done in the last eight years have shown technical tests for each issue [8,9,10,11,12,13,14,15,16,17,18]. The cryogenic laser interferometer of the CLIO project is a proto-type of the LCGT to integrate these technical products and to demonstrate the thermal noise reduction at the displacement level of 1×10^{-19} m $\cdot$ Hz$^{-1/2}$ around 100 Hz. The construction of the cryogenic laser interferometer vacuum system, which were left at room temperature, started in the end of 2003. In parallel, cryostats and ultra-low vibration pulse tube type cryocoolers were developed to satisfy the targeted cryogenic temperature of 8 K at an inner shield of the cryostat and low cold head vibration that is comparable with the low seismic noise level of $10^{-9}f^{-2}$ m $\cdot$ Hz^{-1} in the Kamioka mine. In April 28, 2006, we successfully operated the CLIO whose three sapphire mirrors were cooled around 20K as a gravitational wave detector. We will present the recent progress of CLIO.

2. Why did we choose the KAMIOKA mine?

The CLIO project is sited in the KAMIOKA mine. There are two merits to operate the interferometer at Kamioka mine. One is that the seismic noise level at the Kamioka mine is 1/100 lower than at the city area. The seismic noise is one of origins to degrade the detector's sensitivity. We can obtain better sensitivity in the low frequency band by placing the interferometer at the Kamioka mine. The other is that the temperature and humidity at the Kamioka mine is stable. Variations of temperature and humidity would cause a mechanical drift. Temperature variation is less than 0.1 degree per a day and humidity variation is less than 1% per a week. Those low variations are desirable for a consecutive observation run.

3. Cryogenic Laser Interferometer

3.1. *Optical Setup*

The optical setup and the control system of the cryogenic laser interferometer is explained in the Figure 2 of Ref.19. First, the mode cleaner (MC), which consists of a triangle cavity whose mirrors are suspended by double pendulum seismic noise isolation systems, has been constructed. The MC cavity length is set to be 9.5 m. Two phase modulations (PM) are applied before the MC. A 12.7 MHz-PM is used for the MC control, while a 15.803

MHz-PM, is used to obtain the length signal of the 100-m FP-cavity control by a Pound-Drever-Hall method. After the output beam of the MC is reflected at a steering mirror and bounces at two mode matching mirrors, which have a 43 m convex and a 30 m concave curvature, are suspended by double pendulum suspension systems, the beam is introduced into a 50/50 power beam splitter. A half portion of the beam is introduced into a set of polarized beam splitter, and a half wave plate and a quarter-wave plate, which are also suspended by double pendulum suspension systems, in each arm. The CLIO is operated as a gravitational wave detector using a locked Fabry-Perot control scheme.

3.2. *Cryostats and Cryocoolers*

The cryostat has mainly three parts. One is a cryogenic area housed with double radiation shields that are cooled by a low mechanical vibration 2-stage Gifford-McMahon (GM) type pulse-tube (PT) cryocooler at 8 K and 80 K, respectively. The second part is an optical bench that is supported by four metal poles above the cryogenic area at the room temperature. The third is a 5 m-radiation-shield-duct that extends toward the vacuum duct to reduce molecule adsorption on the mirror, and that is cooled by a low vibration 1-stage GM type PT cryocooler. These cryocoolers have vibration isolation stages, which are fixed by less thermal conductive rods at the top plate of a housing vacuum chamber, around each 4K and 40K cold head. Each cold head is thermally linked with some bundles of pure aluminum wires with each vibration isolation stage. The replacement of a rotary valve unit from the PT base to an isolated metal anchor base also contributed to the low vibration performance. The schematic design of the cryocooler have been explained in Figure 4 of Ref.19. The fluctuation level at the vibration reduction stage is 50 nm for the vertical motion and 1 μm for the horizontal notion [20,21]. The cooling test of the cryostats were successfully completed. The targeted temperature of 8 K at the inner radiation shield was achieved within 4.5 days after the start of the cryocoolers.

3.3. *Sapphire mirror suspension in a cryostat*

For the seismic noise isolation, the sapphire mirror is suspended by a six-stages pendulum. Please refer its detail structure, dimensions and cooling performance to the Ref. 22. The first three stages are set in the room temperature area. While, the last three pendulum stages are housed in the cryogenic area. In order to cool a sapphire mirror, a triple thermal link is prepared: one is a pure aluminum wires between the inner radiation shield and the 4th copper stage, another is also a pure aluminum wires between the 4th stage and the 5th stage, and the other is two wrapping sapphire fibers that suspend the sapphire mirror. The seismic noise through these heat links is estimated to be below the target sensitivity. The design temperature of the sapphire mirror is around 20 K. The cooling test of this suspension system has been completed with pure aluminum wires for the sapphire mirror suspension in place of sapphire fibers, and it took 7 days for cooling the mirror at 20 K.

4. Current status

We have succeeded to operate the CLIO whose three mirrors were cooled around 20K. Improvement of the detector sensitivity is in progress. We would aim for the stable observation run in a state of low fake events.

Acknowledgments

This project is supported in part by Grant-in-Aid for Scientific Research of the Ministry of Education, Culture, Sports, Science and Technology.

References

1. Sigg D *Class. Quantum Grav.* **21** S409 (2004)
2. Acernese F *Class. Quantum Grav.* **22** S869 (2005)
3. Grote H *Class. Quantum Grav.* **22** S193 (2005)
4. Ando M and the TAMA300 Collaboration *Class. Quantum Grav.* **22** S881 (2005)
5. Shoemaker D *5th Amaldi Conf. on Gravitational Waves* (Pisa: Tirrenia) (2003)
6. Giazzoto A *Class. Quantum Grav.* **21** S1183 (2004)
7. Kuroda K *Int. J. Mod. Phy.* D **8** 557 (1999)
8. Uchiyama T *Phy. Lett.* A **242** 211 (1998)
9. Miyoki S *Cryogenics* **40** 61 (2000)
10. Miyoki S *Cryogenics* **41** 415 (2001)
11. Uchiyama T *Phy. Lett.* A **261** 5 (1999)
12. Kasahara K *J. Cryo. Soc. Japan* **39** 25 (2004)
13. Tomaru T *Phy. Lett.* A **283** 80 (2001)
14. Tomaru T *Class. Quantum Grav.* **19** 2045 (2002)
15. Tomaru T *Phy. Lett.* A **301** 215 (2002)
16. Yamamoto K *Class. Quantum Grav.* **21** S1075 (2004)
17. Tomaru T *Proc. 28th Inter. Cosmic Ray Conf.* (Tokyo : Universal Academy Press) p 3127 (2003)

18. Haruyama T *Proc. 28th Inter. Cosmic Ray Conf.* (Tokyo : Universal Academy Press) p 3135 (2003)
19. Miyoki S *Class. Quantum Grav.* **21** S1173 (2004)
20. Tomaru T *Cryocoolers* **13** 695 (2005)
21. Li R *Cryocoolers* **13** 703 (2005)
22. Uchiyama T to be published in *Proc. 6th Amaldi Conference.*

DEVELOPMENT OF AN AUTOMATIC BIREFRINGENCE MEASURING DEVICE OF MIRROR SUBSTRATES FOR LCGT

M. TOKUNARI, H. HAYAKAWA,* K. YAMAMOTO, T. UCHIYAMA, S. MIYOKI, M. OHASHI and K. KURODA

Institute for Cosmic Ray Research, The University of Tokyo, 5-1-5, Kashiwanoha, Kashiwa, Chiba, 277-8582, Japan

E-mail: tokunari@icrr.u-tokyo.ac.jp

We developed an automatic measuring device of birefringence inhomogeneity in synthetic sapphire substrates to evaluate their crystal quality suitable for laser interferometric gravitational wave (GW) detectors. The phase retardation was measured with an accuracy of 7×10^{-4} rad and the orientation of the fast axis with an accuracy of 2×10^{-2} rad. The automatic measuring device is useful to check the inhomogeneity of substrates for the LCGT project, which is a next-generation laser interferometer project for GW detection.

1. Introduction

Some interferometers for GW detection have been constructed for the purpose of verifying Einstein's theory of general relativity and creating the GW astronomy. The most accurately predictable signals come from the in-spiral phase of binary neutron stars. However, even if the detectable range is expanded to the Virgo cluster (20 Mpc), the expected event rate is so small — the order of 10^{-3} events per year — that improvements in sensitivity by one order of magnitude are required. LCGT (Large-scale Cryogenic Gravitational-wave Telescope)[1] is one of such advanced GW detectors which is likely to detect these at a rate of several events per year.

LCGT has three features: a 3 km baseline, cryogenic interferometers and an underground site (in the Kamioka mine)[2,3]. The main feature is the cryogenic mirrors to reduce the thermal noise, which is the key to realize the quantum limited sensitivity. To reduce the thermal noise more efficiently,

*Present address: Graduate School of Science, Kyoto University, Sakyo-ku, Kyoto 606-8502, Japan.

LCGT has selected sapphire as the substrates since it has a high mechanical quality factor, a high thermal conductivity at cryogenic temperature[4].

In addition, very high optical quality is required for the substrates. However, the fabrication technology of sapphire to date is not mature enough to consistently produce large-size substrates of adequate quality. Therefore, it is necessary to establish the evaluation method of optical quality of sapphire substrates. Measuring the fluctuation of the birefringence gives information about the crystal quality of the sample at the position where the beam passes inside the sample. For this reason, we have developed an automatic measuring device of the birefringence of a sapphire crystal[5].

Sapphire is a uniaxial crystal, which means there is a light axis, along which a linearly polarized light beam propagates with the same phase speed regardless of its polarization direction. This axis is called c-axis. Although the cylindrical axis of the mirror substrate is normally taken to coincide with the c-axis of the crystal in applications to laser interferometers, there can be small local differences that fluctuate due to a possible intrinsic irregularity or impurity of a practical crystal sample, and also due to extrinsic mechanical and thermal stress.

There are two parameters that characterize the birefringence. One is the phase difference (retardation) (R) between an ordinary ray and an extraordinary ray

$$R = \frac{2\pi d(n_{\mathrm{o}} - n_{\mathrm{e}}')}{\lambda}, \tag{1}$$

where d, n_{o}, n_{e}' and λ are the optical path length, the ordinary refractive index, the extraordinary refractive index and the wavelength of the light in vacuum, respectively. The other is the orientation (ϕ) of the fast axis, which is the polarization direction of the extraordinary ray.

2. Setup of the measuring device

Figure 1 shows the experimental setup of the measuring device. A laser beam passes in succession through a polarizer oriented at 45 degrees from the horizontal axis, a compensator, a half-wave plate, the sapphire sample, a quarter-wave plate, an analyzer crossed with respect to the polarizer (135 degrees), and a photo-detector (PD). Firstly, the polarizer and the analyzer are set to extinguish the light at the PD in the initial setting without the sample. If the sample is inserted after that, a part of the light beam leaks due to its birefringence. The compensator can compensate the phase retardation and extinguish the light again while adjusting the orientation of the half-wave plate to align with the fast axis of the sample. The displacement of the compensator (x_c), which gives the extinction (minimum) point, results in the phase-retardation value (R). The birefringence mapping is obtained by scanning the sapphire sample in two-dimensional directions in a plane orthogonal to the light beam using mechanical stages. The above procedures were automated with LabVIEW, which is a graphical programming tool. It took about four minutes to finish one automatic measurement for one line with LabVIEW.

Since the standard error of x_c was 3×10^{-3} mm, the standard error of R was 7×10^{-4} rad, which is equivalent to 1×10^{-9} in terms of the fluctuation of the relative refractive index for samples of 60 mm thickness. The orientation (ϕ) of the sample was calculated from the adjusted orientation of the half-wave plate. The orientation was determined with an accuracy of 2×10^{-2} rad.

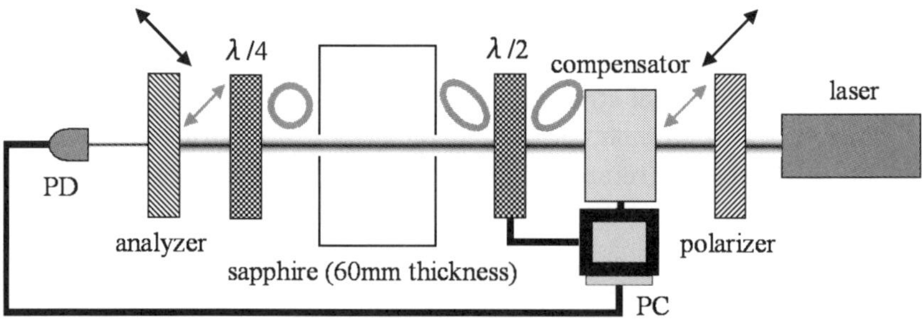

Figure 1. Setup of the birefringence measuring device. Although the polarizer and the analyzer are arranged to extinguish the light passing to the PD without the sample, the inserted sample makes light leak due to its birefringence. The compensator can compensate the phase retardation and extinguish the light again while adjusting the orientation of the half-wave plate to align with the fast axis of the sample. This procedure was controlled by a personal computer (PC).

The procedures were checked by measuring a quarter-wave plate as a sample with known birefringence. The average of measured R data was 1.572 rad ($\sim \pi/2$). This is reasonable because the instrumental error of the quarter-wave plate is about 0.06 rad according to the manufacturer.

3. Result and discussion

We used four high-purity Hemlite sapphire cylinder samples. They were 100 mm in diameter and 60 mm in thickness (samples A, B, C and D).

These samples were made with the Heat Exchanger Method by Crystal Systems Inc.[6].

Figure 2 shows two-dimensional mappings of the phase retardations of the samples B and C. The X-Y plane corresponds to the circular plane of the sample. The reproducibility of a measurement to another, whose interval time was one week, was about 1% for both the phase retardation and the orientation. Layered structure are found in both figures.

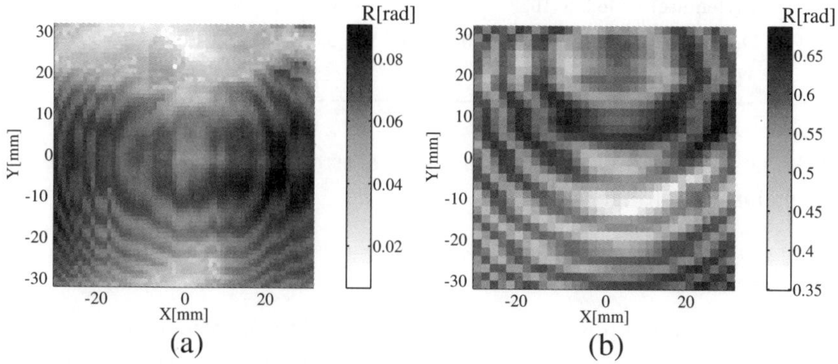

Figure 2. Two-dimensional mappings of the phase retardation: (a) sample B, (b) sample C.

Table 1 gives a summary of the measured data: the average phase retardation, standard deviation of the orientation angle, that of the phase retardation and that of the relative refractive index. For a reference, we included the data of a Hemex-grade sample (sample E) whose properties were measured at the University of Western Australia (UWA)[7]. The average phase retardation represents a gap between the c-axis and the cylindrical axis. The relative refractive index was obtained by using the Eq. (1). The standard deviation of the phase retardation, or the relative refractive index, represents a possible irregularity or inhomogeneity of the crystal sample. It was confirmed that Hemex is of higher grade than Hemlite based on the results that the standard deviations of the relative refractive indices of four Hemlite samples were actually larger by several times than that of Hemex.

In order to check the extrinsic effect of stress-induced birefringence, we rotated sample A by every 90 degrees, and made the same measurements for every 90 degrees. Since the mappings of those measurements rotated according to the sample rotations, it is confirmed that the phase retardation

Table 1. Summary of the measured data with a reference sample: the average phase retardation, standard deviation of the orientation angle, that of the phase retardation and that of the relative refractive index. Samples A, B, C and D are of Hemlite grade. The reference sample E is of Hemex grade.

	$\overline{R}$ rad	$\sigma(\phi)$ rad	$\sigma(R)$ rad	$\sigma(n_o - n'_e)$
A(Hemlite)	19.5×10^{-2}	1.1×10^{-1}	4.2×10^{-2}	11.8×10^{-8}
B(Hemlite)	4.9×10^{-2}	4.5×10^{-1}	2.7×10^{-2}	7.7×10^{-8}
C(Hemlite)	46.9×10^{-2}	0.6×10^{-1}	6.3×10^{-2}	17.7×10^{-8}
D(Hemlite)	30.4×10^{-2}	0.9×10^{-1}	5.1×10^{-2}	14.5×10^{-8}
E(Hemex)	2×10^{-2}	–	1×10^{-2}	2×10^{-8}

of the sample was caused not by mechanical stress-induced birefringence but by intrinsic birefringence.

4. Conclusion

We developed an automatic measuring device of the birefringence of a high-quality sapphire substrate. The phase retardation was measured with an accuracy of 7×10^{-4} rad and the orientation with an accuracy of 2×10^{-2} rad. The measured standard deviations ranged from 0.7×10^{-7} to 1.8×10^{-7} in terms of the relative refractive index with a reproducibility of 1% along lines in parallel with the cylindrical axis of the sample. The automatic birefringence measuring device is useful to check the inhomogeneity of LCGT substrates.

Acknowledgements

This study was supported by a Grant-in-Aid for Scientific Research on Priority Areas by the Ministry of Education, Culture, Sports, Science and Technology.

References

1. K. Kuroda *et al.*, *Class. Quantum Grav.* **20**, S871 (2003).
2. S. Sato *et al.*, *Phys. Rev.* **D69**, 102005 (2004).
3. M. Ohashi *et al*, *Class. Quantum Grav.* **20**, S599 (2003).
4. T. Uchiyama *et al*, *Phys. Lett.* **A261**, 5 (1999).
5. M. Tokunari, H. Hayakawa, K. Yamamoto, T. Uchiyama, S. Miyoki, M. Ohashi and K. Kuroda, *J. Phys.: Conference Series* **32**, 432 (2006).

6. C. P. Khattak, F. Schmid and M. B. Smith, *Windows and Dome Technologies and Materials* V ed R. W. Tustison, Proc. SPIE 3060, 250 (1997).
7. F. Benabid, M. Notcutt, L. Ju and D. G. Blair, *Phys. Lett.* **A237**, 337 (1998).

QUASAR LUMINOSITY FUNCTION FROM RECENT OBSERVATIONS

KAZUHIDE ICHIKAWA

Institute for Cosmic Ray Research, University of Tokyo, Kashiwa 277 8582, Japan

E-mail: kazuhide@icrr.u-tokyo.ac.jp

We provide a fitting function for quasar luminosity function based on several sets of recent optical observations including SDSS DR3. From that luminosity function, together with black hole mass function, we estimate radiation efficiency of quasars. We also discuss implications from X-ray observations.

1. Introduction

Quasar engines are considered to be massive black holes in the centers of galaxies and one of the important ways to gain insight into such mechanism is to consider quasar energy budget. We can estimate the rate of radiation of energy by quasars from quasar luminosity function and the accumulation of mass in the quasar engines as black holes from black hole mass function. Their ratio, the radiation efficiency ϵ, tells us about some properties of the central black holes. Very simple estimation is given by Fukugita and Peebles[1] to be $\epsilon \sim 0.02$.

In this study, we updated quasar part. We (i) updated bolometric correction with new AGN spectra, (ii) updated optical luminosity function with SDSS DR3 data, (iii) considered contributions from optically unidentified quasars inferred from soft X-ray luminosity function and/or harder X-ray background. Our estimation of average radiation efficiency is $0.02 \lesssim \epsilon \lesssim 0.2$. This study is an extension of the work of K. Oguro for his master's thesis.

2. Bolometric corrections

We connected observations at various wavelength to derive bolometric corrections for AGNs. For energy up to optical to UV, we adopt Elvis *et al.*[3] which is consistent with more recent data of Vanden Berk *et al.*[4] and Telfer *et al.*[5]. From EUV, we adopt spectral index $\alpha_{EUV} = -1.76$ from Telfer *et al.*[5] For soft X-ray, we adopt the EUV value as $\alpha_s = \alpha_{EUV}$ up to 0.5 keV and $\alpha_s = -1.23$ from Almaini *et al.*[6] at higher energies. For hard X-ray region, we consider three types of AGNs which are assumed to have different spectrum only at that range. We define x-quasar (x-seyfert) to be AGN with soft X-ray luminosity more (less) than $10^{44.3}$ erg/s. We define type 2 x-seyfert to be AGN with column density larger than 10^{21} cm^{-2}. For x-seyfert, the spectrum is adopted from Zdziarski *et al.*[7] and spectra of x-quasar and type 2 x-seyfert are inferred with model of Comastri *et al.*[8]

The bolometric corrections for B band (at 4450 Å) are 11.4, 11.2 and 11.3 respectively for x-seyfert, x-quasar and type 2 x-seyfert. For soft X-ray (0.5-2 keV) correction factors are 76.6, 75.1 and, for hard X-ray (2-10 keV), they are 76.1, and 66.9, 69.5 and 67.1.

3. Optical luminosity function

We connected several sets of quasar data at various redshifts z: $0.4 < z < 2.1$ from Croom *et al.*[9], $2.2 < z < 4.6$ from Warren *et al.*[10] and $3.6 < z < 5.0$ from Fan *et al.*[11]. We find that these luminosity functions are well described by the following double power-law with hybrid-luminosity-density evolution:

$$\phi(M_B, z)dM_B dV = \frac{\phi(z)dM_B dV}{10^{-0.4(a-1)(M_B-M(z))} + 10^{-0.4(b-1)(M_B-M(z))}}, \quad (1)$$

where

$$\phi(z) = \phi_0 \times [1 + \exp\{c(z - z_c)\}]^{-1}, \quad (2)$$

$$M(z) = M_c - 1.086k\tau(z) \quad (\Leftrightarrow L(z) = L_c(0)\exp(k\tau(z))). \quad (3)$$

Here, $\tau(z)$ is the look-back time,

$$\tau(z) = \int_0^z \frac{dz'}{(1+z')\sqrt{\Omega_m(1+z')^3 + \Omega_\Lambda}}. \quad (4)$$

The best fit parameters are $a = 3.4 \pm 0.2$, $b = 1.25 \pm 0.13$, $M_c = -20.90 \pm 0.28$, $k = 6.20 \pm 0.35$, $\phi_0 = (1.3 \pm 0.3) \times 10^{-6}$ Mpc^{-3} mag^{-1}, $c = 2.05 \pm 0.20$, $z_c = 2.70 \pm 0.20$. This is shown in Fig. 1.

We also find that this functional form well describes the newer SDSS quasar luminosity function[12] as shown in Fig. 2 (in this case, we adopt the normalization ϕ_0 to be 1.0×10^{-6} Mpc^{-3} mag^{-1} for a better fit to SDSS data, and this is used for the later calculation).

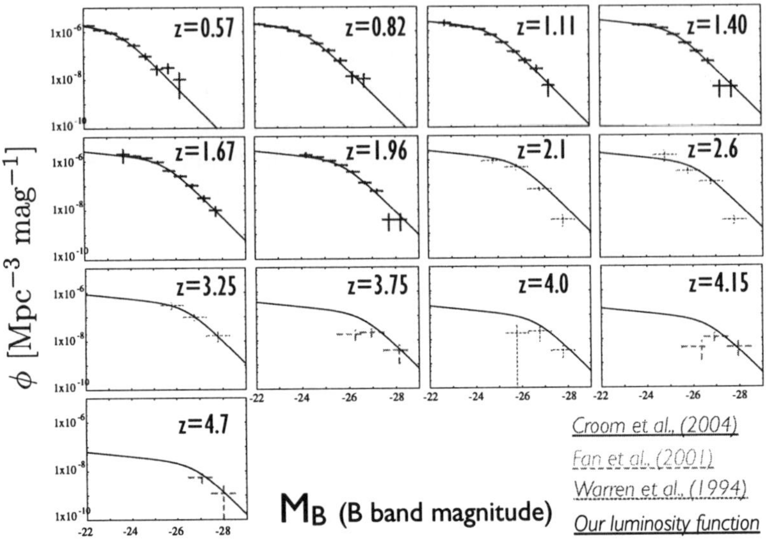

Figure 1. The optical luminosity functions derived from three sets of observations and fitting function eq.(1).

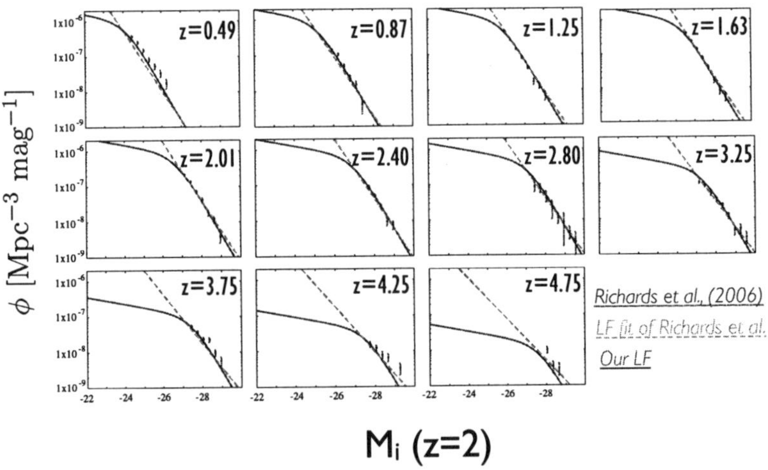

Figure 2. The SDSS luminosity function data, fitting function given in the SDSS paper and our fitting function eq.(1) (renormalized for a better fit).

4. Radiation efficiency

The energy released from quasars is calculated by integrating the luminosity function, for example, as

$$E_{\mathrm{emit}} = \int_0^\infty \frac{dt}{dz} dz \int_{-\infty}^{\infty} L_B C_B \phi(M_B, z) dM_B. \tag{5}$$

Using the luminosity function eq.(1) and average bolometric correction $C_B = 11.3$, we obtain $E_{\rm emit} = (1.5 \pm 0.2) \times 10^{58}$ erg/Mpc^3 or $\Omega_{\rm emit} = (6.2 \pm 1.1) \times 10^{-8}$ (ratio to the critical density).

We note that this is the contribution only from optically identified quasars. Actually, soft X-ray luminosity function (SXLF) obtained by Miyaji *et al.*[13] is described by double power-law with similar power-law index at higher luminosity but larger one at lower luminosity. This can be interpreted that more quasars are identified by the soft X-ray surveys. Using the luminosity function eq.(1) with the lower luminosity power-law index modified to be the SXLF value $b = 1.6 \pm 0.10$[13], we obtain $E_{\rm emit} = (3.4 \pm 0.6) \times 10^{58}$ erg/Mpc^3 or $\Omega_{\rm emit} = (1.4 \pm 0.2) \times 10^{-7}$.

Furthermore, AGNs which can be seen neither by optical nor soft X-ray are considered to exist (type 2 x-seyfert). We can infer their energy release from observation of X-ray background (XRB). We determine number ratio of x-seyfert to type 2 x-seyfert which reproduces the observed XRB to be 4.8. Here, we fixed the distribution for the gas column density around type 2 x-seyfert according to Risaliti *et al.*[14]. Adding the so-inferred contribution from type 2 x-seyfert gives $E_{\rm emit} = (11.8 \pm 2.1) \times 10^{58}$ erg/Mpc^3 or $\Omega_{\rm emit} = (4.8 \pm 0.9) \times 10^{-7}$.

Finally, we derive radiation efficiency from the ratio of $\Omega_{\rm emit}$ to black hole mass density[15,16,17] $\rho_{\rm BH,tot} = (3.2 \pm 0.8) \times 10^5 M_\odot$ Mpc^{-3} or $\Omega_{\rm BH,tot} = (2.4 \pm 0.6) \times 10^{-6}$. We obtain $0.027 \pm 0.008 \leq \epsilon \leq 0.179 \pm 0.050$.

References

1. M. Fukugita and P. J. E. Peebles, *Astrophys. J.* **616**, 643 (2004).
2. K. Oguro, master's thesis (Univ. Tokyo, supervisor: M. Fukugita), (2005).
3. M. Elvis *et al.*, *Astrophys. J. Suppl.* **95**, 1 (1994).
4. D. E. Vanden Berk *et al.*, *Astron. J.* **122**, 549 (2001).
5. R. Telfer *et al.*, *Astrophys. J.* **565**, 773 (2002).
6. O. Almaini *et al.*, *Mon. Not. Roy. Astron. Soc.* **282**, 295 (1996).
7. A. A. Zdziarski *et al.*, *Astrophys. J.* **438**, L63 (1995).
8. A. Comastri *et al.*, *Astron. Astrophys.* **296**, 1 (1995).
9. S. M. Croom *et al.*, *Mon. Not. Roy. Astron. Soc.* **349**, 1397 (2004).
10. S. J. Warren *et al.*, *Astrophys. J.* **421**, 412 (1994).
11. X. Fan *et al.*, *Astron. J.* **121**, 31 (2001).
12. G. T. Richards *et al.*, arXiv:astro-ph/0601434.
13. T. Miyaji, G. Hasinger and M. Schmidt, *Astron. Astrophys.* **353**, 25 (2000).
14. G. Risaliti *et al.*, *Astrophys. J.* **522**, 157 (1999).
15. P. Salucci *et al.*, *Mon. Not. Roy. Astron. Soc.* **307**, 637 (1999).
16. Q. Yu and S. Tremaine, *Mon. Not. Roy. Astron. Soc.* **335**, 965 (2002).
17. M. C. Aller and D. Richstone, *Astron. J.* **124**, 3035 (2002).

Scientific Program

February 22, Wednesday

09:50–10:00 Welcome Address Katsuhiko Sato

1) Highest Energy Universe

10:00–10:20 AGASA Results and the Status of TA

Masaki Fukushima

10:20–10:40 Results from the High-Resolution Fly's Eye (HiRes) Experiment

Chales C. H. Jui

Coffee Break

11:00–11:30 Status of the Pierre Auger Project

Antonio Insolia

11:30–11:55 Astrophysical Origin of the Highest Cosmic Rays

Susumu Inoue

11:55–12:25 Cosmic Rays and Magnetic Fields in Large Scale Structure of the Universe

Hyesung Kang

12:25–12:40 AMBER: UHECR Detection via "Radio Fluorescence"

Benjamin Stokes

Lunch

2) Special Lecture

14:00–15:00 The Sky in Hard X-Rays - INTEGRAL Spacecraft Results

Rashid A. Sunyaev

3) Around the Knee

15:00–15:30 Cosmic Rays at the Knee - Observations and Implications

Thomas K. Gaisser

15:30–15:50 The AMS Detector: A Particle Physics Experiment in Space

Roberto Battiston

Coffee Break

16:20–16:50 From Amanda to IceCube

Francis L. Halzen

16:50–17:20 Ultra High Energy Cosmic Rays and Neutrinos

Peter Mészáros

4) Poster Session 1

17:30–18:10 Poster Abstracts 1 (Poster No.1-14)

18:10–18:50 Poster Session 1

February 23, Thursday

5) GeV Sky

09:00–09:20 Precise Measurement of Low Energy (< TeV) Cosmic-Rays with BESS

Akira Yamamoto

09:20–09:35 The Pamela Experiment on board Resurs-DK1 Satellite: Status and Perspectives

Marco Casolino

09:35–10:05 Particle Acceleration in Kinetic Plasma Processes

Masahiro Hoshino

Coffee Break

10:30–11:00 Observations of High-Energy Phenomena with SWIFT and GLAST

Neil Gehrels

11:00–11:20 Gamma-Ray Burst: Problems delineated by HETE-2 and Other Observations and Giant Flares of SGRs

Nobuyuki Kawai

11:20–11:50 Theoretical Models of High Energy Emission from GRBs

Ehud Nakar

11:50–12:05 Numerical Simulations of Collapsars with Neutrino Heating and Magnetic Fields

Shigehiro Nagataki

12:05–12:25 GRB 050315 and the Uniqueness of GRB Model

Remo Ruffini

Lunch

6) Special Lecture

13:40–14:40 Unveiling the Dark TeV Sky

Trevor C. Weekes

7) TeV Sky 1

14:40–15:00 Recent Results from CANGAROO

Masaki Mori

15:00–15:30 Observations of Galactic Gamma-Ray Sources with H.E.S.S.

David Berge

Coffee Break

15:55–16:25 Particle Acceleration in Supernova remnants and the Resulting Nonthermal EmissionHeinrich

J. Völk

8) Poster Session 2

16:30–17:10 Poster Abstracts 2 (Poster No.15-27)

17:10–17:50 Poster Session 2

18:00–20:00 Banquet

February 24, Friday

9) TeV Sky 2

09:00–09:20 Recent Results from the MAGIC Project and Outlook

Jose A. C. Perez

09:20–09:50 Gamma-Ray Diagnostics of Particle Acceleration

Paolo Coppi

09:50–10:10 All-Sky Survey High Resolution Air-Shower Detector

Makoto Sasaki

10:10–10:25 TeV Gamma-Rays from Old Supernova Remnants

Ryo Yamazaki

10:25–10:40 The CALET Project for Investigating High Energy Universe

Shoji Torii

Coffee Break

10) MeV and keV Sky

11:00–11:30 Suzaku initital results

Tadayuki Takahashi

11:30–11:50 X-Ray Diagnostics of Acceleration Processes

Aya Bamba

11:50–12:20 Supernovae in the Universe

Shoichi Yamada

12:20–12:40 Search for Supernova Neutrinos at Super-Kamiokande

Masayuki Nakahata

12:40–13:10 Neutrino Production in Supernovae

Todd Thompson

Lunch

14:30–15:30 Concluding Remarks

Peter Mészáros

Poster Papers

01 JEM/EUSO: Extreme Universe Space Observatory on JEM/ISS

Toshikazu Ebisuzaki

02 Propagation of Ultra-High Energy Cosmic Rays above 10^{19} eV in a Structured Extragalactic Magnetic Field and Galactic Magnetic Field

Hajime Takami

04 High Energy Neutrino Emission from Gamma-Ray Bursts

Kohta Murase

05 Simulation of Salt Neutrino Detector Performance for Ultra High-Energy Neutrino Detection

Yusuke Watanabe

06 Measurement of Attenuation Length for UHF Radio Wave in Natural Rock Salt Samples Concerning Ultra High-Energy Neutrino Detection

Masami Chiba

07 Particle Acceleration by MHD Turbulence

Jungyeon Cho

08 SU(2)L-triplet Dark Matter and HEAT Anomaly in Cosmic Positron Experiment

Shigeki Matsumoto

11 Cosmic Gamma-Ray Background Anisotropy from Dark Matter Annihilation

Shin'ichiro Ando

12 High Energy Cosmic Rays, Neutrinos, and Photons from Gamma-Ray Bursts

Katsuaki Asano

13 Probing the Cosmic Dark Ages with GeV Gamma-Rays from Very High Redshift Gamma-Ray Bursts

Susumu Inoue

14 Turbulence Transport and Particle Acceleration in Solar Corona

Huirong Yan

15 Optical Measurements for CANGAROO-III

Ryuta Kiuchi

16 Consideration of Cassegrain Imaging Atmospheric Cherenkov Telescopes

Yohei Yukawa

17 Influence of a Non-Dipole Magnetic Field on the Peak Energies of Cyclotron Absorption Lines

Osamu Nishimura

18 Magnetorotational Collapse of Very Massive Stars: Formation of Jets and Black Holes

Yudai Suwa

19 Explosive Nucleosynthesis Induced by Relativistic Jets in Population III Supernovae

Nozomu Tominaga

20 Core-Collapse Very Massive Stars: Evolution, Explosion, and Nucleosynthesis of Population III 500–1000 Solar-Mass Stars

Takuya Ohkubo

21 The Unique Type Ib Supernova 2005bf: A WN Star Explosion Model for Peculiar Light Curves and Spectra

Masaomi Tanaka

22 Time Development of Relativistic Two-Temperature Plasma with Electron-Positron Pair Production

Myonggwan Kim

25 Current Status of CLIO for the Detection of Gravitational Waves

Tomomi Akutsu

26 Development of an Automatic Birefringence Measuring Device of Mirror Substrates for Gravitational Wave Detectors

Masao Tokunari

27 Quasar Luminosity Function from Recent Observations

Kazuhide Ichikawa